AF360887

Carcass and Meat Quality in Ruminants

Carcass and Meat Quality in Ruminants

Editors

Guillermo Ripoll
Begoña Panea

MDPI • Basel • Beijing • Wuhan • Barcelona • Belgrade • Manchester • Tokyo • Cluj • Tianjin

Editors

Guillermo Ripoll
Department of Animal
Science
Agrifood Research and
Technology Centre of Aragon
Zaragoza
Spain

Begoña Panea
Department of Animal
Science
Agrifood Research and
Technology Centre of Aragon
Zaragoza
Spain

Editorial Office
MDPI
St. Alban-Anlage 66
4052 Basel, Switzerland

This is a reprint of articles from the Special Issue published online in the open access journal *Animals* (ISSN 2076-2615) (available at: www.mdpi.com/journal/animals/special_issues/meat_ruminants).

For citation purposes, cite each article independently as indicated on the article page online and as indicated below:

LastName, A.A.; LastName, B.B.; LastName, C.C. Article Title. *Journal Name* **Year**, *Volume Number*, Page Range.

ISBN 978-3-0365-5982-7 (Hbk)
ISBN 978-3-0365-5981-0 (PDF)

Contents

About the Editors

Guillermo Ripoll

Guillermo Ripoll obtained his PhD in "Ingeniería Agraria, Alimentaria, Forestal y del Desarrollo Rural Sostenible" (Universities of Seville and Cordoba, 2020) and Master's in Agrifood Biotecnology (University of Avila, 2016). He was also an Agricultural Engineer (University of Avila, 2014 and Agricultural Technical Engineer, University of Zaragoza, 2000). His researcher ORCID code is 0000-0001-8042-8688. His main research interests concern carcass, meat and meat product quality, including shelf-life, intrinsic and extrinsic quality cues and perception of consumers with a farm to fork approach. His research areas or special interests are color and visual color perception and near-infrared spectroscopy. His research group is "Sustainable Agrosilvopastoal and Food Systems (SAGAS)" and previously was "Calidad y tecnología de la carne-A04".

In terms of scientific activitiy, he has been a part of 47 research projects. He has authored 125 peer-reviewed articles (Q1 52%, D1 25%; h-index: 25). He has published 72 technical papers, 15 book chapters, and presented 150 communications in national (84) and international (76) conferences. Additionally, he is a co-inventor of the software Nodriza (registrered) and the utility model "Procedimiento de predicción de dureza de la carne de ganado vacuno mediante la detección de la mutación genética SNP98535683 (Thr182Ala) de la Calpastatina (CAST) y/o de la mutación genética SNP98545188 (CAST2)" (registered). He has directed 12 MsC and TFG.

He is a member of the Asociación Interprofesional para el Desarrollo Agrario-AIDA and has served as the coordinator of "Product quality" section in the *Jornadas sobre Producción Animal* of AIDA from 2017, as well as the "Red nacional cubana de pequeños rumiantes" (RECUPER), and he is a member of the Editorial Board of *Animals* and *ITEA*.

He has taught students of International Master of Animal Nutrition of CIHEAM and students of Science and Technology of Food grade in the Veterinary Faculty of University of Zaragoza.

Begoña Panea

Begoña Panea graduated in Veterinary Medicine, specializing in Bromatology, Health and Food Technology at the University of Zaragoza (1989), and obtained her PhD in Veterinary Medicine at the University of Zaragoza (2002). She is currently a researcher at the Centro de Investigacion y Tecnologia Agroalimentaria de Aragon (Zaragoza, Spain). She has experience in Animal Production and Meat and Meat Products Science, mainly researching the following themes: production factors that affect the quality of the carcass and meat, product classification and diversification, useful life of meat and meat products, meat products, the sensorial analysis of food, and studies with consumers.

 *animals*

Editorial

A New Insight on Carcass and Meat Quality in Ruminants

Guillermo Ripoll [1,2,]* and **Begoña Panea** [1,2]

1   Animal Science Department, Centro de Investigación y Tecnología Agroalimentaria de Aragón (CITA), Avda. Montañana 930, 50059 Zaragoza, Spain
2   Agrifood Institute of Aragon-IA2 (CITA-Zaragoza University), Avda. Miguel Servet 177, 50013 Zaragoza, Spain
*   Correspondence: gripoll@aragon.es

1. Introduction

Ruminant production systems are very important in many areas of the world and a key aspect of the economy and culture. Ruminants play an important role in low-income areas because they can graze marginal pastures that other species cannot use, or even are agricultural high natural value systems [1]. Moreover, these systems fix the population to unfavorable rural areas and even can prevent forest fires. Although meat consumption is increasing worldwide [2], the production of meat from ruminants is decreasing together with the number of farms, especially in Europe [3]. Production systems are important because they directly affect carcass and meat composition and quality. The interactions between diet, the animal's physiological status, and the environment will impact the yield, composition, quality, and sensorial appeal of its meat products; so, the use of forages, novel feeds, industry by-products, and more sustainable production systems for holistic benefit could modify the value of meat products delivered for human consumption and this must be considered [4].

Food quality is a complex term that includes, in addition to safety, such intrinsic characteristics as appearance, color, texture, and flavor, which are modified by both pre- and post mortem factors. Meat has a short shelf life, and new insights into methods of preservation are gaining interest in preventing deterioration and ensuring the absence of foodborne microorganisms and pathogenic bacteria which lead to meat spoilage. These reactions lead to a loss of the nutritional and sensory qualities of the meat products. For this Special Issue, we are also interested in studies on any of these factors or preservation methods for improving the quality and shelf-life of meat. A quick search with the key words shows around 2500 papers from 2020 to now dealing with this topic. This reveals that carcass and meat quality in ruminants is still a hot topic. Novel strategies can be implemented in production, preparation, storage, and distribution systems to induce qualitative and quantitative changes in meat product composition and to optimize the beneficial properties for human health [5]. On the other hand, meat and meat products contain essential components of the human diet such as protein of high biological value, essential fatty acids, vitamins with high bioavailability, etc. Meat products are nowadays perceived as less healthy and less attractive by consumers, and this makes them more selective in the products they consume, as they are increasingly aware of improving their health through the foods they consume [6]. There is a new trend in the production of healthier meat products to satisfy consumer demands.

This Special Issue of *Animals*, entitled "Carcass and Meat Quality in Ruminants", aims to compile the recent literature with a focus on carcass development, quality, and valorization in addition to meat quality. It includes ten original research articles about various types of meat from ruminant species (bovine [7–11], ovine [12–14], and caprine [15,16]) as well as one review article about strategies to reduce *E. Coli* contamination [17]. These articles, while their aims are different, provide a deep insight on the current topics of carcass and meat science of ruminants.

Citation: Ripoll, G.; Panea, B. A New Insight on Carcass and Meat Quality in Ruminants. *Animals* **2022**, *12*, 3153. https://doi.org/10.3390/ani12223153

Received: 28 October 2022
Accepted: 11 November 2022
Published: 15 November 2022

Publisher's Note: MDPI stays neutral with regard to jurisdictional claims in published maps and institutional affiliations.

2. Summary of Published Papers

Indigenous veld goats (IVG) are a group of specific pure-breed indigenous eco-types represented by the IVG Association, which defines specific standards that a goat must adhere to. These eco-types are characterized by large frames and competitive meat yield, them being animals with disease resistance and adaptability to harsh climates. The paper of van Wyk, Hoffman, Strydom, and Frylinck [15] focuses on the effect of breed and castration on the meat quality of six different muscles to establish quality baselines for IVG eco-types. Various meat quality characteristics of six muscles from large-frame Boer goats and indigenous veld goats were studied. The animals were raised on hay and natural grass, and on a commercial pelleted diet to a live weight of 30–35 kg. All goats were slaughtered at a commercial abattoir and the dressed carcasses were chilled at 4 °C within 1 h post mortem. The muscles were dissected from both sides 24 h post mortem and aged for 1 d and 4 d. Variations in meat characteristics such as ultimate pH, water-holding capacity, drip loss, myofibril fragment length, intramuscular fat, connective tissue characteristics, and Warner–Bratzler shear force were recorded across muscles. Bucks had higher lightness and hue-angle values, whereas wethers had increased redness and chroma values. This study alleviates some misconceptions that exist about the potential quality of "indigenous" goat meat. More muscle meat quality differences were found between sexes than between breeds, while large-frame IVGs consisted of a mixture of the different goat eco-types. In addition, the study further showed that goat muscles have different characteristics from those of other red-meat animals and the muscle baseline data will allow informed decisions to support muscle-specific marketing strategies, which may be used to improve consumer acceptability of chevon.

The aim of the study by Sánchez, Marti, Verdú, González, Font-i-Furnols, and Devant [11] was to characterize three different commercial beef-fattening systems. The fattening systems were intensive Mediterranean fattening programs with different sex, breed, nutrition, and days on feed. Fattening systems were described according to their performance, behavior, and carcass and meat quality when raised simultaneously under the same housing and care conditions. In the authors' words, data generated from this study are the first step for decision making and offer technical information to consider whether raising crossbred Angus bulls can be a good alternative to Holstein bulls in a Mediterranean dairy beef-fattening system. Authors did not find great differences in efficiency, intramuscular fat, or meat tenderness among the three Mediterranean production systems evaluated. As a relevant finding, it was reported that the purchase decision was indicative of an unforeseen impairment in meat quality. In summary, according to the present study, a transition from a production system based on Holstein bulls to crossbred Angus is only reasonable to improve carcass conformation, and only marketing approaches for meat distinction could strengthen this decision.

Mohd Azmi, Mat Amin, Ahmad, Mohd Nor, Meng, Zamri Saad, Abu Bakar, Abdullah, Irawan, Jayanegara, and Abu Hassim [9] examined the effect that a mixture of 4% bypass fat and 26% concentrate supplementations in the buffalo basal diet had on both the carcass characteristics and the proximate and fatty acid composition in three muscles of Murrah cross and swamp buffaloes. Additionally, they studied the profitability of raising buffaloes. The results showed that supplemented bypass fat significantly increased the pre-slaughter weight, hot and cold carcass weights, meat-to-fat ratio, pH, moisture, and crude protein, while the carcass yield and carcass fat percentages were significantly decreased. Furthermore, Murrah cross showed a significantly higher pre-slaughter weight, hot and cold carcass weights, carcass bone percentage, and total fatty acid, but a lower meat-to-bone ratio when compared to swamp buffaloes. Supplementing using bypass fat increased the cost of buffalo feeding but resulted in a higher revenue and net profit. In conclusion, the concentrate and bypass fat supplementations in the buffalo diet could alter the nutrient compositions of buffalo meat without a detrimental effect on carcass characteristics, leading to a higher profit.

The study performed by Bharanidharan, Thirugnanasambantham, Ibidhi, Bang, Jang, Baek, Kim, and Moon [8] focused on the influence of dietary protein level on growth performance, fatty acid composition, and the expression of lipid metabolic genes in intramuscular adipose tissues from 18- to 23-month-old Hanwoo steers, representing the switching point of the lean-to-fat ratio. It was observed that the high-protein diet significantly increased the expression of intramuscular *PPARα* and *LPL* while it did not affect to genes involved in fatty acid uptake, such as *CD36* and *FABP4*, nor lipogenesis, such as *ACACA*, *FASN*, and *SCD*. In addition, it downregulated intramuscular *VLCAD* related to lipogenesis but also *GPAT1*, *DGAT2*, and *SNAP23*, which are involved in fatty acid esterification and adipocyte size. Hanwoo steers fed a high-protein diet at 18–23 months of age resulted in a relatively lower lipid turnover rate than steers fed a low-protein diet, which could be responsible for shortening the feeding period. These results showed a low lipid turnover rate, which could be responsible for shortening the feeding period. Furthermore, Hanwoo steers fed a high-protein diet during this period showed increased intramuscular fatty acid content, oleic acid, and fineness in the marbling texture during later life by downregulating *SNAP23*.

Because aromatic plant distillation residues are being considered with growing interest in a two-fold object, enhancing meat quality by increasing the antioxidant properties and reducing feed prices, Yagoubi, Smeti, Ben Said, Srihi, Mekki, Mahouachi, and Atti [12] studied the effects of rosemary distillation residue incorporation in concentrate associated with two nitrogen sources as a substitute for standard concentrate on lambs' growth, carcass traits, and meat quality. Growth, carcass weights, dressing percentages, and non-carcass component weights were unaffected by the diet. Moreover, regional and tissular compositions and meat physical properties including color were similar irrespective of the diet. However, meat produced by lambs receiving rosemary distillation residue-based concentrate was richer in vitamin E and polyphenol contents than the control lambs. Rosemary by-products may substitute the standard concentrate resulting in similar lamb growth and carcass traits, while improving meat quality by increasing vitamin E content, which could improve its antioxidant power. The results provide evidence that the use of rosemary residues as a cereal substitute up to 30% in concentrate for sheep feeding did not alter animal performances. This smart strategy of using aromatic plant by-products could be effective especially in the Mediterranean region, where this by-product is available in an plentiful amount and is free. The cost per kilogram of meat produced by Barbarine lambs was reduced to 40%. In addition, faba bean (*Vicia Faba*) could be used as a substitute to soybean without affecting carcass nor meat quality; this nitrogen source could potentially be produced, given its production is relatively cheap compared to the nutritional value, to reduce the import of soybean meal, which is still expensive.

Sainfoin is a forage legume with a medium content of proanthocyanidins (PAC), which may affect animal performance and product quality. Therefore, Baila, Lobon, Blanco, Casasus, Ripoll, and Joy [13] studied the effect of PAC from sainfoin fed to dams, using polyethylene glycol as a blocking agent, on the performance and carcass and meat quality of their suckling male lambs. They found that the presence of PAC in the dams' diet did not affect the growth, blood metabolites, and carcass weight and fatness of the suckling lambs but decreased the lightness of caudal fat and increased the weight of the digestive compartments. Regarding the meat characteristics, PAC only decreased polyphenol content. The inclusion of PAC from sainfoin in the dams' diet had no significant effect on the ADG, plasmatic antioxidant activity, and carcass and meat quality of their suckling lambs. Therefore, fresh sainfoin can be fed to ewes during lactation to produce suckling lambs, achieving good performance and meat quality.

The use of pea has been recommended to replace soybean meal in the diet of ruminants, but it may affect meat quality. However, the title of the study of Blanco, Ripoll, Lobon, Bertolin, Casasus, and Joy [14] (The Inclusion of Pea in Concentrates Had Minor Effects on the Meat Quality of Light Lambs) speaks for itself. The aim of this study was to evaluate the effect of the proportion of pea (0%, 10%, 20%, and 30%) in fattening concentrates fed to light lambs for 41 days on carcass color and meat quality. Pea inclusion affected neither the color of the lamb carcasses nor affected most of the parameters of the meat quality. However, the inclusion of pea affected the cholesterol content, and the 20% pea concentrate yielded meat with greater cholesterol contents than the 30% pea concentrate did. The inclusion of pea had minor effects on individual FAs but affected the total saturated fatty acids ($p < 0.01$) and the thrombogenicity index. A greater total saturated fatty acid content was recorded for the 20% pea concentrate than for the rest of the concentrates, and a greater thrombogenicity index was recorded for the 20% concentrate than for the 10% pea concentrate. The results indicated the viability of the inclusion of pea in the fattening concentrate of light lambs without impairing meat quality, with the 30% pea concentrate being the most suitable to reduce the soya dependency.

Since goat milk has a higher value than kid meat in Europe, some farmers rear kids with milk replacers, although some studies have stated that kids raised on natural milk yield higher-quality carcasses. With the aim of enlightening this topic, Ripoll, Alcalde, Argüello, Córdoba, and Panea [16] evaluated the influence of the use of milk replacers on several carcass characteristics of suckling kids from eight Spanish goat breeds. For all studied variables, interactions were found between the rearing system and the breed. In general, the milk replacer increased the head and visceral weights, as well as the length measurements and muscle percentages. Conversely, the natural milk-rearing system increased carcass compactness and resulted in higher fat contents, independent of the deposit. The choice of one or another rearing system should be made according to the needs of the target market.

The amount and distribution of subcutaneous fat are important factors affecting beef carcass quality. The degree of fatness is determined by visual assessments scored on the SEUROP system. New technologies such as the image analysis method have been developed and applied to enhance the accuracy and objectivity of this classification system. In the study by Mendizabal, Ripoll, Urrutia, Insausti, Soret, and Arana [7], 50 young bulls were slaughtered and after slaughter, the carcasses were weighed and a SEUROP system fatness score assigned. A digital picture of the outer surface of the left side of the carcass was taken and the area of fat cover (fat area) was measured using an image analysis system. Commercial cutting of the carcasses was performed 24 h post mortem. The fat trimmed away on cutting (cutting fat) was weighed. A regression analysis was carried out for the carcass cutting fat on the carcass fat area to establish the accuracy of the image analysis system. A greater accuracy was obtained by the image analysis ($R^2 = 0.72$; $p < 0.001$) than from the visual fatness scores ($R^2 = 0.66$; $p < 0.001$). The findings of this study suggest that measuring carcass fat area using an image analysis can be regarded as a suitable indicator of carcass fatness in young bulls of Spanish meat breeds. Furthermore, including this assessment method in the framework of the EU's SEUROP classification system could be worthwhile because it provides an objective measure of carcass fatness. Nevertheless, before applying an image analysis to other breeds or production systems, the method should be tested on the carcasses of fatter animals spanning the broadest possible range of fatness scores and, if it is feasible, spanning the entire interval from 1 to 5.

In autochthonous dairy cattle farms, the production of salami could represent an alternative commercial opportunity. However, the diet of animals can modify the quality of meat and meat products. Therefore, a study was carried out by Alabiso, Maniaci, Giosue, Di Grigoli, and Bonanno [10] to investigate the fatty acid composition of salami made using the meat from grazing or housed young bulls and grazing adult cows of the Cinisara breed. Animal category influenced the FA composition, although the addition of lard mitigated the differences found in fresh meat. The salami from grazing animals showed higher polyunsaturated fatty acid content and a higher level of linoleic acid than that from other animal categories. Salami made from grazing adult cows' meat showed a lower polyunsaturated/saturated fatty acid ratio, but a better *n-6/n-3* ratio compared to housed bulls due to the lower content of linoleic acid. Multivariate analysis showed an important influence of animal category on fatty acid composition due to age, feeding system, and meat fat content of animals, despite the addition of lard.

Finally, closing the Special Issue is a review entitled "Preharvest Management and Postharvest Intervention Strategies to Reduce Escherichia coli Contamination in Goat Meat: A Review by Kannan, Mahapatra and Degala [17]. While researchers have long focused on postharvest intervention strategies to control *E. coli* outbreaks, recent works have also included preharvest methodologies. In goats, these include minimizing animal stress, manipulating the diet a few weeks prior to processing, feeding diets high in tannins, controlling feed deprivation times while preparing for processing, and spray washing goats prior to slaughter. The postharvest intervention methods studied in small ruminant meats have included spray washing using water, organic acids, ozonated water, and electrolyzed water, and the use of ultraviolet (UV) light, pulsed UV-light, sonication, low-voltage electricity, organic oils, and hurdle technologies. These methods show strong antimicrobial activity and are considered environmentally friendly. However, cost-effectiveness, ease of application, and possible negative effects on meat quality characteristics must be carefully considered before adopting any intervention strategy for a given meat-processing operation. As discussed in this review paper, novel pre- and postharvest intervention methods show significant potential for future applications in goat farms and processing plants.

3. Conclusions

This Special Issue on the theme of carcass and meat quality in ruminants has attracted the interest of authors from all over the world, publishing one review and ten original research papers.

Progress has been made in the topic investigated, but it is still necessary to increase the research on it due to the extensiveness of this field and the peculiarities and special problems of the different species of ruminants.

Finally, I would like to conclude that the manuscripts found in this Special Issue have been submitted by internationally recognized research teams. I would like to thank all authors for their contributions.

Author Contributions: Conceptualization, G.R. and B.P.; methodology G.R. and B.P.; writing—original draft preparation, G.R. and B.P.; writing—review and editing, G.R. and B.P. All authors have read and agreed to the published version of the manuscript.

Funding: This research received no external funding.

Conflicts of Interest: The authors declare no conflict of interest.

References

1. EEA: European Environment Agency. *High Nature Value Farmland: Characteristics, Trends and Policy Challenges*; European Environment Agency, European Communities: Copenhagen, Denmark, 2004.
2. Schumann, B.; Schmid, M. Packaging concepts for fresh and processed meat—Recent progresses. *Innov. Food Sci. Emerg. Technol.* **2018**, *47*, 88–100. [CrossRef]
3. Bernués, A.; Ruiz, R.; Olaizola, A.; Villalba, D.; Casasús, I. Sustainability of pasture-based livestock farming systems in the European Mediterranean context: Synergies and trade-offs. *Livest. Sci.* **2011**, *139*, 44–57. [CrossRef]
4. Holman, B.W.B.; Ponnampalam, E.N. Meat Products: From Animal (Farm) to Meal (Fork). *Foods* **2022**, *11*, 933. [CrossRef] [PubMed]
5. Pateiro, M.; Domínguez, R.; Lorenzo, J.M. Recent Research Advances in Meat Products. *Foods* **2021**, *10*, 1303. [CrossRef] [PubMed]
6. Ruiz-Capillas, C.; Herrero, A.M. Novel Strategies for the Development of Healthier Meat and Meat Products and Determination of Their Quality Characteristics. *Foods* **2021**, *10*, 2578. [CrossRef] [PubMed]
7. Mendizabal, J.A.; Ripoll, G.; Urrutia, O.; Insausti, K.; Soret, B.; Arana, A. Predicting Beef Carcass Fatness Using an Image Analysis System. *Animals* **2021**, *11*, 2897. [CrossRef] [PubMed]
8. Bharanidharan, R.; Thirugnanasambantham, K.; Ibidhi, R.; Bang, G.; Jang, S.S.; Baek, Y.C.; Kim, K.H.; Moon, Y.H. Effects of Dietary Protein Concentration on Lipid Metabolism Gene Expression and Fatty Acid Composition in 18–23-Month-Old Hanwoo Steers. *Animals* **2021**, *11*, 3378. [CrossRef] [PubMed]
9. Azmi, A.F.M.; Amin, F.M.; Ahmad, H.; Nor, N.M.; Meng, G.Y.; Saad, M.Z.; Bakar, M.Z.A.; Abdullah, P.; Irawan, A.; Jayanegara, A.; et al. Effects of Bypass Fat on Buffalo Carcass Characteristics, Meat Nutrient Contents and Profitability. *Animals* **2021**, *11*, 3042. [CrossRef] [PubMed]

10. Alabiso, M.; Maniaci, G.; Giosue, C.; Di Grigoli, A.; Bonanno, A. Fatty Acid Composition of Salami Made by Meat from Different Commercial Categories of Indigenous Dairy Cattle. *Animals* **2021**, *11*, 1060. [CrossRef] [PubMed]

11. Sánchez, D.; Marti, S.; Verdú, M.; González, J.; Font-i-Furnols, M.; Devant, M. Characterization of Three Different Mediterranean Beef Fattening Systems: Performance, Behavior, and Carcass and Meat Quality. *Animals* **2022**, *12*, 1960. [CrossRef] [PubMed]

12. Yagoubi, Y.; Smeti, S.; Ben Said, S.; Srihi, H.; Mekki, I.; Mahouachi, M.; Atti, N. Carcass Traits and Meat Quality of Fat-Tailed Lambs Fed Rosemary Residues as a Part of Concentrate. *Animals* **2021**, *11*, 655. [CrossRef] [PubMed]

13. Baila, C.; Lobon, S.; Blanco, M.; Casasus, I.; Ripoll, G.; Joy, M. Sainfoin in the Dams' Diet as a Source of Proanthocyanidins: Effect on the Growth, Carcass and Meat Quality of Their Suckling Lambs. *Animals* **2022**, *12*, 408. [CrossRef] [PubMed]

14. Blanco, M.; Ripoll, G.; Lobon, S.; Bertolin, J.R.; Casasus, I.; Joy, M. The Inclusion of Pea in Concentrates Had Minor Effects on the Meat Quality of Light Lambs. *Animals* **2021**, *11*, 2385. [CrossRef] [PubMed]

15. Van Wyk, G.L.; Hoffman, L.C.; Strydom, P.E.; Frylinck, L. Differences in Meat Quality of Six Muscles Obtained from Southern African Large-Frame Indigenous Veld Goat and Boer Goat Wethers and Bucks. *Animals* **2022**, *12*, 382. [CrossRef] [PubMed]

16. Ripoll, G.; Alcalde, M.J.; Argüello, A.; Córdoba, M.G.; Panea, B. Influence of the use of milk replacers on carcass characteristics of suckling kids from eight spanish goat breeds. *Animals* **2021**, *11*, 3300. [CrossRef] [PubMed]

17. Kannan, G.; Mahapatra, A.K.; Degala, H.L. Preharvest Management and Postharvest Intervention Strategies to Reduce Escherichia coli Contamination in Goat Meat: A Review. *Animals* **2021**, *11*, 2943. [CrossRef] [PubMed]

 *animals*

Review

Preharvest Management and Postharvest Intervention Strategies to Reduce *Escherichia coli* Contamination in Goat Meat: A Review

Govind Kannan *, Ajit K. Mahapatra and Hema L. Degala

Agricultural Research Station, College of Agriculture, Family Sciences and Technology, Fort Valley State University, Fort Valley, GA 31030, USA; mahapatraa@fvsu.edu (A.K.M.); degalah@fvsu.edu (H.L.D.)
* Correspondence: govindak@fvsu.edu; Tel.: +1-478-825-4613

Simple Summary: Goat farms and processing facilities worldwide are primarily small-scale, limited resource operations. Cost-effectiveness and practicality are critical factors to be considered before adopting any pre- and/or post-harvest strategies for pathogen reduction in goat meat. Preharvest management methods in goats that can reduce *Escherichia coli* in meat include minimizing animal stress, selecting diets and feed deprivation times that can reduce fecal shedding of bacteria, and adding tannin-rich feed supplements. In addition, use of appropriate postharvest nonthermal intervention technologies that can reduce microbial loads in carcasses and meat can extend the shelf-life and marketability of goat meat products. Reducing stress prior to slaughter and using nonthermal intervention methods can result in better meat quality and economic returns for producers.

Citation: Kannan, G.; Mahapatra, A.K.; Degala, H.L. Preharvest Management and Postharvest Intervention Strategies to Reduce *Escherichia coli* Contamination in Goat Meat: A Review. *Animals* **2021**, *11*, 2943. https://doi.org/10.3390/ani11102943

Academic Editors: Guillermo Ripoll and Begoña Panea

Received: 8 August 2021
Accepted: 9 October 2021
Published: 12 October 2021

Publisher's Note: MDPI stays neutral with regard to jurisdictional claims in published maps and institutional affiliations.

Abstract: Goat meat is the main source of animal protein in developing countries, particularly in Asia and Africa. Goat meat consumption has also increased in the US in the recent years due to the growing ethnic population. The digestive tract of goat is a natural habitat for *Escherichia coli* organisms. While researchers have long focused on postharvest intervention strategies to control *E. coli* outbreaks, recent works have also included preharvest methodologies. In goats, these include minimizing animal stress, manipulating diet a few weeks prior to processing, feeding diets high in tannins, controlling feed deprivation times while preparing for processing, and spray washing goats prior to slaughter. Postharvest intervention methods studied in small ruminant meats have included spray washing using water, organic acids, ozonated water, and electrolyzed water, and the use of ultraviolet (UV) light, pulsed UV-light, sonication, low-voltage electricity, organic oils, and hurdle technologies. These intervention methods show a strong antimicrobial activity and are considered environmentally friendly. However, cost-effectiveness, ease of application, and possible negative effects on meat quality characteristics must be carefully considered before adopting any intervention strategy for a given meat processing operation. As discussed in this review paper, novel pre- and post-harvest intervention methods show significant potential for future applications in goat farms and processing plants.

Keywords: goat meat; food safety; *E. coli*; preharvest management; postharvest intervention

1. Introduction

Enterohemorrhagic *Escherichia coli* (*E. coli*) is considered one of the most economically important food-borne pathogens. Much of the research has focused on post-slaughter sanitation to improve the safety of meat products, and as a result, various strategies are practiced in meat plants to reduce carcass contamination. In recent years, researchers have also been working on developing intervention strategies in the live animal prior to slaughter to reduce foodborne pathogens. Because fecal shedding is correlated with carcass contamination, reducing bacterial loads in the gastrointestinal tracts of live animals is important in the production of safe and wholesome food products.

It is widely recognized that hygienic risks at slaughterhouses should be assessed in reference to the number of organisms indicative of fecal contamination [1]. The Agricultural Marketing Service of the US Department of Agriculture has established 500 CFU g^{-1} as a critical limit for generic *E. coli* in red meat such as beef [2]. Although *E. coli* organisms are normal inhabitants of the gastrointestinal tracts of ruminants, some pathogenic strains can cause hemorrhagic colitis in humans [3]. *Escherichia coli* O157:H7 is the most common enterohemorrhagic *E. coli* serotype implicated in many outbreaks of bloody diarrhea and the hemolytic–uremic syndrome resulting in kidney failure.

Goat meat is one of the most widely consumed meats in the world, especially in Asia and Africa, and importation of goat meat into the US has steadily increased mainly due to increased demand by ethnic consumers. However, research on intervention strategies to reduce pathogens in goat carcasses, cuts, and products is very limited. Data available on preslaughter intervention strategies to reduce foodborne pathogens in goats are scanty.

Dietary manipulation, feed deprivation duration prior to slaughter, supplementation with high tannin-containing diets, minimizing animal stress, and live animal washing have been studied as possible pre-harvest intervention strategies to reduce *E. coli* in goats. Research on postharvest methods in goats have included treating goat meat with ozonated water, electrolyzed oxidizing water, ultraviolet light, sonication, organic acids, and organic oils, to name a few. In addition, nonthermal hurdle technologies with different treatment-time combinations have also been found to be promising. The body of literature currently available on the various low-cost pre- and post-harvest intervention strategies to reduce *E. coli* populations in goats was reviewed in this work to benefit smallholder farmers and small-scale meat processors and retailers worldwide.

2. Prevalence of *E. coli* in Goats

Several countries from different continents have reported *E. coli* O157:H7 in humans. Although this organism has been isolated from several animal species, ruminant livestock species are regarded as natural carriers of *E. coli*. These food animals typically do not show any clinical signs while shedding *E. coli* through feces. Isolation of *E. coli* O157:H7 from goats was first reported in 1994 from a human outbreak of *E. coli* O157:H7 in United Kingdom [4].

Researchers have reported different prevalence rates in different countries. According to Dulo et al. [5], *E. coli* O157:H7 was present in 2.2% of cecal content samples and 3.2% of carcass swab samples obtained from goats in Somali region of Ethiopia. The researchers suggested that poor hygiene and slaughter practices may cause contamination of meat and human health risks as consumption of raw meat is a common practice in Ethiopia. A comparison of *E. coli* O157:H7 contamination rate among beef, lamb, and chevon samples from slaughter plants and retail outlets in Addis Ababa, Ethiopia, showed that beef was the most frequently contaminated meat, followed by sheep and goat meat [6]. This study also revealed that contamination rates were higher at retail shops than at slaughterhouses for beef (21.9 vs. 4.7%), sheep (10.9 vs. 6.3%), and goat (9.4 vs. 6.3%) carcass and meat samples. Sixty percent of goat meat samples collected from different markets in Dhaka, Bangladesh has been reported to be positive for *E. coli* O157:H7 strains, although no official infections have been reported either due to improper tracking of outbreaks and causative organisms or to possible acquired immunity in the population [7]. Another report from Bangladesh also showed that prevalence of multidrug-resistant *E. coli* was higher in younger goats than older goats. The authors also reported that the prevalence of drug-resistant *E. coli* was higher in goats raised in poor hygienic conditions than those raised in good hygienic conditions, and higher in goats with recent history of transportation [8]. Shiga toxin-producing *E. coli* prevalence in Jordan was found to be greater in intensively reared goats with occasional grazing (65%) than in extensively reared goats with year-round grazing (50%) [9].

Preharvest meat goat management, slaughter and processing, and postharvest carcass and meat handling methods vary among different countries and regions. The *E. coli* counts on goat skin prior to slaughter generally range from 2.2 to 2.5 $\log_{10}$ CFU cm^{-2}, and those on carcasses before washing range from 2.1 to 2.3 log10 CFU cm^{-2} [10,11]. It is not clear to what extent factors such as sex, age, and breed can influence *E. coli* shedding in goats. A generic flow diagram of various steps involved in pre- and post-harvest phases of chevon production is included to illustrate potential intervention points (Figure 1).

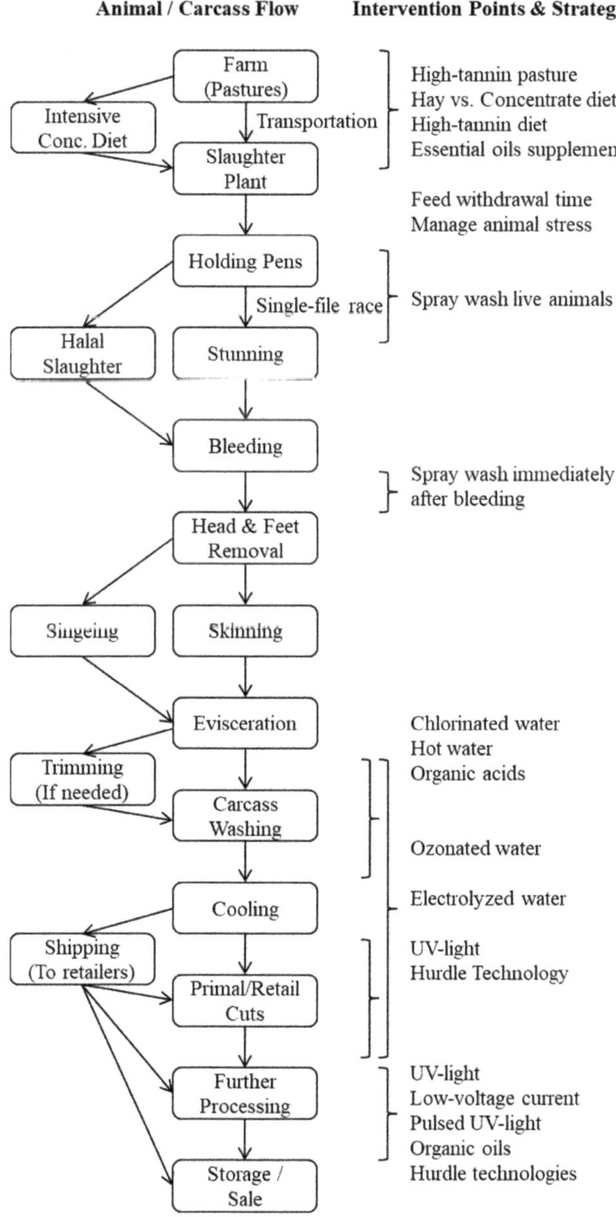

Figure 1. A generic flow diagram of various steps involved in pre- and post-harvest phases of goat meat production to illustrate potential intervention points [12–34].

3. Pre-Harvest Intervention

Studies involving preharvest intervention to reduce *E. coli* population in goats are very limited (Table 1). Before recommending any management method as a viable means of controlling potential transfer of organisms from the gastrointestinal tract to the skin, carcass, and meat, it is important to consider how an intervention strategy can affect animal welfare, productivity, and product quality aspects [35]. Since most goat meat producers around the world are smallholder farmers, it is also important to factor in the economic implications for farmers.

Table 1. Reduction in microbial counts resulting from preharvest intervention strategies in goats.

Intervention Stage	Intervention Method	Sample Type	Microorganism	Reduction	Goat Breed	Reference
Holding pens	Spray washing live goats in single file race prior to processing with potable water (0.4–0.8 ppm chlorine; approx. 12 L of water at 15–18 °C per animal) for 1 min.	Skin swab samples	*Generic E. coli* *Aerobic plate count*	No change $0.8 \log_{10}$ CFU cm^{-2}	Spanish	[10]
Farm/holding pens	Feed deprivation for 27 h prior to processing (compared to no feed deprivation).	Carcass swab samples	*Generic E. coli* *Aerobic plate count*	$0.8 \log_{10}$ CFU cm^{-2} $1.0 \log_{10}$ CFU cm^{-2}	Boer × Spanish	[36]
Farm/holding pens	Feed deprivation (in general)	-	-	Reduces gut fill and fecal contamination of carcasses	-	[21]
Farm/holding pens	Feed deprivation for 24 h prior to processing (compared to 12 h feed deprivation)	Carcass swab samples	*Generic E. coli* *Aerobic plate count*	No change $0.5 \log_{10}$ CFU cm^{-2}	Kiko × Spanish	[11]
Farm	Feeding hay diet for 4 days prior to harvesting (compared to concentrate feeding).	Rectal samples	*Generic E. coli* *Total coliform* *Enterobacteriaceae*	$2.4 \log_{10}$ CFU g^{-1} $2.6 \log_{10}$ CFU g^{-1} $3.1 \log_{10}$ CFU g^{-1}	Kiko × Spanish	[13]
Farm	Feeding alfalfa hay diet for 90 days prior to processing (compared to concentrate feeding).	Rectal samples	*Generic E. coli*	$1.8 \log_{10}$ CFU g^{-1}	Boer × Spanish	[37]
Farm	Feeding ground sericea lespedeza (*Lespedeza cuneata*) for 14 weeks.	Fecal samples	*Generic E. coli* *Total plate count*	No change $1.6 \log_{10}$ CFU g^{-1}	Kiko × Spanish	[38]
Farm	Feeding brown seaweed (*Ascophyllum nodosum*) supplement for 14 days.	Rumen samples	*Generic E. coli*	$1.4 \log_{10}$ CFU g^{-1}	Boer × Spanish	[16]
During processing	Spray washing skin for 1 min. using potable water followed by 1 min. with chlorinated water (50 mg L^{-1} hypochlorite) immediately after bleeding.	Skin swab samples	*Aerobic plate count*	$1.0 \log_{10}$ CFU cm^{-2}	Boer × Spanish	[16]

3.1. Meat Goat Management and Productivity

Small ruminant management practices that increase animal stress may alter the normal course of conversion of muscle to meat, leading to inferior meat quality [39]. Chronic stress in animals can negatively affect animal growth, weight gain, and immune function, which in turn can lead to poorer health, carcass yield, and economic returns [40]. Animal management during the 24-h period prior to slaughter is crucial not only for animal welfare reasons, but also for profitability. Poor preslaughter handling practices in goats can increase dehydration and live weight shrinkage and decrease carcass yield and meat quality [21], in addition to increasing the chances of fecal contamination of skin and carcasses [41]. Preparation of goats for slaughter generally involves loading onto a trailer, transportation, unloading, feed deprivation, exposure to novel environments, noise and vibration, disruption of social groups, and changes in temperature and humidity, all of which will have to be carefully considered and managed appropriately to minimize negative effects [40,42].

Some researchers who evaluated preharvest intervention methods to reduce fecal contamination have also evaluated other aspects of economic importance. For example, Kannan et al. [10] reported that skin bacterial counts can be significantly reduced in goats by preslaughter spray washing without increasing animal stress.

3.2. Dietary Regimens and E. coli Populations

A considerable amount of data is available on the effect of diet on fecal shedding of generic *E. coli* in cattle and sheep; however, studies on goats are very limited. Finishing beef cattle and lambs in the US are often fed grain rations to improve productivity. Meat goats are raised primarily on pastures with grain supplements in some cases. Dietary starch is protected from ruminal microbial degradation by a protein called zein [3]. Because there is less pancreatic amylase activity in the small intestines of ruminants, much of the dietary starch reaches the cecum and colon, where it undergoes a secondary microbial fermentation [43]. Fermentation of starch by bacteria, including *E. coli*, in the cecum and colon produce volatile fatty acids that could reduce the pH of the colonic digesta and inhibit *E. coli* [3]. According to Gutta et al. [13], the pH values of the rumen contents of hay-fed small ruminant animals were higher (7.08) than concentrate-fed animals (6.43). A similar effect was also noticed in colon contents, with pH values of 7.02 and 6.56 in hay-fed and concentrate-fed animals, respectively. Regardless of these unfavorable conditions, research shows *E. coli* grows in the intestinal tract of cattle fed high-grain rations. A suggested explanation for higher *E. coli* counts despite lower pH is the fermentation of the easily assimilable carbohydrate portion of concentrate diets in the rumen, so that substances that facilitate microbial growth likely become easily available. Moreover, Gutta et al. [13] observed that in goats and sheep, *E. coli*, coliform, *Enterobacteriaceae*, and total plate counts were more associated with pH in the colon than in the rumen. The authors observed a negative correlation tendency between *E. coli* numbers and colon pH. A similar effect was reported by Scott et al. [44] in cattle, which indicates that fermentation of starch in the colon produces nutrients necessary for both bacterial growth and pH decline. Acid resistance in *E. coli* organisms in the colon is greater when animals are fed a concentrate diet rather than a hay diet [45]. Russell et al. [46] speculated that by modifying the concentration of undissociated volatile fatty acids (VFA), colonic pH may play a role in regulating the resistance of *E. coli* to low pH values. It is not clear if the negative correlation tendency between *E. coli* counts and colon pH observed by Gutta et al. [13] was due to the acid resistance of organisms in the colon. The site of persistence of *E. coli* in adult ruminant animals appears to be the colon [47].

Shedding of *E. coli* by ruminants is influenced by the diet [48]. There are conflicting reports on the effects of forage and concentrate diets on fecal shedding and colonization by *E. coli* in the gastrointestinal tract of ruminants. Gutta et al. [13] reported that concentrate-fed goats and sheep had higher *E. coli* (6.44 vs. 4.01 ± 0.468 $\log_{10}$ CFU g^{-1}), total coliform (6.74 vs. 4.16 ± 0.469 $\log_{10}$ CFU g^{-1}), *Enterobacteriaceae* (6.93 vs. 3.83 ± 0.651 $\log_{10}$ CFU g^{-1}), and total plate counts (7.79 vs. 7.28 ± 0.170 $\log_{10}$ CFU g^{-1}) in the rectum than the hay-fed animals. The authors suggested that microbial loads in the gastrointestinal tract of goats and sheep can be reduced by feeding hay for four days before slaughter, although diet did not have any effect on skin bacterial contamination in goats and sheep [11]. To determine the effect of diet on gastrointestinal tract bacterial populations in goats, Lee et al. [37] conducted a feeding trial involving three dietary treatments: 90-day alfalfa hay alone, 90-day concentrate alone (18% crude protein), or 45-day alfalfa hay diet followed by 45-day concentrate. The authors found that diet did not have any effect on rumen *E. coli* population; however, diet had a significant effect on rectal *E. coli* counts with the population being lower in hay-fed goats compared with concentrate-fed goats.

A change in ruminant diet from concentrate to hay in the days before slaughter has been shown to reduce fecal shedding of bacteria, although this effect was also not consistent among studies. Shifting cattle from a high grain (90% corn/soybean) diet to 100% timothy hay resulted in a significant reduction of generic *E. coli* in feces [45]. Gregory et al. [14] reported that switching cattle from pasture to hay 48 h prior to slaughter significantly reduced the *E. coli* burden throughout the gut. The authors further speculated that the increased intestinal *Enterococci* populations due to hay feeding can inhibit *E. coli* populations. However, a diet change from alfalfa pellet to poor quality forage has been shown to increase *E. coli* O157:H7 fecal shedding in experimentally infected sheep [49]. Sheep shifted from a 50:50 corn/alfalfa ration to low quality hay shed greater populations of *E. coli* O157:H7 than animals fed exclusively on the corn/alfalfa ration [48]. These discrepancies in results from different studies on diet switching from concentrate to hay reveal that other factors such as breed, age, season, feed quality, and consumption could also be involved in determining *E. coli* shedding in ruminants.

3.3. Feed Withdrawal and E. coli Populations

It is a common practice to withhold feed in meat animals during periods of preslaughter transportation and holding, to minimize the difficulties of handling overfilled guts during evisceration and incidences of gastrointestinal tract rupture and soiling of hide/skin. Since the primary sources of carcass contamination with enterogenic pathogens are the hide and gastrointestinal tract, feed withdrawal may be an important step in reducing the bacterial load of feces excreted during the preslaughter period.

Conflicting reports have been published on the effects of feed deprivation on fecal shedding of *E. coli*. Gutta et al. [13] reported that feed deprivation for 24 h increased *E. coli*, total coliform counts, and *Enterobacteriaceae* counts in the rumen of Kiko × Spanish goats and Dorset × Suffolk sheep, compared with 12 h deprivation and with no significant change in the pH of rumen liquor. However, no difference in *E. coli* or total coliform counts were observed on sheep and goat (Kiko × Spanish) carcasses as a result of feed deprivation time [11]. In contrast, experimentally infected adult sheep showed no increase in fecal shedding of *E. coli* O157:H7 during feed withdrawal [48]. According to Vanguru et al. [36], Boer × Spanish goats subjected to either 0, 9, 18, or 27 h of feed deprivation prior to slaughter showed that the 27-h feed deprivation group had higher rumen pH (6.95) than those at 0 h (6.23) or 9 h (6.46) feed deprivation, although there were no differences in the microbial counts of rumen or fecal samples among the groups. The authors concluded that feed deprivation time alone up to 27 h may not significantly influence gut, skin, or carcass microbial loads.

Low pH is not favorable for growth of naturally occurring *E. coli*; however, the organisms can survive low pH and start growing again if the pH becomes favorable [46,50]. The effect of fasting on *E. coli* growth in the ruminant digestive tract is mediated by a reduction of VFA concentration and the subsequent increase in the pH of the digestive tract contents [51]. *Escherichia coli* grew best in rumen liquor in vitro when the concentration of VFA was less than 25 mM and the pH was 7.2, and the growth of *E. coli* was completely arrested when VFA concentration was greater than 75 mM. Also, a linear decrease in *E. coli* growth as pH declined, and zero growth at a pH 6.0, were observed [51].

The inconsistencies in results from different studies on the effects of diet and feed deprivation on gut *E. coli* populations and fecal shedding could be due to various other factors such as environment, season, breed, age, previous feeding regime, and bacterial strain [41]. Bacterial shedding by ruminants is very complex and can be affected by various interrelated factors, such as extent of attachment to intestinal mucosa, ease of detachment from mucosa, distribution and growth in solid-liquid phases of gut contents, and passage rates through different parts of the gastrointestinal tract [52,53]. The type of feed can affect the consistency of gastrointestinal tract contents and passage rate. Grazing on young pasture often results in diarrhea in ruminants, and the abbreviated stay of fecal contents in the colon also prevents reabsorption of water, which could result in purging and potentially dirty hindquarters in ruminants [21].

3.4. Tannin-Containing Feed Supplements

A preharvest diet containing brown seaweed (*Ascophyllum nodosum*) extract has been reported to reduce fecal shedding of *E. coli* and other enteric bacteria. Supplementation of brown seaweed extract in a conventional grain diet reduced *E. coli* counts both in fecal samples and hide swipes in feedlot Angus steers [15]. Enterohemorrhagic *E. coli* and *Salmonella spp.* populations were significantly reduced by brown seaweed extract supplementation in feedlot steers [54]. A brown seaweed extract-containing diet has been shown to result in enhanced antioxidant status and immune function in live animals, and improved meat quality characteristics and food safety [16,55–57].

Among the several biologically active compounds present in *A. nodosum*, phlorotannins, a group of polyphenols, have been shown to possess marked antibacterial activity. The mode of action has not been understood fully yet, although there is evidence that the phlorotannins could be at least partially responsible for its antibacterial property [58]. The authors reported that phlorotannins isolated from *A. nodosum* had significant antimicrobial activity that affected several rumen bacteria. The antibacterial effects of terrestrial tannins are attributed to several mechanisms, including inhibition of oxidative phosphorylation and extracellular microbial enzymes, dysfunction of cell membranes, and deprivation of substrate metal ions and minerals [59]. Since the mechanism involved in the antibacterial activity of phlorotannins is not fully elucidated, Wang et al. [58] speculated that the antimicrobial mechanisms could be similar in both phlorotannins and terrestrial tannins. In addition, the authors demonstrated, using transmission electron microscopy, that phlorotannins can inhibit bacterial growth by interfering in the cell membrane functions. There is also evidence that phlorotannins possess more potent antibacterial activities compared with hydrolysable terrestrial tannins or condensed tannins, probably due to the greater number of hydroxyl groups [58] and the degree of phloroglucinol polymerization [60] in phlorotannins.

Lee et al. [38] fed Kiko × Spanish goats either ground sericea lespedeza (SL; *Lespedeza cuneata* (Dum-Cours) G. Don) or bermudagrass (BG) hay at 75% of daily intake with a corn-based supplement (25%) for 14 weeks before slaughter to see the effect on fecal *E. coli* shedding. They reported that high dietary condensed tannins in SL increased *E. coli* and total coliform populations in the rumen without affecting the populations in the rectum. The sericea hay-fed goats had higher rumen pH due to lower production of total VFA, a condition that favored *E. coli* and total coliform growth in the rumen, although such an effect was absent in the rectum. However, the authors found that the total plate count in the rectum samples from goats fed sericea was lower than that from goats fed bermudagrass hay. In another study, Mechineni et al. [12] fed Spanish goats either BG, SL, or a combined SL + BG diet for eight weeks. Half of the goats from each paddock were subjected to a 3 h transportation stress. The results indicated that dietary treatment did not affect gastrointestinal tract, skin, and carcass microbial populations or meat quality. Transportation stress also had no significant effect on gastrointestinal tract, skin, and carcass microbial populations or on meat quality.

3.5. Essential Oil-Containing Feed Supplements

Dietary supplementation of essential oils derived from plants such as thyme, rosemary, and sage have been evaluated in ruminants for their antimicrobial and rumen modifying activities. These compounds are secondary metabolites that often contain terpenoids and phenylpropanoids [61].

The antibacterial effects of essential oils are primarily due to their hydrophobic properties, which enable them to disrupt the bacterial cell wall and mitochondria by partitioning lipids. This results in changes in membrane integrity, ion transport processes, and cell osmotic pressure [62]. Although essential oils are more effective against gram-positive bacteria, low molecular weight molecules of essential oils can get through the cell wall of gram-negative bacteria via diffusion through membrane proteins, thus disrupting the membrane integrity [63]. The bacterial cells can offset these effects using ionic pumps; however, the high energy cost involved results in the slowing down of ATP synthesis and microbial growth, and the eventual death of bacteria cells [64]. In addition to bacterial cell membrane disruption, essential oils have also been reported to coagulate cell contents by protein denaturation [65], interact with proteins through hydrophobic interactions [66] and through other functional groups [66,67], and inhibit enzymatic activity [68].

The antimicrobial properties of the herb thyme (*Thymus vulgaris*) are well researched. In vitro studies have shown that the essential oils of thyme inhibit *E. coli* growth [17] and modify rumen VFA concentrations [69]. The primary constituents of thyme essential oils are thymol and carvacrol [70]. Thyme also contains caffeic acid, which has antibacterial properties [71]. Rosemary and thyme essential oils can inhibit certain rumen bacteria involved in biohydrogenation due to a high polyphenol content, resulting in an increase in polyunsaturated fatty acids in rumen digesta [18]. However, the results on the effects of these compounds on ruminal fermentation have not been consistent.

3.6. Spray Washing

Small ruminants with dirty skin or hair/wool presented for slaughter can pose food safety risks, since the degree of visible contamination on the hide or skin has been shown to affect the contamination levels of the resultant carcass. The skin of a live animal becomes contaminated with both pathogenic and nonpathogenic microorganisms derived from a wide range of sources such as feces, soil, water, and vegetation [72]. Animals can spread the contamination to other cleaner animals during transport and holding, either directly via body contact, indirectly via contact with contaminated floors, or both [73]. The presence of dirt or feces on animals can potentially lead to the transfer of microbes to the carcass and meat and to the slaughter equipment.

Methods to deal with excessively dirty animals during the preslaughter period vary in different countries, and these may include isolation of these animals as unfit for slaughter in their present condition, provision of a clean-up period by allowing to graze pasture, or inclusion of a swim-washing or spray-washing step prior to slaughter [21]. However, spray washing live animals prior to slaughter is not permitted in some countries, although studies have indicated this treatment reduces hide/skin fecal contamination and microbial loads [10,22].

Kannan et al. [10] found that the aerobic plate counts on the skin were the same for both spray-washed (1 min with potable water) and unwashed Spanish goats, but the counts were significantly less in the treated group when sampled after washing. However, skin *E. coli* counts did not decrease significantly due to spray-washing treatment. Spray washing also did not influence carcass *E. coli* or aerobic plate counts. Several factors can influence the antimicrobial effect of spray-washing, including duration of the spray-washing treatment, pressure of the spray nozzle, and whether or not the water contains any antibacterial agents. Byrne et al. [22] observed that a 3-min spray-washing treatment in cattle yielded better results compared with a 1-min washing treatment in reducing hide *E. coli* populations. It is likely longer durations and increased water pressure during spray- washing can increase discomfort and stress in live goats. A study by Kannan et al. [16] showed that applying a two-step spray-washing treatment in Boer × Spanish goats after exsanguination that comprised of 1 min washing with potable water (to remove fecal material) followed by 1 min with chlorinated water (to kill bacteria) decreased both skin *E. coli* and aerobic plate counts. In practical situations, this strategy would allow increasing of the duration and pressure of washing, if needed, since the washing treatment is imposed after bleeding.

3.7. Animal Behavior and Physiology

Minimizing stress in goats during loading, transportation, unloading and holding prior to slaughter is very important in reducing meat hygiene risks. It is important to consider goat behavior and physiology in response to a preharvest intervention method before recommending it as a viable pathogen reduction technology. Defecation is a frequently noticed stress response in cattle. Stress combined with light physical exercise can accelerate gastrointestinal tract emptying, while vigorous exercise slows gastric emptying [21]. Ruminants infected with pathogenic bacteria can harbor the organisms in the cecum, where they multiply. Stress can speed up emptying of the cecum into the colon and increase the rate of excretion, which can in turn increase the likelihood of contamination of the hide/skin of pathogen-free animals and increasing the food safety risk in the slaughter plant [21,74].

Several studies have evaluated stress levels in goats in response to preslaughter management practices such as transportation, feed deprivation, social isolation, and spray washing. Loading meat goats onto a transport trailer increases stress as indicated by increasing plasma cortisol and glucose concentrations. A 2½-hour transportation combined with an 18-h feed deprivation increases plasma cortisol concentrations in goats [40]. However, when goats are feed-deprived without transportation, cortisol concentrations do not increase [42]. While feed deprivation can facilitate evisceration without rupturing the digestive tract during the slaughter process, it can also contribute to live weight and carcass shrinkage losses and result in economic losses to the producer.

Social isolation of goats when they are moved through the single file race with individual compartments just prior to slaughter can increase stress. Isolation of goats from their social group for 15 min increased plasma cortisol concentrations; however, stress levels did not increase in socially isolated goats that could maintain visual contact with other goats [42]. Any intervention method used to treat excessively dirty animals prior to slaughter such as spray washing is likely to involve social isolation and further increase stress levels in goats. However, spray washing alone does not significantly increase stress in goats compared with unwashed controls if the water pressure used is carefully controlled to prevent any visible discomfort in the animals [10].

An important factor that can influence the efficacy of spray washing in small ruminants is the length of hair/wool and smoothness of the animal's coat. Kannan et al. [11] observed that under identical preslaughter management conditions, sheep may be more prone to skin bacterial contamination than goats. This is probably because the goat breeds used for meat production in the US generally have smoother and thinner coats. In addition, the season of the year and the behavior of the animals in the holding pens can influence the extent of skin contamination and the efficacy of spray-washing treatment. Sheep tend to spend more time lying down on the concrete floor of the holding pens compared to goats in winter, probably because goats with thin smooth coats cannot tolerate cold concrete floors for extended periods [11]. Since animals held in holding pens continue to defecate, they can pick up fecal materials from the contaminated floor when they lie down, the extent of which can be greater when animals have rough long coats [11].

4. Postharvest Intervention

Post-harvest intervention can be at any point in the processing line of carcasses after dressing is completed or in the product movement path before they reach consumers (Figure 1). This can involve applying intervention technology to whole carcasses, primal cuts, retail cuts, or ready-to-cook cuts such as bone-in cubed or boneless cubed meat. Goat meat is usually marketed as fresh meat, and marketing further-processed goat meat is extremely rare in the US. Postharvest intervention techniques can be broadly classified into thermal and nonthermal methods. Thermal techniques are effective in reducing pathogen counts on meat; however, they could adversely affect the quality characteristics of fresh meat due to heat-induced changes in muscle structure, composition, and biochemistry. Researchers have focused on nonthermal novel intervention methods that are effective in reducing pathogen counts (Table 2) without having any noticeable negative effects on fresh meat quality due to temperature increase [30]. It is essential to evaluate the effects of nonthermal technology used on fresh meat properties, such as appearance, color, odor, and lipid oxidation, and on cooked meat properties, such as texture, juiciness, and flavor. In addition, some low-cost technologies can leave chemical residues in meat, in addition to negatively affecting the flavor of meat. Technologies that cause significant negative effects on fresh meat quality that are easily discernable by consumers cannot be considered as viable intervention methods, even if they are efficient in reducing microbial numbers. Several studies assessed the effectiveness of these nonthermal technologies on poultry, red meat, and sea foods. The results of these studies show that these novel nonthermal technologies can be potentially applied to any meat, regardless of animal species, and therefore could be easily adopted in small-scale goat meat processing units. In the following sections, those methods that qualify as nonthermal, low-cost technologies with practical applications in goat meat processing are reviewed.

4.1. Organic Acids

Spray treatment of food animal carcasses using various decontaminant solutions has been extensively studied; however, studies in chevon carcasses have been limited. Goats have very limited subcutaneous fat deposition and more visceral fat accumulation. As a result, a dressed goat carcass is typically devoid of fat coverage compared to the extensive subcutaneous fat seen in beef or lamb carcass. Therefore, spray-washing parameters specific to goat carcasses may be required for practical applications. Spray-washing treatments studied to decontaminate food animal carcasses have included ozonated water, chlorinated water, trisodium phosphate, purified water, hot water, and organic and other acid solutions, to name a few. Adding chlorine to water used for spray washing appears to have little added advantage over water alone, based on research results [25]. Organic acids (lactic, acetic, and citric) were found to be very effective for broad-spectrum decontamination of meat carcasses and were proven to be better than many other compounds studied. An advantage of using organic acids over other intervention strategies is that residual antimicrobial activity is seen over extended storage time, although carcass decontamination may not necessarily improve the safety of meat cuts [26]. Among the organic acids, lactic acid is preferred due to its nonirritant characteristic, in addition to its decontamination properties.

Table 2. Reduction in microbial counts resulting from postharvest intervention strategies in goats.

Intervention Stage	Intervention Method	Sample Type	Microorganism	Reduction	Reference
Postharvest	2.5% acetic acid spray for 10 sec. using low-pressure hand sprayer	Carcasses	*Generic E. coli*	1.18 log10 CFU cm^{-2}	[75]
Postharvest	Pulsed dc square wave electricity with 10, 20, or 30 mA cm^{-2} current intensities.	Inoculated goat meat samples	*E. coli* O157:H7	8.0 log$_{10}$ CFU mL^{-1}	[33]
Further processing	Spraying acidic electrolyzed water for 12 min.	Inoculated boneless goat meat samples	*E. coli* K12	1.2 log$_{10}$ CFU mL^{-1}	[76]
Further processing	Spraying alkaline electrolyzed water for 12 min.	Inoculated boneless goat meat samples	*E. coli* K12	0.9 log$_{10}$ CFU mL^{-1}	[76]

Table 2. *Cont.*

Intervention Stage	Intervention Method	Sample Type	Microorganism	Reduction	Reference
Further processing	Applying UV-C * for 12 min. at 200 µW cm^{-2}	Inoculated boneless goat meat samples	*E. coli* K12	1.2 log$_{10}$ CFU mL^{-1}	[30]
Further processing	Spreading 1% lemongrass oil on the surface for 8 min.	Inoculated boneless goat meat samples	*E. coli* K12	2.1 log$_{10}$ CFU mL^{-1}	[30]
Further processing	Dipping in ozonated water for 12 min.; pH 6.8, ORP ** (mV) 562.75, ozone concentration 0.68 mg L^{-1}.	Inoculated boneless goat meat samples	*E. coli* K12	0.4 log$_{10}$ CFU mL^{-1}	[31]
Further processing	Hurdle Technology: Dipping in ozonated water for 6 min. (pH 6.8, ORP (mV) 562.75, ozone concentration 0.68 mg L^{-1}) followed by dipping in acidic electrolyzed water for 6 min.	Inoculated boneless goat meat samples	*E. coli* K12	0.86 log$_{10}$ CFU mL^{-1}	[31]
Further processing	Hurdle Technology: Spreading 1% lemongrass oil on the surface for 1 min followed by applying UV-C for 1 min. at 200 µW cm^{-2}	Inoculated boneless goat meat samples	*E. coli* K12	6.6 log$_{10}$ CFU mL^{-1}	[30]

* UV-C: ultraviolet-light; ** ORP: oxidation reduction potential.

The decontamination efficacy of organic acid spray washing depends on various factors, including strength of acid, solution temperature and contact time, spray pressure, and acid adaptation of organisms [27,28]. A 1.5 to 2.5% solution of food grade organic acid is recommended as ideal for carcass spraying. Microbial population reductions of 1 to 4 log CFU cm^{-2} have been reported with the use of varying concentrations [77]. High spray pressure has the potential of damaging the carcass surface and allowing the organisms to penetrate into the carcass, which could contaminate the meat [78]. Higher water temperature can negatively affect carcass surface color, although Dorsa [25] reported that a water temperature range of 70–96 °C did not permanently affect carcass appearance. Studies in goat and sheep carcasses have shown that a spray-wash treatment with a 2% lactic acid and 1.5% acetic + 1.5% propionic acid combination can reduce total viable counts from 0.52 to 1.16 log units with minimal changes in meat color and odor scores [79], and can extend shelf life to 8–11 days, compared to a shelf life of 3 days in untreated samples. However, lactic acid and acetic acid treatment application directly on meat cuts may result in permanent adverse changes in sensory properties [26].

4.2. Ozonated Water

Ozone is an effective antimicrobial for the treatment of meat due to several advantages, including its reactivity, penetrability, and spontaneous decomposition into a nontoxic product. Several researchers have assessed the antimicrobial efficacy of ozone, as it decomposes to oxygen continuously without leaving any residue in meat [29,31]. After production, ozone water must be used within 15 min for the decontamination of foods [80] as its half-life at room temperature is short due to instability in aqueous solutions [81]. The antimicrobial activity of ozonated water depends on temperature and pH, and on the presence of dissolved compounds such as sugar, minerals, surfactants, and organic matter [82]. Although several studies reported use of ozone in gaseous or aqueous form on fruits, vegetables, and greens, studies in meat, especially in red meat, are limited (Table 3). Degala et al. [31] reported that ozonated water (pH 6.80, oxygen-reduction potential 562.75 mV, ozone concentration 0.68 mg L^{-1}) treatment of chevon samples (20 $\pm$ 1 g) for 2, 4, 6, 8, 10, or 12 min resulted in an initial reduction of *E. coli* O157:H7 by 0.19 $\log_{10}$ CFU mL^{-1} after 2 min treatment, and by 0.52 $\log_{10}$ CFU mL^{-1} after 10 min treatment. However, the possible discoloration of meat at higher ozone concentrations, and ozone's rapid degradation to oxygen may limit its application in the meat industry.

4.3. Electrolyzed Water

Electrolyzed water, now used as a novel nonthermal technology for inactivating microorganisms in the food industry, was first invented in Japan, and since the 1980s, it has been used as a medical product. The strong antimicrobial activity of this environment-friendly technology is due to its pH, oxidation-reduction potential, and chlorine content [32,83]. Electrolyzed water can be produced on site by passing a diluted salt solution through an electrolytic cell that contains anode and cathode electrodes separated by a bipolar membrane. Electrolysis of sodium chloride (NaCl) solution yields Na^+ and Cl^- ions [31,83] and results in acidic electrolyzed water (pH range 2.0–4.0; oxidation-reduction potential > 1000 mV) and alkaline electrolyzed water (pH range 10–11.5; oxidation-reduction potential 800–900 mV).

Electrolyzed water has been used as a decontaminant in a variety of food products such as poultry carcasses, eggshells, salmon fillets, and frozen shrimp. However, studies on its use in red meats are reported for pork and beef, while very little is known about its antimicrobial efficiency in goat meat (Table 3). Arya et al. [76] evaluated the antimicrobial effects of both acidic electrolyzed water and alkaline electrolyzed water on goat meat by spraying samples for different time periods from 2 to 12 min using a household electrolyzed water generator. The authors observed 1.22 and 0.96 $\log_{10}$ CFU mL^{-1} reductions in *E. coli* K12 in 12-min acidic and alkaline electrolyzed water treatments, respectively. Although this technology is very effective in decontaminating fresh meat, it can leave chemical residues that can potentially affect meat color and flavor, since it is a chlorine-based technology. Because of its antimicrobial effectiveness, low cost, and ease of operation, electrolyzed water treatment of goat meat may be a viable intervention strategy for small and very small meat processors and retailers [76]. This technology can also be used in conjunction with other nonthermal technologies to further enhance the safety of goat meat and other muscle foods.

Table 3. Examples of studies on reducing *E. coli* populations in other red meats.

Meat Type	Intervention Method	Microorganism	Reduction	Reference
Beef carcasses	Water wash + aqueous ozone was sprayed on the surface of inoculated sample surface	*E. coli* O157:H7	No effect	[84]
Beef trimmings	1% ozonated water treatment for 7 or 15 min and effect was observed for 0 to 7 days.	*E. coli*	0.64 to 1.05 log CFU g^{-1}	[85]
Pork	Meat samples were treated with LcEW * (pH: 6.8; ORP: 700 mV; Chlorine concentration: 100.1 ppm) for 5 min at 23 °C	*E. coli* O157:H7	1.7 log CFU g^{-1}	[32]
Beef	Beef samples were treated with AEW ** (pH: 2.3–2.7; ORP ***: 110–1200 mV; Chlorine concentration: 50 ppm) for 3 min at 23 °C	*E. coli* O157:H7	1.6 log CFU g^{-1}	[86]

Table 3. *Cont.*

Meat Type	Intervention Method	Microorganism	Reduction	Reference
Ground beef	Ground beef samples were treated with 0.5% and 1% citral (essential oil extract of *Melissa officinalis* leaf) for 30 s and stored at 4 °C	*E. coli* cocktail	0.5–1.0 log CFU g^{-1}	[87]
Beef patties	Samples were treated with 0.2% of ginger and basilica essential oils	*E. coli*	Significantly lower *E. coli* numbers compared to control	[88]

* LcEW: low-concentrated electrolyzed water; ** AEW: acidic electrolyzed water; *** ORP: oxidation reduction potential.

4.4. Ultraviolet Light

Ultraviolet (UV) light is another nonthermal method approved for surface decontamination of food. Both continuous UV light and pulsed UV light are used in the food industry to inactivate pathogens in foods. Ultraviolet light falls in the wavelength range between 100 and 400 nm in the electromagnetic spectrum and is categorized as long wave (UV-A, 320–400 nm), medium wave (UV-B, 280–320 nm), and short wave (UV-C, 200–280 nm). The bactericidal effect of UV light is dependent upon its dose; the best antibacterial effect is between 245 and 285 nm. The ultraviolet dose can be calculated using the formula:

$$D = I \times t$$

where:
D = Ultraviolet light dose, μJ cm^{-2}
I = Intensity, μW cm^{-2}
t = Exposure time

Ultraviolet light treatment damages the genetic material and cell membrane of bacteria, thus inactivating the organisms. In addition, the release of free radicals during UV light treatment can also damage the cell membrane, enzymes, and nucleic acids of microorganisms. The antimicrobial efficacy of UV light depends on the optical properties of the medium, color, transparency, and the number of soluble and suspended solids [89]. Ultraviolet treatment has been reported to be more effective against gram negative bacteria than gram positive bacteria, yeasts, molds, and viruses, thus UV light may be suitable for inactivating pathogenic *E. coli* in meat.

Pulsed UV light is advantageous compared to other thermal and chemical methods due to its antibacterial efficiency without leaving chemical residues on meat and without causing changes in meat flavor, texture, and nutrient contents [90]. The limitations of pulsed UV light treatment include its ability to inactivate microorganisms present only on the surface of meat. Pulsed UV light induces photochemical, photophysical and/or photothermal damage, resulting in the inactivation or death of bacterial cells [90]. The photophysical inactivation effect of pulsed UV light on bacteria is attributed to the disruption of the cell membrane and components, and the photothermal inactivation effect is due to temporary intracellular heating caused by the absorption of photons [89].

Various foods are processed using UV light at different intensities and treatment times to inactivate pathogenic microorganisms, including *E. coli* O157:H7, as well as to inactivate certain enzymes that promote oxidation and deterioration in foods (Table 3). Degala et al. [30] evaluated the antimicrobial effect of UV light on goat meat by applying intensities of 100 and 200 μW cm^{-2}, with treatment times ranging from 2 to12 min, yielding energy dosages of 0.2–2.4 mJ cm^{-2}. The authors found that log reduction of *E. coli* K12 significantly increased when UV light intensity increased from 100 to 200 μW cm^{-2}, with a maximum reduction of 1.18 log$_{10}$ CFU mL^{-1} at an intensity of 200 μW cm^{-2} for 12 min. However, increasing the treatment time from 2 min to 12 min did not significantly influence the log reduction. The authors also reported that superficial lipids and proteins in meat with strong UV light-absorbing properties can interfere with the antimicrobial activity of UV light on meat [30]. Bryant et al. [34] assessed the efficiency of pulsed UV-light in inactivating *E. coli* K12 on goat meat and beef surfaces. In this study, the meat samples were placed in the pulsed UV-light sterilization chamber at three distances (4.47, 8.28 and 12.09 cm) from the light source and treated for different time periods from 5 to 60 s. The authors observed a maximum log reduction of 1.66 and 1.74 CFU mL^{-1} on goat meat and beef, respectively, at 4.47 cm distance for 60 s.

4.5. Sonication

Application of ultrasonic waves as an intervention technology to reduce pathogens in foods is generally considered as safe, nontoxic, and environmentally friendly. Ultrasound waves can be classified, based on frequency ranges, into two categories: (i) high power ultrasound (frequency range 20 to 100 kHz) and (ii) low power ultrasound (frequency of 100 kHz or above) [91]. High power ultrasound is regarded suitable for use in food processing and preservation [92].

Ultrasonic waves cause intracellular cavitation and thinning that increase the permeability of the cell membrane [93]. Ultrasound waves create negative pressure, resulting in breakage of the cell wall through a series of compression cycles caused by cavitation bubbles that pass through the solution [92]. When these cavitation bubbles collapse, hydroxyl radicals are produced, which recombine to form hydrogen peroxide and molecular hydrogen resulting in the damaging of DNA and thinning of the cell membrane [94]. The antimicrobial efficacy of ultrasound waves, however, depends upon the frequency and amplitude of the ultrasound waves, the temperature and viscosity of the liquid medium, the shape and size of the microorganisms, and the type of cell wall and its physiological state [95]. Ultrasound has been assessed for its antimicrobial efficiency in different types of meat and meat products; however, to the best of our knowledge its use in goat meat has not been reported.

Caraveo et al. [96] observed a 3 log$_{10}$ CFU cm^{-2} reduction in enterobacteria and mesophilic aerobic and psychrophilic bacteria in beef treated with 40 kHz frequency and 11 W cm^{-2} intensity for 60–90 min. In addition to inactivating microorganisms in meat, ultrasonic waves have also been shown to improve texture, margination, water-holding capacity, and cooking-yield of meat [92]. Ultrasound waves have also been used to improve the tenderness and water-holding capacity of beef [97]. The antimicrobial efficacy of ultrasound with a frequency of 40 kHz and intensity of 2.5 W cm^{-2} in distilled water and lactic acid solutions for 3 and 6 min against *S. anatum*, *E. coli*, *Proteus* species and *P. fluorescens* on chicken wing surfaces was investigated [98]. These authors reported a 1.0 log$_{10}$ CFU cm^{-2} reduction in the number of microorganisms on skin surface in water for 3 min, and a greater reduction in bacteria when the time was extended to 6 min. Sonication in the lactic acid aqueous solution for 3 min resulted in more than 1.0 log$_{10}$ CFU cm^{-2} reduction, and after 6 min the reduction exceeded 1.5 log$_{10}$ CFU cm^{-2}.

4.6. Low-Voltage Direct Electric Current

The use of low-intensity currents to inactivate bacteria on meat has received some research attention in recent years due to its efficiency, low cost, and ease of application. The mechanisms involved in the inactivation of microorganisms by low-voltage electric current are not completely understood. It is believed that the physical activity of electric currents may inactivate *E. coli*, possibly by disrupting the bacterial cell membrane. It has been reported that the electric current affects the cell membrane orientation and thereby the cell viability [99]. When NaCl is used as a medium to apply low level current, both NaCl and the low-level current can act synergistically in inactivating microorganisms. When electrolyzed, chlorides are converted to chlorine gas, which can play a significant role in inactivation of *E. coli* [100].

Different species of microbes can be inactivated using different intensities of direct electric current [100]. Even a low micro amperage can be effective in reducing the number of microorganisms such as *E. coli* and inhibiting their growth [101].

Saif et al. [33] applied different intensities (10, 20 and 30 mA/cm^2) of dc square wave electric signal for different durations (2, 8 and 32 min) to goat meat samples inoculated with *E. coli* O157:H7 and surface-coated with a thin film of 0.15 M NaCl, and found that all three intensities of current were effective in inactivating bacterial cells at a treatment duration of 32 min. The researchers also found that decreasing treatment duration decreased log reduction of *E. coli*, and frequencies of $\geq$1 kHz and duty cycles of $\geq$50% accelerated inactivation of the bacteria at a current intensity of 20 mA cm^{-2}. Mahapatra et al. [102] used a low-voltage electric inactivation system to apply low voltage dc current to beef samples inoculated with *E. coli* O157:H7 and surface-coated with a thin film of 0.15 M NaCl solution, and observed that increases in current intensity, frequency, duty cycle, and treatment duration increased percent reduction in *E. coli*. However, this study also revealed that the application of a low intensity current can potentially affect sensory properties, such as color and tenderness. Localized heating due to low intensity current can cause discoloration and increase Warner–Bratzler hardness values in meat [102].

4.7. Organic Essential Oils

Plants produce secondary metabolites known as essential oils that act as defensive mechanisms against microorganisms [103]. Several essential oils are used in meat preservation due to their antimicrobial and antioxidant activities. Their antimicrobial activity depends on several factors, including chemical structure, pH, temperature, and oxygen level. Although essential oils are generally recognized as safe, achieving high reductions in microbial counts may require higher concentrations or increased treatment time, both of which can have negative effects on the quality characteristics of food [104]. Several authors have studied the antimicrobial effects of essential oils in combination with other intervention technologies in meat and have found significant reductions in bacterial numbers with minimal effects on the quality characteristics of fresh meat. A few researchers have studied the efficacy of essential oils in decontaminating beef (Table 3); however, their uses in other red meats have not been adequately explored.

Degala et al. [30] studied the effects of three different concentrations of lemon grass oil (0.25%, 0.5%, 1.0%) treatment for different time periods from 2 to 12 min on *E. coli* K12 counts on inoculated goat meat and found that the reductions significantly increased with increasing concentrations of lemon grass oil. The authors observed a reduction of 2.16 log$_{10}$ CFU mL^{-1} with 1% lemongrass oil treatment. Other researchers have reported reductions below detection level with a 1.5% lemon grass oil treatment on minced beef and cooked beef patties [105,106]. Although there is clear evidence that increasing treatment intensity using increased lemongrass oil concentration results in better log reductions, Degala et al. [30] found that increasing treatment time from 2 to 12 min did not result in a significant increase in log reductions. Therefore, the researchers recommended lemongrass oil treatment of 1% concentration for 2 min for extending the microbial shelf life of goat meat. Lemongrass (1.56%) has been reported to significantly reduce *L. monocytogenes* in ground beef [107].

4.8. Hurdle Technologies

Although several technologies are capable of inactivating microorganisms when used individually, they may cause negative effects on the sensory properties of fresh meat, particularly when used in higher treatment strengths. When two or more processes are combined, the microbial shelf life of meat can be increased using the synergistic effects of lower individual treatment intensities and minimum energy input [108]. Combination treatments are likely to have zero, or minimal, negative effects on the sensory properties of meat, such as any texture, color, or flavor deterioration due to lipid oxidation. Furthermore, combination treatments can combat microbial stress adaptation associated with lower (sub-lethal) treatment intensities [109].

Several combination techniques on meat have been studied that involved organic oils, ozonated water, electrolyzed water, ultrasonic waves, and UV light, to name a few. Hurdle treatment of lemongrass oil with cold nitrogen plasma enhanced the antimicrobial activity of lemongrass on pork and minimized the potential negative effects on the sensory properties of the meat [110]. Degala et al. [30] reported that a hurdle treatment comprising of 1% lemongrass oil plus UV light with 200 μW cm^{-2} for 2 min resulted in a synergistic *E. coli* K12 reduction of 6.66 $\log_{10}$ CFU mL^{-1} (below detection levels). Ozonated water was used along with electrolyzed water to reduce contamination of cattle hide prior to slaughter [111]. Degala et al. [31] observed an *E. coli* K12 reduction of 0.86 $\log_{10}$ CFU mL^{-1} on goat meat with a 12 min combination treatment comprising ozonated water and acidic electrolyzed water. This study also revealed that the combination treatment was significantly more effective than ozonated water treatment alone based on log reductions.

5. Combination of Pre- and Post-Harvest Interventions

Although individual intervention strategies have proven to be effective in reducing pathogens on goat carcasses, a combination of both pre- and post-harvest strategies may be of value in further ensuring food safety in goat carcasses. For example, a dietary manipulation during the weeks prior to slaughter to reduce the number of *E. coli* in the gastrointestinal tracts of goats, followed by postmortem spray washing to decontaminate the skin or carcass, has been demonstrated to be useful. Kannan et al. [16] reported that brown seaweed extract supplementation for two weeks prior to slaughter, combined with chlorinated spray wash during processing, can be used as a viable decontamination strategy in goat processing. Spray washing reduced the total number of bacteria on the skin of goats. Although dietary treatment did not influence skin or dressed carcass bacterial loads, rumen *E. coli* counts were significantly lower in the seaweed extract-supplemented group while rumen pH and VFA concentrations were unaffected. The authors further suggested that the two-step pathogen reduction strategy can be easily implemented in very small meat processing plants with minimum modifications in the existing Hazard Analysis Critical Control Points (HACCP) plans. Ba et al. [23] used organic acid spray treatments, both immediately prior to slaughter (live animal hide) and immediately after slaughter (carcass surface) to decontaminate carcasses. The two-step treatment effectively reduced multiple bacterial species, including *E. coli*, by 2–5 log units, without negatively affecting the meat quality.

6. Recommendations

Farmers can adopt certain meat goat management measures that can minimize gastrointestinal *E. coli* populations in addition to improving animal wellbeing and productivity. Goat-handling methods that do not increase stress are of great value in improving animal performance and product quality. Feeding goats on a hay diet a few days prior to shipping them to the processing facility can reduce gut *E. coli* counts. Feeding with high-condensed tannin-containing forages, such as sericea lespedeza in any form (ground, pellet, or hay), may reduce the numbers of certain types of bacteria. Farmers can also consider feed supplements such brown seaweed extract and essential oils derived from herbal plants in meat goat diets, due to their overall positive effect on antioxidant activity, immune function, and product quality characteristics, in addition to reducing gastrointestinal *E. coli* populations.

Processing plant operators must coordinate the overall feed deprivation time meat goats are subjected to prior to processing. Ideally, overnight feed deprivation is recommended, as extended feed deprivation times can increase stress in goats and shorter feed deprivation durations may not result in adequate emptying of gut contents. At the industrial level, spray washing carcasses with organic acids is a proven method to reduce pathogens. Spraying electrolyzed water is another method that could be implemented in processing operations of any scale. At further processing levels, the use of organic essential oils and UV light may be practicable methods depending on the prevailing regulations in different countries. Hurdle technologies can also be adopted for further processed products, although the use of these methods may depend on the type of product, conditions of packaging and storage, and the intended length of shelf life.

7. Conclusions

Goat meat production and processing methods vary from country to country in the developing world, although in most cases goats are raised free range on brush or unestablished pastures and processed following the halal method. Concentrate feeding for meat goats is not common, except in some larger establishments where they are finished on concentrate supplements. In many developing countries, slaughter and meat hygiene practices, as well as regulations and inspections, are still not well established or followed. Small-scale butcher shops without refrigeration facilities, and chevon being sold as bone-in cubed meat are common sights in underdeveloped countries, particularly in the rural areas. However, processing facilities that cater to export goat meat markets likely maintain international hygiene standards and requirements. Reducing initial bacterial contamination in goat carcasses during the slaughter process becomes crucial in controlling the microbial quality of resultant products in regions where there are frequent interruptions in the cold chain.

Any pathogen reduction intervention strategy, pre- or post-harvest, must not compromise animal welfare, live and carcass weights, meat quality characteristics, and other variables of economic importance. For example, an easy-to-apply preharvest management method of hay feeding and shorter feed deprivation period could decrease gut bacterial counts in goats, with no significant effect on their physiological status. Any post-harvest method that increases the temperature of the carcass or cuts can potentially affect the appearance, color, tenderness, flavor, and lipid and pigment oxidation. The nonthermal methods discussed in this paper, irrespective of meat type, could be viable pathogen reduction methods if applied according to recommendations. However, most technologies available are effective for surface decontamination of meat with minimum penetrability.

The cost-effectiveness and practicality of any pathogen reduction strategy are critical in goat meat production, since the majority of goat processing plants worldwide are small-scale, limited-resource operations. The carcass skin-washing step can be added to the operation with ease with virtually no additional cost to the slaughter facility, but with the advantage of reducing biological hazards in goat carcasses. The additional cost incurred by the producer or processor are offset by the other benefits derived. For example, brown seaweed supplementation may increase goat meat production cost, but it has added benefits, such as better disease resistance in animals, increased antioxidant status, color stability, and shelf life, in addition to reduced fecal shedding of pathogens. Therefore, the application of an intervention strategy may depend on the scale of operation of a production or processing establishment.

Author Contributions: Conceptualization and original draft preparation, G.K.; writing, reviewing and editing, A.K.M. and H.L.D. All authors have read and agreed to the published version of the manuscript.

Funding: This work received no external funding.

Informed Consent Statement: Not applicable.

Data Availability Statement: All data referenced in this manuscript are already published.

Conflicts of Interest: The authors declare no conflict of interest.

References

1. Harris, L.J.; Stiles, M.E. Reliability of *Escherichia coli* counts for vacuum packaged ground beef. *J. Food Prot.* **1992**, *55*, 266–270. [CrossRef]
2. Agricultural Marketing Service-USDA. Microbial Testing of AMS Purchased Meat, Poultry and Egg Commodities. 2013. Available online: http://www.ams.usda.gov/resources/microbial-testing (accessed on 1 September 2021).
3. Callaway, T.R.; Elder, R.O.; Keen, J.E.; Anderson, R.C.; Nisbe, D.J. Forage feeding to reduce preharvest *Escherichia coli* populations in cattle, a review. *J. Dairy Sci.* **2003**, *86*, 852–860. [CrossRef]
4. Shukla, R.; Slack, R.; George, A.; Cheasty, T.; Rowe, B.; Scutter, J. *Escherichia coli* O157 infection associated with a farm visitor center. *Commun. Dis. Rep. CDR Rev.* **1995**, *5*, R86–R90. [PubMed]

5. Dulo, F.; Feleke, A.; Szonyi, B.; Fries, R.; Bauman, M.P.O.; Grace, D. Isolation of multidrug-resistant *Escherichia coli* 0157 from goats in the Somali Region of Ethiopia: A cross-sectional, abattoir-based study. *PLoS ONE* **2015**, *10*, e0142905. [CrossRef] [PubMed]

6. Bekele, T.; Zeede, G.; Tefera, G.; Feleke, A.; Zerom, K. *Escherichia coli* O157:H7 in raw meat in Addis Ababa, Ethiopia: Prevalence at an abattoir and retailers and antimicrobial susceptibility. *Int. J. Food Contam.* **2014**, *1*, 4. [CrossRef]

7. Tabashsum, Z.; Nazneen, M.; Ahsan, C.R.; Bari, M.L.; Yasmin, M. Isolation and characterization of *Escherichia coli* O157:H7 in raw goat meat in Dhaka city using conventional and molecular based technique. In Proceedings of the 2nd AFSSA Conference on Food Safety and Food Security, Dong Nai Province, Vietnam, 15–18 August 2014; pp. 89–95.

8. Islam, K.; Ahad, A.; Barua, M.; Islam, A.; Chakma, S.; Dorji, C.; Uddin, M.A.; Islam, S.; Lutful Ahasan, A.S.M. Isolation and epidemiology of multidrug resistant *Escherichia coli* from goats in Coxs Bazar, Bangladesh. *J. Adv. Vet. Anim. Res.* **2016**, *3*, 166–172. [CrossRef]

9. Novotna, R.; Alexa, P.; Hamrik, J.; Madanat, A.; Smola, J.; Cizek, A. Isolation and characterization Shiga toxin-producing *Escherichia coli* from sheep and goats in Jordan with evidence of multiresistant serotype O157:H7. *Vet. Med.—Czech.* **2005**, *50*, 111–118. [CrossRef]

10. Kannan, G.; Jenkins, A.K.; Eega, K.R.; Kouakou, B.; McCommon, G.W. Preslaughter spray-washing effects on physiological stress responses and skin and carcass microbial counts in goats. *Small Rumin. Res.* **2007**, *67*, 14–19. [CrossRef]

11. Kannan, G.; Gutta, V.R.; Lee, J.H.; Kouakou, B.; Getz, W.R.; McCommon, G.W. Preslaughter diet management in sheep and goats: Effects on physiological responses and microbial loads on skin and carcass. *J. Anim. Sci. Biotechnol.* **2014**, *5*, 42–51. [CrossRef]

12. Mechineni, A.; Kommuru, D.S.; Terrill, T.H.; Kouakou, B.; Lee, J.H.; Gujja, S.; Burke, J.M.; Kannan, G. Forage type and transportation stress effects on gut microbial counts and meat quality in goats. *Can. J. Anim. Sci.* **2021**, *101*, 126–133. [CrossRef]

13. Gutta, V.R.; Kannan, G.; Lee, J.H.; Kouakou, B.; Getz, W.R. Influences of short-term pre-slaughter dietary manipulation in sheep and goats on pH and microbial loads of gastrointestinal tract. *Small Rumin. Res.* **2009**, *81*, 21–28. [CrossRef]

14. Gregory, N.G.; Jacobson, L.H.; Nagle, T.A.; Muirhead, R.W.; Leroux, G.J. Effect of preslaughter feeding system on weight loss, gut bacteria, and the physico-chemical properties of digesta in cattle. *N. Z. J. Agric. Res.* **2000**, *43*, 351–361. [CrossRef]

15. Behrends, L.L.; Blanton, J.R.; Miller, M.F.; Pond, K.R.; Allen, V.G. Tasco supplementation in feedlot cattle: Effects on pathogen load. *J. Anim. Sci.* **2000**, *78*, 106.

16. Kannan, G.; Lee, J.H.; Kouakou, B.; Terrill, T.H. Reduction of microbial contamination in meat goats using dietary brown seaweed (*Ascophyllum nodosum*) supplementation and chlorinated wash. *Can. J. Anim. Sci.* **2019**, *99*, 570–577. [CrossRef]

17. Al-Kassie, G.A.M. Influence of two plant extracts derived from thyme and cinnamon on broiler performance. *Pak. Vet. J.* **2009**, *29*, 169–173.

18. Smeti, S.; Hajji, H.; Mekki, I.; Mahouachi, M.; Atti, N. Effects of dose and administration form of rosemary essential oils on meat quality and fatty acid profile of lamb. *Small Rumin. Res.* **2018**, *158*, 62–68. [CrossRef]

19. Sheridan, J.J. Sources of contamination during slaughter and measures for control. *J. Food Saf.* **1998**, *18*, 321–339. [CrossRef]

20. Stevens, M.P.; van Diemen, P.M.; Dziva, F.; Jones, P.W.; Wallis, T. Options for the control of enterohaemorrhagic *Escherichia coli* in ruminants. *Microbiology* **2002**, *148*, 3767–3778. [CrossRef]

21. Gregory, N.G. Livestock Presentation and Welfare before Slaughter. In *Animal Welfare and Meat Science*; Gregory, N.G., Grandin, T., Eds.; CABI Publishing: New York, NY, USA, 1998; pp. 15–41.

22. Byrne, C.M.; Bolton, D.J.; Sheridan, J.J.; McDowell, D.A.; Blair, I.S. The effects of preslaughter washing on the reduction of *Escherichia coli* O157:H7 transfer from cattle hides to carcasses during slaughter. *Lett. Appl. Microbiol.* **2000**, *30*, 142–145. [CrossRef]

23. Ba, H.V.; Seo, H.W.; Pil-Nam, S.; Kim, Y.S.; Park, B.Y.; Moon, S.S.; Kang, S.J.; Choi, Y.M.; Kim, J.H. The effects of pre- and post-slaughter spray application with organic acids on microbial population reductions on beef carcasses. *Meat Sci.* **2018**, *137*, 16–23. [CrossRef]

24. James, C.; Thornton, J.; Ketteringham, L.; James, S. Effect of steam condensation, hot water or chlorinated water immersion on bacterial numbers and quality of lamb carcass. *J. Food Eng.* **2000**, *43*, 219–225. [CrossRef]

25. Dorsa, W.J. New and established carcass decontamination procedures commonly used in the beef-processing industry. *J. Food Prot.* **1997**, *60*, 1146–1151. [CrossRef] [PubMed]

26. Smulders, F.J.M.; Greer, G.G. Integrating microbial decontamination with organic acids in HACCP programme for muscle foods: Prospects and controversies. *Int. J. Food Microbiol.* **1998**, *44*, 149–169. [CrossRef]

27. Cutter, C.N.; Siragusa, G.R. Efficacy of organic acids against *Escherichia coli* O157:H7 attached to beef carcass tissue using a pilot scale carcass washer. *J. Food Prot.* **1994**, *57*, 97–103. [CrossRef] [PubMed]

28. Berry, E.D.; Cutter, C.N. Effects of acid adaptation of *Escherichia coli* O157:H7 on efficacy of acetic acid spray washes to decontaminate beef carcass tissue. *Appl. Environ. Microbiol.* **2000**, *66*, 1493–1498. [CrossRef] [PubMed]

29. Mahapatra, A.K.; Muthukumarappan, K.; Julson, J.L. Application of ozone, bacteriocins and irradiation in food processing: A review. *Crit. Rev. Food Sci. Nutr.* **2005**, *45*, 447–461. [CrossRef]

30. Degala, H.L.; Mahapatra, A.K.; Demirci, A.; Kannan, G. Evaluation of non-thermal hurdle technology for ultraviolet-light to inactivate *Escherichia coli* K12 on goat meat surfaces. *Food Control* **2018**, *90*, 113–120. [CrossRef]

31. Degala, H.L.; Scott, J.R.; Nakkiran, P.; Mahapatra, A.K.; Kannan, G. Inactivation of *E. coli* O157:H7 on goat meat surface using ozonated water alone and in combination with electrolyzed oxidizing water. In *American Society of Agricultural and Biological Engineers*; Paper No. 2462209; ASABE: St. Joseph, MI, USA, 2016. [CrossRef]

32. Arya, R.; Bryant, M.; Degala, H.L.; Mahapatra, A.K.; Kannan, G. Effectiveness of a low-cost household electrolyzed water generator in reducing the populations of *Escherichia coli* K12 on inoculated beef, chevon, and pork surfaces. *J. Food Process. Preserv.* **2018**, *42*, e13636. [CrossRef]

33. Saif, S.M.H.; Lan, Y.; Williams, L.L.; Joshee, L.; Wang, S. Reduction of *Escherichia coli* O157:H7 on goat meat surface with pulsed dc square wave signal. *J. Food Eng.* **2006**, *77*, 281–288. [CrossRef]

34. Bryant, M.; Degala, H.L.; Mahapatra, A.K.; Gosukonda, R.M.; Kannan, G. Inactivation of *Escherichia coli* K12 by pulsed UV-light on goat meat and beef: Microbial responses and modeling. *Int. J. Food Sci. Technol.* **2020**, *56*, 563–572. [CrossRef]

35. Jacobson, H.L.; Tanya, A.N.; Gregory, N.G.; Bell, R.G.; Roux, G.L.; Haines, J.M. Effect of feeding pasture-finished cattle different conserved forages on *Escherichia coli* in the rumen and faeces. *Meat Sci.* **2002**, *62*, 93–106. [CrossRef]

36. Vanguru, M.; Lee, J.H.; Kouakou, B.; Terrill, T.H.; Kannan, G. Effect of feed deprivation time on bacterial contamination of skin and carcass in meat goats. *Trop. Subtrop. Agroecosyst.* **2009**, *11*, 259–261.

37. Lee, J.H.; Koaukou, B.; Kannan, G. Influences of dietary regimens on microbial content in gastrointestinal tracts of meat goats. *Livestock Sci.* **2009**, *125*, 249–253. [CrossRef]

38. Lee, J.H.; Vanguru, M.; Kannan, G.; Moore, D.A.; Terrill, T.H.; Kouakou, B. Influence of dietary condensed tannins from sericea lespedeza on bacterial loads in gastrointestinal tracts of meat goats. *Livestock Sci.* **2009**, *126*, 314–317. [CrossRef]

39. Kadim, I.T.; Mahagoub, O.; AlKindi, A.Y.; Al-Marzooqi, W.; Al-Sakri, N.M.; Almaneay, M.; Mahmoud, I.Y. Effect of transportation at high ambient temperatures on physiological responses, carcass, and meat quality characteristics in two age groups of Omani sheep. *Asian-Australas. J. Anim. Sci.* **2007**, *20*, 124–431. [CrossRef]

40. Kannan, G.; Terrill, T.H.; Kouakou, B.; Gazal, O.S.; Gelaye, S.; Amoah, E.A.; Samake, S. Transportation of goats: Effect on physiological stress responses and live weight loss. *J. Anim. Sci.* **2000**, *78*, 1450–1457. [CrossRef]

41. Reid, C.A.; Avery, S.M.; Warriss, P.; Buncic, S. The effect of feed withdrawal on *Escherichia coli* shedding in beef cattle. *Food Control* **2002**, *13*, 393–398. [CrossRef]

42. Kannan, G.; Terrill, T.H.; Kouakou, B.; Gelaye, S.; Amoah, E. Simulated Preslaughter holding and isolation effects on stress responses and live weight shrinkage in meat goats. *J. Anim. Sci.* **2002**, *80*, 1771–1780. [CrossRef]

43. Huntington, G.B. Starch utilization by ruminants: From basics to the bunk. *J. Anim. Sci.* **1997**, *75*, 852–867. [CrossRef]

44. Scott, T.; Wilson, C.; Bailey, D.; Klopfenstein, T.; Milton, T.; Moxley, R.; Smith, D.; Gray, J.; Hungerford, L. Influence of diet on total and acid resistant *E. coli* and colonic pH. *Nebraska Beef Rep.* **2000**, *389*, 39–41.

45. Diez-Gonzalez, F.; Callaway, T.R.; Kizoulis, M.G.; Russell, J.B. Grain feeding and the dissemination of acid resistant *Escherichia coli* from cattle. *Science* **1998**, *281*, 1666–1668. [CrossRef] [PubMed]

46. Russell, J.B.; Diez-Gonzalez, F.; Jarvis, G.N. Invited review: Effects of diet shifts on *Escherichia coli* in cattle. *J. Dairy Sci.* **2000**, *83*, 863–873. [CrossRef]

47. Grauke, L.J.; Kudva, I.T.; Yoon, J.W.; Hunt, C.W.; Williams, C.J.; Hovde, C.J. Gastrointestinal tract location of *Escherichia coli* O157:H7 in ruminants. *Appl. Environ. Microbiol.* **2002**, 2269–2277. [CrossRef] [PubMed]

48. Kudva, I.T.; Hunt, C.W.; Williams, C.J.; Nance, U.M.; Hovde, C.J. Evaluation of dietary influences on *Escherichia coli* O157:H7 shedding by sheep. *Appl. Environ. Microbiol.* **1997**, *63*, 3878–3886. [CrossRef]

49. Kudva, I.T.; Hatfield, P.G.; Hovde, C.J. Effect of diet on the shedding of *Escherichia coli* O157:H7 shedding in a sheep model. *Appl. Environ. Microbiol.* **1995**, *61*, 1363–1370. [CrossRef] [PubMed]

50. Lin, J.; Smith, M.P.; Chapin, K.C.; Baik, H.S.; Bennett, G.N.; Foster, J.W. Mechanisms of acid resistance in enterohemorrhagic *Escherichia coli*. *Appl. Environ. Microbiol.* **1996**, *62*, 3094–3100. [CrossRef]

51. Rasmussen, M.A.; Cray, W.C.; Casey, T.A.; Whip, S.C. Rumen contents as a reservoir of enterohemorrhagic *Escherichia coli*. *FEMS Microbiol. Lett.* **1993**, *114*, 79–84. [CrossRef] [PubMed]

52. Chart, H. VTEC enteropathogenicity. *J. Appl. Microbiol.* **2000**, *88*, 12S–23S. [CrossRef]

53. McWilliam Leitch, E.C.; Duncan, S.H.; Stanley, K.N.; Stewart, C.S. Dietary effects on the microbiological safety of food. *Proc. Nutr. Soc.* **2001**, *60*, 247–255. [CrossRef]

54. Barham, A.R.; Barham, B.L.; Blanton, J.R., Jr.; Allen, V.G.; Pond, K.R.; Miller, M.F. Effects of Tasco®14 on prevalence levels of enterohemorragic *Escherichia coli* and *Salmonella* spp. in feedlot steers. *J. Anim. Sci.* **2001**, *79*, 257.

55. Allen, V.G.; Pond, K.R.; Saker, K.E.; Fontenot, J.P.; Bagley, C.P.; Ivy, R.L.; Evans, R.R.; Schmidt, R.E.; Fike, J.H.; Zhang, X.; et al. Tasco: Influence of a brown seaweed on antioxidants in forages and livestock—A review. *J. Anim. Sci.* **2001**, *79*, E21–E31. [CrossRef]

56. Saker, K.E.; Fike, J.H.; Veit, H.; Ward, D.L. Brown seaweed- (Tasco™) treated conserved forage enhances antioxidant status and immune function in heat-stressed wether lambs. *J. Anim. Physiol. Anim. Nutr.* **2004**, *88*, 122–130. [CrossRef]

57. Montgomery, J.L.; Allen, V.G.; Pond, K.R.; Miller, M.F.; Wester, D.B.; Brown, C.P.; Evans, R.; Bagley, C.P.; Ivy, R.L.; Fontenot, J.P. Tasco-Forage: IV. Influence of a seaweed extract applied to tall fescue pastures on sensory characteristics, shelf-life, and vitamin E status in feedlot-finished steers. *J. Anim. Sci.* **2001**, *79*, 884–894. [CrossRef]

58. Wang, Y.; Xu, Z.; Bach, S.J.; McAllister, T.A. Sensitivity of *Escherichia coli* to seaweed (*Ascophyllum nodosum*) phlorotannins and terrestrial tannins. Asian-Australasian. *J. Anim. Sci.* **2009**, *22*, 238–245. [CrossRef]

59. Scalbert, A. Antimicrobial properties of tannins: Review article number 63. *Phytochemistry* **1991**, *30*, 3875–3883. [CrossRef]

60. Nagayama, K.; Iwamura, Y.; Shibata, T.; Hirayama, I.; Nakamura, T. Bactericidal activity of phlorotannins from the brown alga *Ecklonia kurome*. *J. Antimicrob. Chemother.* **2002**, *50*, 889–893. [CrossRef]

61. Casamiglia, S.; Busquet, M.; Cardoza, P.; Castillejos, L.; Ferrett, A. Essential oils as modifiers of rumen microbial fermentation: A review. *J. Dairy Sci.* **2017**, *90*, 2580–2595. [CrossRef]
62. Simitzis, P.E. Enrichment of animal diets with essential oils—A great perspective in improving animal performance and quality characteristics of the derived products. *Medicines* **2017**, *4*, 35. [CrossRef]
63. Dorman, H.J.D.; Deans, S.G. Antimicrobial agents from plants: Antibacterial activity of plant volatile oils. *J. Appl. Microbiol.* **2000**, *88*, 308–316. [CrossRef]
64. Burt, S. Essential oils: Their antibacterial properties and potential applications in foods—A review. *Int. J. Food Microbiol.* **2004**, *94*, 223–253. [CrossRef]
65. Gustofson, R.H.; Bowen, R.E. Antibiotic use in animal agriculture. *J. Appl. Microbiol.* **1997**, *83*, 531–541. [CrossRef]
66. Prescott, L.M.; Harley, J.P.; Klein, D.A. Control de microorganismos por agentes fisicos y quimicos. In *Microbiologia*; McGraw-Hill-Interamericana de España: Madrid, Spain, 2004; pp. 145–162.
67. Ouattara, B.; Simard, R.E.; Holley, R.A.; Piette, G.J.-P.; Bégin, A. Antibacterial activity of selected fatty acids and essential oils against six meat spoilage organisms. *Int. J. Food Microbiol.* **1997**, *37*, 155–162. [CrossRef]
68. Wendakoon, C.N.; Sakaguchi, M. Inhibition of amino acid decarboxylase activity of *Enterobacter aerogenes* by active components in spice. *J. Food Prot.* **1995**, *58*, 280–283. [CrossRef]
69. Evans, J.D.; Martin, S.A. Effects of thymol on ruminal microorganisms. *Curr. Microbiol.* **2000**, *41*, 336–340. [CrossRef]
70. Sengül, T.; Yurtseven, S.; Cetin, M.; Kocyigit, A.; Sögüt, B. Effect of thyme (*T. vulgaris*) extracts on fattening performance, some blood parameters, oxidative stress and DNA damage in Japanese quails. *J. Anim. Feed Sci.* **2008**, *17*, 608–620. [CrossRef]
71. Cowan, M.M. Plant products as antimicrobial agents. *Clin. Microbiol. Rev.* **1999**, *12*, 564–582. [CrossRef]
72. McEvoy, J.M.; Doherty, A.M.; Finnerty, M.; Sheridan, J.J.; McGuire, L.; Blair, I.S.; MacDowell, D.A.; Harrington, D. The relationship between hide cleanliness and bacterial numbers on beef carcasses at a commercial abattoir. *Lett. Appl. Microbiol.* **2000**, *30*, 390–395. [CrossRef]
73. Reid, C.A.; Small, A.; Avery, S.M.; Buncic, S. Presence of food-borne pathogens on cattle hides. *Food Control* **2002**, *13*, 411–415. [CrossRef]
74. Berends, B.R.; Urlings, H.A.P.; Snijders, J.M.A.; van Knapen, F. Identification and quantification of risk factors in animal management and transport regarding Salmonella spp. in pigs. *Int. J. Food Microbiol.* **1996**, *30*, 37–53. [CrossRef]
75. Wudie, A.; Zewude, G.; Ali, T.; Jibat, T. Study on effect of acetic acid spray on *Escherichia coli* load and meat pH at an export abattoir, Modjo, Ethiopia. *Int. J. Appl. Res. Vet. Med.* **2013**, *11*, 130–136.
76. Ding, T.; Rahman, S.M.E.; Purev, U.; Oh, D.-H. Modelling of *Escherichia coli* O157:H7 growth at various storage temperatures on beef treated with electrolyzed oxidizing water. *J. Food Eng.* **2010**, *97*, 497–503. [CrossRef]
77. Dickson, J.S.; Anderson, M.E. Microbial decontamination of food animal carcasses by washing and sanitizing systems: A review. *J. Food Prot.* **1992**, *55*, 133–140. [CrossRef]
78. De Zuniga, A.G.; Anderson, M.E.; Marshall, R.T.; Iannotti, E.L. A model system for studying the penetration of microorganisms into meat. *J. Food Prot.* **1991**, *54*, 256–258. [CrossRef]
79. Dubal, Z.B.; Paturkar, A.M.; Waskar, V.S.; Zende, R.J.; Latha, C.; Rawool, D.B.; Kadam, M.M. Effect of food grade organic acids on inoculated *S. aureus*, *L. monocytogenes*, *E. coli*, and *S. typhimurium* in sheep/goat meat stored at refrigeration temperature. *Meat Sci.* **2004**, *66*, 817–821. [CrossRef]
80. Hung, Y.C. Reducing microbial safety risk on blueberries through innovative washing technologies. In *Final Report, Grant Code: SRSFC Project # 2010-15*; Department of Food Science and Technology, University of Georgia: Griffin, GA, USA, 2015.
81. Khadre, M.A.; Yousef, A.E.; Kim, J.G. Microbial aspects of ozone applications in food: A review. *J. Food Sci.* **2001**, *66*, 1242–1252. [CrossRef]
82. Chawla, A.S. Application of Ozonated Water Technology for Improving Quality and Safety of Peeled Shrimp Meat. Master's Thesis, Department of Food Science, Louisiana State University, Baton Rouge, LA, USA, 2006.
83. Haiti, S.; Mandal, S.; Minz, P.S.; Vij, S.; Khetra, Y.; Singh, B.P. Electrolyzed oxidizing water (EOW): Non-thermal approach for decontamination of food borne microorganisms in food industry. *Food Sci. Nutr.* **2012**, *3*, 760–768. [CrossRef]
84. Castillo, A.; McKenzie, K.S.; Lucia, L.M.; Acuff, G.R. Ozone treatment for reduction of *Escherichia coli* O157:H7 and *Salmonella* serotype *Typhimurium* on beef carcass surfaces. *J. Food Protect.* **2003**, *66*, 775–779. [CrossRef] [PubMed]
85. Stivarius, M.R.; Pohlman, F.W.; McElyea, K.S.; Apple, J.K. Microbial, instrumental color and sensory color and odor characteristics of ground beef produced from beef trimmings treated with ozone or chlorine dioxide. *Meat Sci.* **2002**, *60*, 299–305. [CrossRef]
86. Rahman, S.M.E.; Wang, J.; Oh, D.-H. Synergistic effect of low concentration electrolyzed water and calcium lactate to ensure microbial safety, shelf life and sensory quality of fresh pork. *Food Control* **2013**, *30*, 176–183. [CrossRef]
87. Chien, S.-Y.; Sheen, S.; Sommers, C.; Sheen, L.-Y. Combination effect of high-pressure processing and essential oil (*Melissa officinalis* Extracts) or their constituents for the inactivation of *Escherichia coli* in ground beef. *Food Bioproc. Technol.* **2019**, *12*, 359–370. [CrossRef]
88. Dzudie, T.; Kouebou, C.P.; Essia-Ngang, J.J.; Mbofung, C.M.F. Lipid sources and essential oils effects on quality and stability of beef patties. *J. Food Eng.* **2004**, *65*, 67–72. [CrossRef]
89. Gayán, E.; Monfort, S.; Álvarez, I.; Codón, S. UV-C inactivation of Escherichia coli at different temperatures. *Innov. Food Sci. Emerg. Technol.* **2011**, *12*, 531–541. [CrossRef]

90. Keener, L.; Krishnamurthy, K. Shedding light on food safety: Applications of pulsed light processing. *Food Saf. Mag.* **2014**, *20*, 28–33.
91. Sango, M.D.; Abela, D.; McElhatton, A.; Valdramidis, V.P. Assisted ultrasound applications for the production of safe foods. *J. Appl. Microbiol.* **2014**, *116*, 1067–1083. [CrossRef] [PubMed]
92. Turantas, F.; Kilic, G.B.; Kilic, B. Ultrasound in the meat industry: General applications and decontamination efficiency. *Int. J. Food Microbiol.* **2015**, *198*, 59–69. [CrossRef] [PubMed]
93. Bilek, S.E.; Turantas, F. Decontamination efficiency of high power ultrasound in the fruit and vegetable industry, a review. *Int. J. Food Microbiol.* **2013**, *166*, 155–162. [CrossRef] [PubMed]
94. Kumar, M.A.; Mason, T.J. Sonochemistry. In *Kirk-Othmer Encycl. Chem. Technol. yOn-Line*; Wiley Interscience: Hoboken, NJ, USA, 2007.
95. Rastogi, R.P.; Kumar, A.; Tyagi, M.B.; Sinha, R.P. Molecular mechanisms of ultraviolet radiation-induced DNA damage and repair. *J. Nucleic Acids* **2010**, *2010*, 592980. [CrossRef] [PubMed]
96. Caraveo, O.; Alarcon-Rojo, A.D.; Renteria, A.; Santellano, E.; Paniwnky, L. Physicochemical and microbiological characteristics of beef treated with high-intensity ultrasound and stored at 4 °C. *J. Sci. Food Agric.* **2014**, *95*, 2487–2493. [CrossRef]
97. Dolatowski, Z.J.; Stasiak, D.M.; Giemza, S. Effects of sonication on properties of reduced pH meat. *Pol. J. Food Nutr. Sci.* **2001**, *10*, 192–196.
98. Kordowska-Wiater, M.; Stasiak, D.M. Effect of ultrasound on survival of gram-negative bacteria on chicken skin surface. *Bull. Vet. Inst. Pulawy.* **2011**, *55*, 207–210.
99. Jackman, A.S.; Maini, G.; Sharman, A.K.; Knowles, C.J. The effects of direct electric current on viability and metabolisms of acidophilic bacteria. *Enzyme Microb. Technol.* **1999**, *24*, 316–324. [CrossRef]
100. Birbir, M.; Hacioglu, H.; Birbir, Y.; Altug, G. Inactivation of *Escherichia coli* by alternative electric current in rivers discharged into sea. *J. Electrost.* **2009**, *67*, 640–645. [CrossRef]
101. Davis, C.P.; Weinberg, S.; Anderson, M.D.; Rao, G.M.; Warren, M.M. Effects of microamperage, medium, and bacterial concentration on iontophoretic killing of bacteria in fluid. *Antimicrob. Agents Chemother.* **1989**, *33*, 442–447. [CrossRef]
102. Mahapatra, A.K.; Harris, D.L.; Nguyen, C.N.; Kannan, G. Reduction of *Escherichia coli* O157:H7 on beef surfaces using low-voltage direct electric current and the impact on sensory properties. *J. Electrost.* **2011**, *69*, 30–35. [CrossRef]
103. Hyldgaard, M.; Mygind, T.; Meyer, R.L. Essential oils in food preservation: Mode of action, synergies, and interactions with food matrix components. *Front. Microbiol.* **2012**, *3*, 1–12. [CrossRef]
104. Wood, O.B.; Bruhn, C.M. Position of the American dietetic association: Food irradiation. *J. Am. Diet Assoc.* **2000**, *100*, 246–253. [CrossRef]
105. Salem, A.M.; Amin, R.A.; Afifi, G.S.A. Studies on antimicrobial and antioxidant efficiency of some essential oils in minced beef. *J. Am. Sci.* **2010**, *6*, 691–700.
106. Rounds, L.; Havens, C.M.; Feinstein, Y.; Friedman, M.; Ravishankar, S. Plant extracts, spices, and essential oils inactivate *Escherichia coli* O157:H7 and reduce formation of potentially carcinogenic heterocyclic amines in cooked beef patties. *J. Agric. Food Chem.* **2012**, *60*, 3792–3799. [CrossRef] [PubMed]
107. Oliveira, T.L.C.; Cardosa, M.G.; Soares, R.A.; Ramos, E.M.; Piccoli, R.H.; Tebaldi, V.M.R. Inhibitory activity of *Syzygium aromaticum* and *Cymbopogon citratus* (DC) *Stapf.* Essential oils against *Listeria monocytogenes* inoculated in bovine ground meat. *Braz. J. Microbiol.* **2013**, *44*, 357–365. [CrossRef] [PubMed]
108. Ross, A.I.V.; Griffiths, M.W.; Mittal, G.S.; Deeth, H.C. Combining non-thermal technologies to control foodborne microorganisms. *Int. J. Food Microbiol.* **2003**, *89*, 125–138. [CrossRef]
109. Yousef, A.E. Efficacy and Limitations of Non-Thermal Preservation Technologies. In *Institute of Food Technologists (IFT) Annual Meeting Book of Abstracts*; Session 9-1; Institute of Food Technologists: Chicago, IL, USA, 2001.
110. Cui, H.; Wu, J.; Li, C.; Lin, L. Promoting anti-listeria activity of lemongrass oil on pork loin by cold nitrogen plasma assist. *J. Food Saf.* **2017**, *37*, e12316. [CrossRef]
111. Bosilevac, J.M.; Shackelford, S.D.; Bricht, D.M.; Koohmaraie, M. Efficacy of ozonated and electrolyzed oxidative waters to decontaminate hides of cattle before slaughter. *J. Food Prot.* **2005**, *68*, 1393–1398. [CrossRef] [PubMed]

Article

Sainfoin in the Dams' Diet as a Source of Proanthocyanidins: Effect on the Growth, Carcass and Meat Quality of Their Suckling Lambs

Clàudia Baila , Sandra Lobón , Mireia Blanco , Isabel Casasús , Guillermo Ripoll and Margalida Joy *

Departamento de Ciencia Animal, Centro de Investigación y Tecnología Agroalimentaria de Aragón (CITA), Instituto Agroalimentario de Aragón–IA2 (CITA-Universidad de Zaragoza), Avda. Montañana 930, 50059 Zaragoza, Spain; cbaila@cita-aragon.es (C.B.); slobon@cita-aragon.es (S.L.); mblanco@cita-aragon.es (M.B.); icasasus@cita-aragon.es (I.C.); gripoll@cita-aragon.es (G.R.)
* Correspondence: mjoy@cita-aragon.es; Tel.: +34-976716442

Simple Summary: Several studies point out that the use of local forage legumes, such as sainfoin, can be appropriate for feeding sheep autochthonous breeds, with additional benefits also for the soil. Besides, sainfoin has a medium content of proanthocyanidins (PAC), also known as condensed tannins, the effects of which have been studied in fattening lambs but seldom on suckling lambs. The aim of the study was to evaluate the effect of PAC of sainfoin fed to dams on the productive traits, weight of the digestive organs, and on carcass and meat quality of their suckling lambs. The inclusion of PAC from sainfoin in the dam diet did not produce detrimental changes on the growth and carcass and meat characteristics of their suckling lambs. Therefore, sainfoin can be fed to ewes during lactation to produce suckling lambs, achieving good performances and meat quality.

Abstract: Sainfoin (*Onobrychis viciifolia*) is a forage legume with a medium content of proanthocyanidins (PAC), which may affect animal performance and product quality. The objective of the present study was to assess the effect of PAC from sainfoin fed to dams, using polyethylene glycol (PEG) as a blocking agent, on the performance and carcass and meat quality of their suckling male lambs. After lambing, twenty lactating dams were fed fresh sainfoin *ad libitum* plus 200 g per day of barley; ten were orally dosed with water (Sainfoin), and ten were dosed orally with a water dilution of 100 g PEG (Sainfoin + PEG). Their lambs (4.1 ± 0.64 kg at birth) suckled *ad libitum* until they reached the target slaughter weight of 10–12 kg. The presence of PAC in the dams' diet did not affect the growth, blood metabolites and carcass weight and fatness of the suckling lambs but decreased the lightness of caudal fat ($p < 0.05$) and increased the weight of the digestive compartments ($p < 0.05$). Regarding the meat characteristics, PAC only decreased polyphenols content ($p < 0.05$). In conclusion, the presence of PAC in the dams' diet had not significant effects on the performance and product quality of their suckling lambs.

Keywords: *Onobrychis viciifolia*; condensed tannins; performance; plasma metabolites; meat color

Citation: Baila, C.; Lobón, S.; Blanco, M.; Casasús, I.; Ripoll, G.; Joy, M. Sainfoin in the Dams' Diet as a Source of Proanthocyanidins: Effect on the Growth, Carcass and Meat Quality of Their Suckling Lambs. *Animals* **2022**, *12*, 408. https://doi.org/10.3390/ani12040408

Academic Editor: Adriana Bonanno

Received: 12 January 2022
Accepted: 7 February 2022
Published: 9 February 2022

Publisher's Note: MDPI stays neutral with regard to jurisdictional claims in published maps and institutional affiliations.

1. Introduction

Nowadays, there is increasing social pressure for livestock production systems to minimize their negative environmental impacts, and to reduce the inclusion of feed components that compete for land use with human food crops [1]. In the last decades, the European Union has encouraged the use of local legumes for animal feeding in order to reduce the dependency on soybean meal, and to benefit from their positive environmental effects [2,3]. Among legume forages, sainfoin (*Onobrychis viciifolia*) has proven to be an excellent forage to be fed during lactation in ewes [4]. Furthermore, consumers are increasingly aware of the importance of food quality on human health, which has increased the demand for products obtained from forage-fed animals as they are considered healthier than those obtained from concentrate-fed diets [4,5], as well as more respectful with animal welfare.

In the Mediterranean area, the traditional production of suckling lambs —slaughtered at 10–12 kg body weight (BW)—is based on a system in which dams are fed diets mainly composed by straw, cereals and byproducts, and lambs are fed exclusively on their dams' milk. In this framework, the inclusion of high proportions of fresh forage in the diet of the ewes could be an interesting alternative. In fact, grazing sainfoin during lactation improved the meat quality of light lambs even after a finishing period on concentrates, when compared to lambs reared with dams grazing alfalfa [6]. This could be due to a possible synergy between dietary proanthocyanidins (PAC) present in sainfoin and other antioxidant components in the muscle [7] or the milk [8].

Previous research concerning the effect of PAC on lamb performance is not conclusive, as some studies reported that weight increased [9,10], decreased [5] or did not change [11]. Regarding meat quality, there is also no consensus about the relation between PAC in lamb diets and the color parameters and heme pigments contents [12,13]. On the other hand, improvements in the antioxidant capacity of tissues due to the action of PAC have been observed [14,15]. Therefore, this great variability of results may depend on molecular weight, structure and degree of polymerization of PAC, as well as on the type of diet and animal studied [16].

Most of the studies regarding the inclusion of PAC have been carried out in fattening lambs with limited research on suckling lambs, the diet of which is based almost exclusively in milk. Therefore, to study the effect of PAC on the meat of suckling lambs, the source of PAC has to be included in their dam's diet, and results cannot be extrapolated from those obtained in fattening lambs. Hence, the aim was to evaluate the effect of PAC of fresh sainfoin fed to dams on productive parameters, weight of the digestive organs, and carcass and meat quality of suckling lambs.

2. Materials and Methods

2.1. Experimental Site

The experimental procedures (CEEA, 2017–07), which were in compliance with the guidelines of the Directive 2010/63/EU of the European Parliament and of the Council of 22 September on the protection of animals used for experimental purposes, were approved by The Animal Ethics Committee of the Centro de Investigación y Tecnología Agroalimentaria de Aragón (CITA).

2.2. Animal Management and Experimental Design

The experiment was conducted in the facilities of CITA in Zaragoza, Spain —41°3′ N, 0°47′ W and 216 m above sea level— in spring 2019, during 28 days. All the methodology carried out during the experiment has been explained in detail in a previous study [17]. Briefly, after lambing twenty multiparous Rasa Aragonesa ewes with their male lambs were assigned into two homogeneous groups according to ewe body weight (BW; 61 ± 6.2 kg), body condition score (BCS; 3.3 ± 0.57), lambing date (April 6 ± 0.1 d) and lamb body weight at birth (4.1 ± 0.64 kg). All dams were fed fresh sainfoin (*Onobrychis viciifolia* cv Reznos) —dry matter (DM): 213 g/kg; crude protein (CP): 116 g/kg DM; neutral detergent fiber (NDF): 369 g/kg DM; acid detergent fiber (ADF): 264 g/kg DM; total PAC: 38.8 g eq. PAC sainfoin/kg DM—, water and mineral blocks ad libitum and 200 g/head/day of barley —DM: 912 g/kg; NDF: 250 g/kg DM; ADF: 87 g/kg DM; CP: 95 g/kg DM— distributed in two meals. Before each meal, ten ewes were drenched with 100 mL of water (Sainfoin), whereas ten ewes were orally dosed with 100 mL of polyethylene glycol (PEG) solution (50 g of PEG 4000/100 mL; Sainfoin + PEG to inactivate the effects of PAC. Each pair of dam-lamb was placed in an individual pen (2.2 m²). Lambs exclusively suckled their dams ad libitum until they reached the target slaughter weight of 10–12 kg BW.

The detailed chemical composition of feedstuffs and milk has been reported in a previous study [17]. The dry matter intake (1879.5 ± 281.3 g DM/d) and the milk yield and chemical composition was similar between groups —milk yield: 1.25 L/d; crude fat: 6.54%, CP: 5.02%, lactose: 5.28%—, except for the polyphenols (42.3 vs. 51.8 µg eq. [gallic acid]/g fresh sample, for Sainfoin and Sainfoin + PEG, respectively) and urea (275 vs. 338 mg/L, for Sainfoin and Sainfoin + PEG, respectively).

2.3. Measurements and Sampling Procedures

Lambs were weighed weekly at 8:00 h with an electronic scale (0.1 kg precision) to calculate the average daily gain (ADG). Blood samples were obtained the day of slaughter, from the jugular vein into heparin tubes (Vaccuette, Madrid, Spain). Samples were immediately centrifuged—$3000\times g$ for 15 min at 4 °C—and stored at -20 °C until the metabolites analyses were performed.

When lambs reached the target weight of 10–12 kg BW, they were stunned by a captive bolt pistol and exsanguinated in the experimental abattoir of the Research Centre, using standard commercial procedures and according to Council Regulation (EC) N° 1099/2009. The contents of the digestive tract corresponding to the sections of reticulum-rumen, omasum-abomasum and duodenum-jejunum were extracted and weighed. Then, the empty digestive compartments were weighed. Hot carcass weight (HCW) was recorded without head and offal. After 24 h chilling at 4 °C in total darkness, the cold carcass weight (CCW) was obtained. The dressing percentage was calculated as:

$$\frac{\text{HCW}}{\text{slaughter weight}} \times 100 \tag{1}$$

and the carcass shrinkage was calculated as:

$$\left(\frac{\text{HCW} - \text{CCW}}{\text{HCW}}\right) \times 100 \tag{2}$$

The fatness degree of the carcasses was determined following the Community Scale for Classification of Carcasses of Ovine Animals and of Light Lambs [18] and scored from 1 (1−, very low) to 4 (4+, very high) following the scale of 1 (low), 2 (slight), 3 (average), and 4 (high). Caudal subcutaneous fat color was measured on tail root using a Minolta CM-2006 d spectrophotometer (Konica Minolta Holdings, Inc., Osaka, Japan), registering lightness (L*), redness (a*), and yellowness (b*), which were used to calculate hue angle (h_{ab}), and chroma (C^*_{ab}). The absolute value of the summation of the translated spectrum (SUM) was calculated as:

$$SUM = \left[\left(\frac{TR_{450}}{2}\right) + TR_{460} + TR_{470} + TR_{480} + TR_{490} + TR_{500} + \left(\frac{TR_{510}}{2}\right)\right] \cdot 10 \tag{3}$$

where TR_i was the reflectance value at i nm. An extensive explanation of the baselines of the method is exposed in Prache and Theriez [19].

After that, the carcass was carefully split longitudinally into the two half carcasses and the *longissimus thoracis et lumborum* (*LTL*) muscles of both sides were collected. Perirenal fat deposit was extracted and weighed.

2.4. Meat Quality

The *LTL* muscles from 4th–6th lumbar vertebrae of the left side were used to measure the pH with a pH-meter equipped with a Crison 507 penetrating electrode (Crison Instruments, S.A., Barcelona, Spain) and to estimate the chemical composition by NIRs (FoodScan™ 2, Foss Analytics, Hilleroed, Denmark). From 6th to 13th thoracic *vertebrae* from both sides were sliced into 2.5 cm-hick samples, the left ones were assigned to days 0, 2 and 7 of display and the right ones for days of display 5 and 9. The slices were placed in trays, wrapped with oxygen permeable polyvinyl chloride film, and kept in darkness at 4 °C until being measured for color and heme pigment estimations. *LTL* color was measured as had been explained above, while heme pigments were measured as described in Lobón, Blanco, Sanz, Ripoll, Bertolín and Joy [4]. The 0-d samples were allowed to bloom also in darkness at 4 °C for 1 h before being measured. After that, meat was freeze-dried and vacuum-stored in total darkness at −80 °C until the analysis of polyphenols and a 2,2-azinobis-3-ethylbensothiazoline-6-sulfonic acid (ABTS) assay, which were determined according to Leal, et al. [20] and Vázquez, et al. [21], respectively.

2.5. Plasma Analysis

Plasma concentrations of creatinine and urea (kinetic methods) were analyzed with an automatic analyzer (GernonStar, RAL/TRANSASIA, Dabhel, India). The methodology to determine antioxidant activity of plasma regarding to polyphenols concentration, super-oxide dismutase (SOD) and ABTS and the method of the determination of lipid oxidation (measured as malondialdehyde; MDA) are described in Baila, et al. [17].

2.6. Statistical Analysis

Data were analysed with the SAS [22] using the lamb as the experimental unit. The productive traits—animal performances and plasmatic metabolites—, carcass characteristics and meat chemical composition of the suckling lambs were analyzed through an variance analysis with a general linear model, with the presence of PAC as the fixed effect. Color and heme pigments of *LTL* muscle were analysed with mixed models—MIXED procedure—with presence of PAC, time of display and their interactions as fixed effects and the lamb as the random effect. The degrees of freedom were adjusted with the Kenward-Rodger correction. Results were reported as least square means and their associated standard errors of the means, and the Tukey correction was applied for pair-wise comparisons. The effects were considered significant at $p < 0.05$.

3. Results

3.1. Lamb Performance and Plasma Metabolites

The productive traits of the suckling lambs are shown in Table 1. The presence of PAC in the dams' diet did not affect any productive trait studied, such as average daily gain, age and weight at slaughter. Regarding the plasma metabolites at slaughter, the presence of PAC in the dams' diet did not affect creatinine and urea plasmatic concentrations of the suckling lambs.

Table 1. Effect of the presence of proanthocyanidins (PAC) in the dams' diet [1] on the performance, plasma metabolites and antioxidant (AO) status of their suckling lambs.

Item	Sainfoin	Sainfoin + PEG	s.e.m [2]	*p*-Value
Birth weight, kg	4.0	4.2	0.29	0.59
Average daily gain, g/d	272	283	21.0	0.63
Slaughter age, d	28.1	25.7	2.38	0.33
Slaughter weight, kg	11.6	11.1	0.33	0.14
Plasma metabolites				
Creatinine, μmol/L	52.2	55.4	7.28	0.67
Urea, mmol/L	4.86	5.16	0.619	0.63
Polyphenols, eq. [gallic acid] mg/mL	1.62	1.70	1.160	0.49
Antioxidant status				
Superoxide dismutase (SOD), U/mL	0.68	0.63	0.115	0.65
Total AO capacity–ABTS [3]	5.32	5.79	0.294	0.13
Lipid oxidation, μM MDA [4]	9.69	9.51	0.699	0.31

[1] Sainfoin: lambs whose dams were fed ad libitum sainfoin + 200 g/d barley; Sainfoin + PEG: lambs whose ewes were fed ad libitum sainfoin + 200 g/d barley + 100 g/d polyethylene glycol (PEG); [2] standard error of the mean; [3] 2,2-azinobis-(3-ethylbensothiazoline)-6-sulfonic acid, μmol eq. [TROLOX]/mL; [4] malondialdehyde.

In the same line, the plasma polyphenols concentration and the antioxidant activity measured as SOD, ABTS and MDA were not affected by the presence of PAC in the dams' diet ($p > 0.05$).

3.2. Digestive Compartments and Carcass Traits

The presence of PAC in the dams' diet significantly increased the weight of the content in the reticulum-rumen ($p < 0.05$) and, concomitantly, in the forestomach ($p < 0.01$) and increased the weight of the digestive compartments ($p < 0.05$), except for the omasum-abomasum (Table 2). Nevertheless, most of the carcass characteristics were not affected by the treatment (Table 2). The color of subcutaneous fat of the suckling lambs was similar between groups, except for lightness, which was decreased with the presence of PAC in the dams' diet ($p < 0.01$).

Table 2. Effect of the presence of proanthocyanidins (PAC) in the dams' diet [1] on the weights of the digestive compartments, the carcass characteristics and the color of caudal fat deposits of their suckling lambs.

Item	Sainfoin	Sainfoin + PEG	s.e.m [2]	*p*-Value
Weight of digestive content, g fresh matter (FM)				
Reticulum-rumen	284	192	39.5	0.033
Omasum-abomasum	185	161	41.6	0.57
Forestomach	469	353	47.5	0.011
Duodenum-jejunum	69	46	18.4	0.23
Weight of digestive compartments, g FM				
Reticulum-rumen	126	78	16.3	0.009
Omasum-abomasum	95	73	14.8	0.15
Forestomach	221	151	17.2	0.001
Duodenum-jejunum	171	132	13.8	0.011
Carcass traits				
Hot carcass weight, kg	7.78	7.64	0.264	0.61
Cold carcass weight, kg	6.19	6.15	0.238	0.87
Dressing percentage, %	55.6	57.2	1.03	0.13
Carcass shrinkage, %	3.98	2.99	0.595	0.12
Fatness score, 1–4 scale	2.10	2.15	0.103	0.74
Perirenal fat weight, g	128	129	24.7	0.98
Color of caudal fat deposits				
Lightness (L*)	68.8	71.4	0.48	0.015
Redness (a*)	2.6	2.5	0.31	0.89
Yellowness (b*)	12.3	11.9	0.45	0.66
Hue angle (h_{ab})	78.2	78.9	0.16	0.77
Chroma (C^*_{ab})	12.6	12.2	0.49	0.71
SUM [3]	102	110	13.1	0.75

[1] Sainfoin: lambs whose dams were fed ad libitum sainfoin + 200 g/d barley; Sainfoin + PEG: lambs whose ewes were fed ad libitum sainfoin + 200 g/d barley + 100 g/d polyethylene glycol (PEG); [2] standard error of the mean; [3] estimator of carotenoids.

3.3. Meat Quality

There were no differences between groups in the pH values of *LTL* at 24 h *post-mortem*, DM, CP, intramuscular fat, total collagen and ash contents (Table 3). The polyphenol content of meat significantly decreased with the presence of PAC in the dams' diet ($p < 0.05$) but did not affect the total antioxidant capacity estimated by ABTS content.

Table 3. Effect of the presence of proanthocyanidins (PAC) in the dams' diet [1] on the pH, chemical composition and total antioxidant (AO) capacity of the meat of their suckling lambs.

Item	Sainfoin	Sainfoin + PEG	s.e.m [2]	*p*-Value
pH $_{24\,h}$	5.5	5.53	0.032	0.46
Dry matter, % fresh matter (FM)	21.0	21.0	0.07	0.85
Crude protein, % FM	21.0	21.3	0.28	0.17
Intramuscular fat, % FM	2.36	2.19	0.124	0.18
Total collagen, % FM	0.77	0.77	0.153	0.99
Ash, % FM	1.81	1.78	0.103	0.75
Polyphenols, µg eq gallic acid/g FM	71.0	81.5	4.17	0.021
Total AO capacity-ABTS [3]	0.44	0.45	0.035	0.76

[1] Sainfoin: lambs whose dams were fed ad libitum sainfoin + 200 g/d barley; Sainfoin + PEG: lambs whose ewes were fed ad libitum sainfoin + 200 g/d barley + 100 g/d polyethylene glycol (PEG); [2] standard error of the mean; [3] 2,2-azinobis-(3-ethylbensothiazoline)-6-sulfonic acid, µmol eq. [TROLOX]/g FM.

No significant interactions were observed between the presence of PAC in the dams' diet and the time of display of meat. The *LTL* color (Figure 1) was not modified by the presence of PAC on the dams' diet, but it was affected by the time of display ($p < 0.05$). All color variables increased from day 0 to day 2 ($p < 0.05$), and thereafter L* and b* remained steady, whereas C*$_{ab}$ and a* increased until day 5, remaining unchanged onwards ($p < 0.001$). Similarly to the color parameters, the presence of PAC in the dams' diet had no significant effect on heme pigments (Figure 2), but the day of display had a significant effect ($p < 0.05$). Metmyoglobin and oxymyoglobin showed a similar evolution over time, increasing until day 5 ($p < 0.001$), whereas deoxymyoglobin followed the inverse pattern, decreasing until 5 day ($p < 0.001$) and remaining steady thereafter.

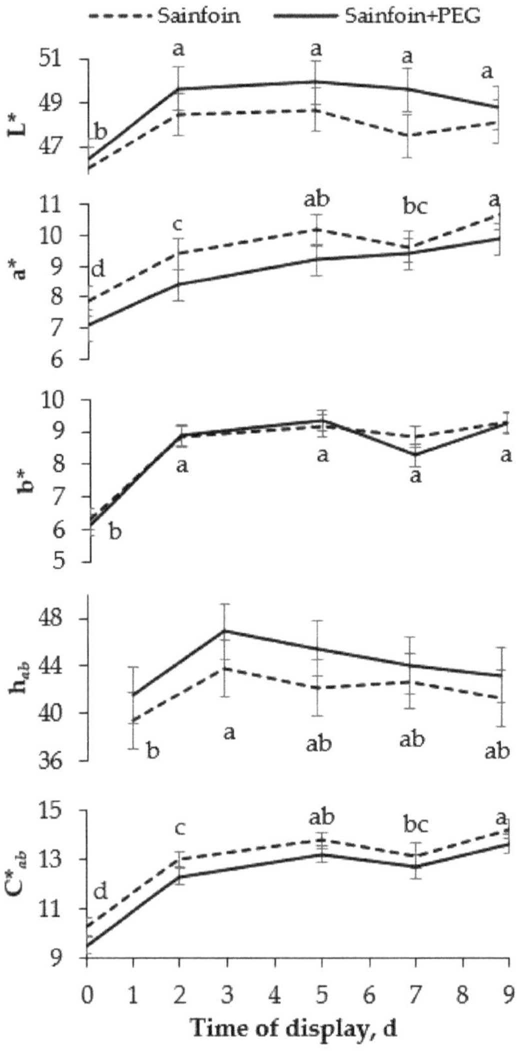

Figure 1. Evolution of instrumental color [lightness (L*), redness (a*), yellowness (b*), hue angle (h$_{ab}$), and chroma (C*$_{ab}$)] of meat of suckling lambs according to the presence of proanthocyanidins (PAC) in their dams' diet. Within a parameter, different letters mean differences at $p < 0.05$ among days. Vertical bars indicate the standard error of the mean.

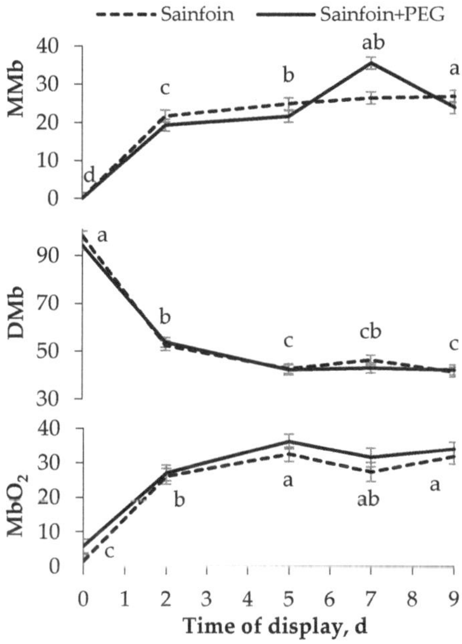

Figure 2. Evolution of heme pigments [metmyoglobin (MMb), deoxymyoglobin (DMb), and oxymyoglobin (MbO$_2$)] of meat of suckling lambs according to the presence of proanthocyanidins (PAC) in their dams' diet. Within a parameter, different letters mean differences at $p < 0.05$ among days. Vertical bars indicate the standard error of the mean.

4. Discussion

The lack of effect of the presence of PAC from sainfoin in the diet of lactating ewes on the weight gains and carcass characteristics of their suckling lambs in the current study can be related to the similar milk production and quality of the dams, as previously shown. Milk yield and composition are the main factors responsible for suckling lamb growth [23], the protein intake being the most determinant [24]. A previous study [10] reported a higher ADG in suckling lambs whose dams received polyphenols of grape seed extract, but not when the supplementation was given directly to fattening lambs. Therefore, it is possible that polyphenols elicit a more pronounced effect during the suckling period compared to the post-weaning phase, as shown in lambs supplemented with grape pomace [25]. In a meta-analysis, it was reported that the performance of weaned lambs was not modified when the concentrations of PAC in their diet ranged between 16 and 25 g PAC/kg DM [26].

The growth of the suckling lambs of this study was comparable to that observed when dams were fed concentrates indoors, and higher than those obtained with dams fed on pasture during lactation [27]. Therefore, in this study, a diet based mainly on fresh sainfoin with only a 10% of supplementation was sufficient to achieve good performances in rearing a male lamb of this autochthonous breed. This alternative feeding management could be advisable to allow diversification of the production system and to increase system resilience in an unfavorable situation from a meteorological, social and economic point of view [28].

The similar plasma urea concentration at slaughter of lambs of both treatments was unexpected, because dams from the Sainfoin group had lower urea concentrations both in the plasma and the milk [17]. The Sainfoin + PEG suckling lambs ingested a greater quantity of urea from their dam's milk, but it was not reflected in their plasma concentration, which suggests that the protein metabolism could be different between groups. The plasma creatinine concentration in suckling lambs was similar between treatments, indicating a similar metabolism of muscle mass [29], so that the PAC of dams' diet had no effect on the use of amino acids of their lambs by reducing the ruminal degradation of dietary protein [16].

The antioxidant effect of PAC is well known [14,15], and some studies even report the transfer of dietary phenolic compounds from the milk to the meat of the suckling lamb, where they act as antioxidants [30]. However, in this study it was not observed in dams' milk, which was reflected in the similar plasma antioxidant activity of lambs from both treatments. The metabolism of PAC along the digestive tract is complex [31], and the fact that the milk ingested by the lambs has already been processed in the mammary gland complicates the mechanism even further. In addition, Leparmarai, et al. [8] suggested that most of the phenolic compounds ingested by sheep were catabolized before reaching the milk or blood, so that the actual amount received by the lambs through the milk could be very low.

The carcass characteristics, dressing percentage and fat cover was similar in the suckling lambs of both experimental groups. Differences observed in carcass weight, fatness score and perirenal fat weight are usually related to the quality of feeding sources [32,33], dry matter intake [5,34,35], and/or age and weight at slaughter [33], all of which were similar in the present study. On average, the dressing percentage obtained in the suckling lambs was greater than the observed in fattened lambs, due to their lower development of the digestive tract of the former [5,33].

Unexpectedly, the reticulum-rumen and forestomach contents and the empty digestive compartments were heavier in lambs of the Sainfoin group. The higher weight of the digestive tract is usually associated with a worst dressing percentage [36]. In this case, the greater weight of the digestive organs of the Sainfoin group lambs produced a numerically lower dressing percentage, although the differences were not statistically significant. On the other hand, the digestive content is closely related with the intake. Since the milk intake of the lambs of both treatments was similar, the cause of this result remains unclear.

Yellowness and SUM (an estimator of carotenoids) in animal fat and meat are mainly influenced by the carotenoid pigments coming from feedstuffs [5,37]. The lack of differences observed in these parameters between treatments is ascribed to the similar sainfoin intake of their dams, and the consequent similar intake of secondary compounds. The lower L* value of caudal fat in Sainfoin lambs disagrees with the results observed by Rivaroli et al. [35], who concluded that the mechanisms of PAC modifying the L* of caudal fat deposits are unclear. Brainard [38] developed a method to estimate if instrumental color differences (expressed as ΔE_{ab}^{*}) are perceptible by human vision. In relation to this, Carrasco, et al. [39] reported that differences of color (ΔE_{ab}^{*}) of caudal fat lower than 5.2 between carcasses were imperceptible. In the present study, the difference between both treatments was 2.6; therefore, the practical implications were minimal. Fat L* values in suckling lambs of Churra Tensina, a similar local breed, ranged between 69 and 72 when their dams received a diet based on fresh forage or hay [40]. On the other hand, the color parameters were similar to those obtained by Lobón et al. [41] in fattened lambs previously raised by dams grazing sainfoin, despite their post-weaning finishing period and, consequently, heavier slaughter weight (23.4 kg BW).

Regarding the meat quality, the pH value of meat was within the normal range, close to 5.50 [27]. The presence of PAC in the dams' diet did not have any effect on the chemical composition of the meat of their suckling lambs, as Gómez-Cortés et al. [42] reported when grape pomace was included as a source of PAC in the diet of lactating ewes. The content of fat and protein in the meat were similar to those obtained in suckling lambs of dams fed with pasture [27] or supplemented with polyphenols [10].

In this study, the content of polyphenol in the meat of suckling lambs reflects the concentration in the milk of their dams. Proanthocyanidins and polyphenols are characterized by providing a great antioxidant power [14] and studies show an improvement of antioxidant status which increases the shelf life of lamb meat [7,43]. However, this result was not reflected on the antioxidant capacity, probably because it was only measured on the first day post slaughter, when the oxidation process had hardly started.

In relation to heminic pigments in *LTL*, Vieira et al. [30] observed a reduction of MMb in the meat of suckling lambs whose dams were fed grape pomace (containing PAC). However, this reduction was observed from day 10 of meat storage, so the time of the present study could have been insufficient to show effects. Contrary to the present results of color, Vasta, et al. [44] in a review pointed out that the meat of lambs fed with PAC was lighter than that of their counterparts fed with PEG, concluding that the mechanisms of action of PAC on meat color are still unclear. Besides in the abovementioned review most of studies involved weaned lambs, whereas here we studied suckling lambs fed exclusively with maternal milk.

5. Conclusions

The inclusion of PAC from sainfoin in the dams' diet had no significant effect on the ADG, plasmatic antioxidant activity, and carcass and meat quality of their suckling lambs. Therefore, fresh sainfoin can be fed to ewes during lactation to produce suckling lambs, achieving good performances and meat quality.

Author Contributions: Conceptualization, M.J., S.L., I.C. and M.B.; formal analysis, C.B. and S.L.; chemical analysis, C.B. and G.R.; investigation, M.J., S.L., I.C., M.B. and C.B., writing-original draft preparation, C.B., review and editing, M.J., M B., I.C., G.R. and S.L.; project administration, M.J.; funding acquisition, M.J. All authors have read and agreed to the published version of the manuscript

Funding: Financial support for this project was provided by the Spanish Ministry of Economy and Competitiveness (RTA2017-00008-C02-01) and the Government of Aragón (Grant Research Group Funds, Group A17_20R). C. Baila had a predoctoral grant from AEI (PRE2018-086670).

Institutional Review Board Statement The animal study protocol was approved by the Ethics Committee of the Centro de Investigación y Tecnología Agroalimentaria de Aragón (CITA) (CEEA, 2017–07).

Informed Consent Statement: Not applicable.

Data Availability Statement: The datasets analyzed in the present study are available from the corresponding author on reasonable request.

Acknowledgments: Special thanks are due to the CITA-Department of Animal, to the Servicio General de Apoyo a la Investigación-SAI, Universidad de Zaragoza and to J.R. Bertolín, M.A. Legua and A. Domínguez for helping with the laboratory analyses.

Conflicts of Interest: The authors declare no conflict of interest.

References

1. Mottet, A.; de Haan, C.; Falcucci, A.; Tempio, G.; Opio, C.; Gerber, P. Livestock: On our plates or eating at our table? A new analysis of the feed/food debate. *Glob. Food Secur.* **2017**, *14*, 1–8. [CrossRef]
2. Calabrò, S.; Tudisco, R.; Balestrieri, A.; Piccolo, G.; Infascelli, F.; Cutrignelli, M.I. Fermentation characteristics of different grain legumes cultivars with the in vitro gas production technique. *Ital. J. Anim. Sci.* **2009**, *8*, 280–282. [CrossRef]
3. Guyader, J.; Janzen, H.H.; Kroebel, R.; Beauchemin, K.A. Forage use to improve environmental sustainability of ruminant production. *J. Anim. Sci.* **2016**, *94*, 3147–3158. [CrossRef] [PubMed]
4. Lobón, S.; Blanco, M.; Sanz, A.; Ripoll, G.; Bertolín, J.; Joy, M. Meat quality of light lambs is more affected by the dam's feeding system during lactation than by the inclusion of quebracho in the fattening concentrate. *J. Anim. Sci.* **2017**, *95*, 4998–5011. [CrossRef] [PubMed]
5. Girard, M.; Dohme-Meier, F.; Silacci, P.; Ampuero Kragten, S.; Kreuzer, M.; Bee, G. Forage legumes rich in condensed tannins may increase n-3 fatty acid levels and sensory quality of lamb meat. *J. Sci. Food Agric.* **2015**, *96*, 1923–1933. [CrossRef] [PubMed]
6. Lobón, S.; Joy, M.; Sanz, A.; Álvarez-Rodríguez, J.; Blanco, M. The fatty acid composition of ewe milk or suckling lamb meat can be used to discriminate between ewes fed different diets. *Anim. Prod. Sci.* **2019**, *59*, 1108–1118. [CrossRef]
7. Valenti, B.; Natalello, A.; Vasta, V.; Campidonico, L.; Roscini, V.; Mattioli, S.; Pauselli, M.; Priolo, A.; Lanza, M.; Luciano, G. Effect of different dietary tannin extracts on lamb growth performances and meat oxidative stability: Comparison between mimosa, chestnut and tara. *Animal* **2018**, *13*, 435–443. [CrossRef]
8. Leparmarai, P.T.; Sinz, S.; Kunz, C.; Liesegang, A.; Ortmann, S.; Kreuzer, M.; Marquardt, S. Transfer of total phenols from a grapeseed-supplemented diet to dairy sheep and goat milk, and effects on performance and milk quality. *J. Anim. Sci.* **2019**, *97*, 1840–1851. [CrossRef]

9. Valderrábano, J.; Calvete, C.; Uriarte, J. Effect of feeding bioactive forages on infection and subsequent development of Haemonchus contortus in lamb faeces. *Vet. Parasitol.* **2010**, *172*, 89–94. [CrossRef]

10. Giller, K.; Sinz, S.; Messadene-Chelali, J.; Marquardt, S. Maternal and direct dietary polyphenol supplementation affect growth, carcass and meat quality of sheep and goats. *Animal* **2021**, *15*, 100333. [CrossRef]

11. Peng, K.; Shirley, D.C.; Xu, Z.; Huang, Q.; McAllister, T.A.; Chaves, A.V.; Acharya, S.; Liu, C.; Wang, S.; Wang, Y. Effect of purple prairie clover (Dalea purpurea Vent.) hay and its condensed tannins on growth performance, wool growth, nutrient digestibility, blood metabolites and ruminal fermentation in lambs fed total mixed rations. *Anim. Feed Sci. Technol.* **2016**, *222*, 100–110. [CrossRef]

12. Luciano, G.; Monahan, F.; Vasta, V.; Biondi, L.; Lanza, M.; Priolo, A. Dietary tannins improve lamb meat colour stability. *Meat Sci.* **2009**, *81*, 120–125. [CrossRef]

13. Seoni, E.; Battacone, G.; Silacci, P.; Kragten, S.A.; Chelali, J.M.; Dohme-Meier, F.; Bee, G. Effect of condensed tannins from Birdsfoot trefoil and dietary protein level on growth performance, carcass composition and meat quality of ram lambs. *Small Rumin. Res.* **2018**, *169*, 118–126. [CrossRef]

14. Soobrattee, M.A.; Neergheen, V.S.; Luximon-Ramma, A.; Aruoma, O.I.; Bahorun, T. Phenolics as potential antioxidant therapeutic agents: Mechanism and actions. *Mutat. Res.-Fundam. Mol. Mech. Mutagen.* **2005**, *579*, 200–213. [CrossRef] [PubMed]

15. López-Andrés, P.; Luciano, G.; Vasta, V.; Gibson, T.M.; Biondi, L.; Priolo, A.; Mueller-Harvey, I. Dietary quebracho tannins are not absorbed, but increase the antioxidant capacity of liver and plasma in sheep. *Br. J. Nutr.* **2013**, *110*, 632–639. [CrossRef]

16. Piluzza, G.; Sulas, L.; Bullitta, S. Tannins in forage plants and their role in animal husbandry and environmental sustainability: A review. *Grass Forage Sci.* **2014**, *69*, 32–48. [CrossRef]

17. Baila, C.; Joy, M.; Blanco, M.; Casasús, I.; Bertolín, J.R.; Lobón, S. Effects of feeding sainfoin proanthocyanidins to lactating ewes on intake, milk production and plasma metabolites. *Animal* **2022**, *16*, 100438. [CrossRef]

18. *EEC. 461/93*; Community Scale for the Classification of Carcasses of Light Lambs. Directorate-General for Agriculture and Rural Development (European Commission): Brussels, Belgium, 1993.

19. Prache, S.; Theriez, M. Traceability of lamb production systems: Carotenoids in plasma and adipose tissue. *Anim. Sci.* **1999**, *69*, 29–36. [CrossRef]

20. Leal, L.N.; Jordán, M.J.; Bello, J.M.; Otal, J.; den Hartog, L.A.; Hendriks, W.H.; Martín-Tereso, J. Dietary supplementation of 11 different plant extracts on the antioxidant capacity of blood and selected tissues in lightweight lambs. *J. Sci. Food Agric.* **2019**, *99*, 4296–4303. [CrossRef] [PubMed]

21. Vázquez, C.V.; Rojas, M.G.V.; Ramírez, C.A.; Chávez-Servín, J.L.; García-Gasca, T.; Martínez, R.A.F.; García, O.P.; Rosado, J.L.; López-Sabater, C.M.; Castellote, A.I. Total phenolic compounds in milk from different species. Design of an extraction technique for quantification using the Folin–Ciocalteu method. *Food Chem.* **2015**, *176*, 480–486. [CrossRef]

22. *SAS Statistical Software*, v.9.3; EE.UU.; SAS Institute Inc.: Cary, NC, USA, 2012.

23. Gallardo, B.; Gómez-Cortés, P.; Mantecón, A.; Juárez, M.; Manso, T.; de la Fuente, M. Effects of olive and fish oil Ca soaps in ewe diets on milk fat and muscle and subcutaneous tissue fatty-acid profiles of suckling lambs. *Animal* **2014**, *8*, 1178–1190. [CrossRef] [PubMed]

24. Gargouri, A.; Caja, G.; Casals, R.; Mezghani, I. Lactational evaluation of effects of calcium soap of fatty acids on dairy ewes. *Small Rumin. Res.* **2006**, *66*, 1–10. [CrossRef]

25. Kafantaris, I.; Kotsampasi, B.; Christodoulou, V.; Makri, S.; Stagos, D.; Gerasopoulos, K.; Petrotos, K.; Goulas, P.; Kouretas, D. Effects of dietary grape pomace supplementation on performance, carcass traits and meat quality of lambs. *In Vivo* **2018**, *32*, 807–812. [CrossRef] [PubMed]

26. Álvarez-Rodríguez, J.; Serrano-Pérez, B.; Villalba, D.; Joy, M.; Bertolín, J.R.; Molina, E. ¿Afectan los taninos condensados de la dieta a los resultados productivos, la composición de ácidos grasos y el color de la carne de cordero? *Itea-Inf. Tec. Econ. Ag.* **2020**, *116*, 116–130. [CrossRef]

27. Fusaro, I.; Giammarco, M.; Chincarini, M.; Odintsov Vaintrub, M.; Palmonari, A.; Mammi, L.M.E.; Formigoni, A.; di Giuseppe, L.; Vignola, G. Effect of ewe diet on milk and muscle fatty acid composition of suckling lambs of the protected geographical origin abbacchio romano. *Animals* **2020**, *10*, 25. [CrossRef]

28. Ripoll-Bosch, R. *Diagnóstico de los Sistemas Ganaderos Ovinos en Áreas Desfavorecidas: Caracterización Productiva de la raza Ojinegra de Teruel, Análisis Integrado de Sostenibilidad y Evaluación de la Huella de Carbono*; Universidad de Zaragoza: Zaragoza, Spain, 2013.

29. Hegarty, R.; McFarlane, J.R.; Banks, R.; Harden, S. Association of plasma metabolites and hormones with the growth and composition of lambs as affected by nutrition and sire genetics. *Aust. J. Agric. Res.* **2006**, *57*, 683–690. [CrossRef]

30. Vieira, C.; Guerra-Rivas, C.; Martínez, B.; Rubio, B.; Manso, T. Effects of grape pomace supplementation on the diet of lactating ewes as compared to vitamin E on the meat shelf life of suckling lambs. *Meat Sci.* **2022**, *184*, 108666. [CrossRef]

31. Quijada, J.; Drake, C.; Gaudin, E.; El-Korso, R.; Hoste, H.; Mueller-Harvey, I. Condensed tannin changes along the digestive tract in lambs fed with sainfoin pellets or hazelnut skins. *J. Agric. Food Chem.* **2018**, *66*, 2136–2142. [CrossRef]

32. Karnezos, T.; Matches, A. Fatty acid composition of dock-fat from lambs grazing alfalfa, sainfoin, wheatgrass. *J. Food Qual.* **1993**, *16*, 409–415. [CrossRef]

33. Bonanno, A.; di Miceli, G.; di Grigoli, A.; Frenda, A.; Tornambè, G.; Giambalvo, D.; Amato, G. Effects of feeding green forage of sulla (Hedysarum coronarium L.) on lamb growth and carcass and meat quality. *Animal* **2011**, *5*, 148–154. [CrossRef]

34. Copani, G.; Niderkorn, V.; Anglard, F.; Quereuil, A.; Ginane, C. Silages containing bioactive forage legumes: A promising protein-rich feed source for growing lambs. *Grass Forage Sci.* **2016**, *71*, 622–631. [CrossRef]
35. Rivaroli, D.; Prunier, A.; Meteau, K.; do Prado, I.N.; Prache, S. Tannin-rich sainfoin pellet supplementation reduces fat volatile indoles content and delays digestive parasitism in lambs grazing alfalfa. *Animal* **2019**, *13*, 1883–1890. [CrossRef]
36. Díaz, M.; Velasco, S.; Caneque, V.; Lauzurica, S.; de Huidobro, F.R.; Pérez, C.; González, J.; Manzanares, C. Use of concentrate or pasture for fattening lambs and its effect on carcass and meat quality. *Small Rumin. Res.* **2002**, *43*, 257–268. [CrossRef]
37. Prache, S.; Aurousseau, B.; Theriez, M.; Renerre, M. Discoloration of sheep carcass subcutaneous fat. *INRA Prod. Anim.* **1990**, *3*, 275–285. [CrossRef]
38. Brainard, D.H. Color appearance and color difference specification. In *The Science of Color*, 2nd ed.; Shevell, S.K., Ed.; Elsevier: Oxford, UK, 2003; p. 340.
39. Carrasco, S.; Panea, B.; Ripoll, G.; Sanz, A.; Joy, M. Influence of feeding systems on cortisol levels, fat colour and instrumental meat quality in light lambs. *Meat Sci.* **2009**, *83*, 50–56. [CrossRef]
40. Joy, M.; Sanz, A.; Ripoll, G.; Panea, B.; Ripoll-Bosch, R.; Blasco, I.; Alvarez-Rodriguez, J. Does forage type (grazing vs. hay) fed to ewes before and after lambing affect suckling lambs performance, meat quality and consumer purchase intention? *Small Rumin. Res.* **2012**, *104*, 1–9.
41. Lobón, S.; Blanco, M.; Sanz, A.; Ripoll, G.; Joy, M. Effects of feeding strategies during lactation and the inclusion of quebracho in the fattening on performance and carcass traits in light lambs. *J. Sci. Food Agric.* **2019**, *99*, 457–463. [CrossRef] [PubMed]
42. Gómez-Cortés, P.; Guerra-Rivas, C.; Gallardo, B.; Lavín, P.; Mantecón, A.; de la Fuente, M.; Manso, T. Grape pomace in ewes diet: Effects on meat quality and the fatty acid profile of their suckling lambs. *Food Res. Int.* **2018**, *113*, 36–42. [CrossRef]
43. Vasta, V.; Luciano, G. The effects of dietary consumption of plants secondary compounds on small ruminants' products quality. *Small Rumin. Res.* **2011**, *101*, 150–159. [CrossRef]
44. Vasta, V.; Nudda, A.; Cannas, A.; Lanza, M.; Priolo, A. Alternative feed resources and their effects on the quality of meat and milk from small ruminants. *Anim. Feed Sci. Technol.* **2008**, *147*, 223–246. [CrossRef]

Article

Influence of the Use of Milk Replacers on Carcass Characteristics of Suckling Kids from Eight Spanish Goat Breeds

Guillermo Ripoll [1,2,]*, María Jesús Alcalde [3], Anastasio Argüello [4], María Guía Córdoba [5] and Begoña Panea [1,2]

1. Centro de Investigación y Tecnología Agroalimentaria de Aragón (CITA), Avenida de Montañana 930, 50059 Zaragoza, Spain; bpanea@cita-aragon.es
2. Instituto Agroalimentario de Aragón (IA2 CITA-Universidad de Zaragoza), 50013 Zaragoza, Spain
3. Departamento de Ciencias Agroforestales, Universidad de Sevilla, 41013 Sevilla, Spain; aldea@us.es
4. Departamento de Patología y Producción Animal y Ciencia y Tecnología de los Alimentos, Universidad de Las Palmas de Gran Canaria, 35416 Las Palmas, Spain; tacho@ulpgc.es
5. Nutrición y Bromatología, Instituto Universitario de Investigación de Recursos Agrarios (INURA), Escuela de Ingeniería Agrarias, Universidad de Extremadura, Avda. Adolfo Suarez s/n, 06007 Badajoz, Spain; mdeguia@unex.es
* Correspondence: gripoll@cita-aragon.es

Simple Summary: Goats are important species due to their contributions to the development of rural areas. Spain has one of the largest goat populations in Europe; however, literature on goat carcasses is very scarce and, therefore, it is crucial to study the specific productive potential of each breed. Previous studies by our team on other Spanish goat breeds have shown some interactions between breed and rearing systems and, consequently, further analysis is necessary. This paper aims to contribute to the general knowledge on the subject.

Abstract: Since goat milk has a higher value than kid meat in Europe, some farmers rear kids with milk replacers, although some studies have stated that kids raised on natural milk yield higher-quality carcasses. Our previous studies showed some interactions between breed and rearing system on carcass and meat quality. This study evaluated the influence of the use of milk replacers on several carcass characteristics of suckling kids from eight Spanish goat breeds (Florida, Cabra del Guadarrama, Majorera, Palmera, Payoya, Retinta, Tinerfeña, and Verata). A total of 246 kids fed milk replacer (MR) or natural milk (NM) were evaluated. Carcass, head, viscera, and kidney fat weights, as well as several carcass measurements (round perimeter, forelimb width, carcass length, forelimb length, and carcass compactness index), were registered. Forelimbs were dissected to study tissue composition. For all studied variables, interactions were found between rearing system and breed. In general, the MR rearing system increased the head and visceral weights, as well as the length measurements and muscle percentages. Conversely, the NM rearing system increased carcass compactness and resulted in higher fat contents, independent of the deposit. The choice of one or another rearing system should be made according to the needs of the target market.

Keywords: rearing system; tissue composition; breed

Citation: Ripoll, G.; Alcalde, M.J.; Argüello, A.; Córdoba, M.G.; Panea, B. Influence of the Use of Milk Replacers on Carcass Characteristics of Suckling Kids from Eight Spanish Goat Breeds. *Animals* **2021**, *11*, 3300. https://doi.org/10.3390/ani11113300

Academic Editor: Monica Isabella Cutrignelli

Received: 21 October 2021
Accepted: 17 November 2021
Published: 18 November 2021

Publisher's Note: MDPI stays neutral with regard to jurisdictional claims in published maps and institutional affiliations.

1. Introduction

Goats are important species due to their contributions to the development of rural areas [1], provisioning meat and milk, which are among the most valuable services of livestock [2]. Spain ranks second in the European Union in terms of number of goats, producing 20% of the goat milk and 10.9% of the kid meat in the European Union [3]. In addition, the sale of suckling kids makes up 20% of the total income per goat on the dairy farm [4], and 80% of this kid meat originates from the suckling kid category (*cabrito*) [5], whilst the other 20% of meat comes from adult goats that are no longer in dairy production.

These suckling kids have a live weight of 6–13 kg and a carcass weight of 3.5–7 kg and are perceived by consumers to be a high-quality meat [6].

Nevertheless, because goat meat has an insignificant position in terms of overall farm production, little attention has been given to it; hence, literature on goat carcass and meat quality is very scarce in comparison with other species. Therefore, it is crucial to study the specific productive potential of each breed.

Eighty-eight percent of European Union goats are raised extensively and slaughtered as kids, with carcass weights between 5 and 11 kg [7]. When kid goats are reared with their dams, the availability of milk for cheese production is decreased. In addition, the milk from goats with feeding kids has less fat and protein than milk from lactating goats without suckling kids [8]. Therefore, some goat farmers remove the kids from their dams at a very young age and rear them with milk replacers. Milk replacers specifically formulated for kids can result in high daily weight gain, but some authors [9–11] have pointed out that suckling kids better metabolize nutrients from natural milk, yielding higher-quality carcasses.

Previous studies by our team on other Spanish goat breeds have shown that different carcass and meat quality traits improved when animals were fed natural milk, although some interactions between breed and rearing system were found [10,12–15]; consequently, further analysis is necessary. The aim of this work was to study the influence of the use of milk replacers on several carcass characteristics of suckling kids from several Spanish goat breeds.

2. Materials and Methods

2.1. Animals

All procedures were conducted according to the guidelines of Directive 2010/63/EU on the protection of animals used for experimental and other scientific purposes (EU, 2010). Suckling male kids from 8 Spanish goat breeds were used: Florida, Cabra del Guadarrama, Majorera, Palmera, Payoya, Retinta, Tinerfeña, and Verata. The different breeds used have different productive purposes. Palmera and Payoya are clearly oriented to milk production, whereas Guadarrama and Retinta are reared mainly to produce meat, with milk production being a secondary purpose. The others breeds are mainly reared for milk production, with meat production being a secondary purpose. For additional information about breed characteristics and productions, see the Official Catalogue from the Spanish Agricultural Ministry, available at https://www.mapa.gob.es/es/ganaderia/temas/zootecnia/razas-ganaderas/razas/catalogo-razas/ (accessed on 1 January 2021).

All kids were evenly reared at two (FL, MA, PL, PY, and TI) or three farms (GU, RE, and VE) per breed in their respective local areas. Each farm reared approximately half of the kids of one breed into each rearing system. On each farm, kids were randomly selected from those born within a range of 10 days. The age of the dam was not considered, although most of them were between the 4th and 6th lactation. All kids were born from a single parturition and were raised with milk replacers (MR) or natural milk from the dams (NM). Kids of the MR rearing system were fed colostrum for the first 2 days and had free access to milk replacer 24 h a day, which was sucked from a teat connected to a unit for feeding a liquid diet. Commercial milk replacers were reconstituted at 17% (w/v) and given warm (40 °C). The main ingredients were skimmed milk ($\approx$60%) and whey. The chemical composition of milk replacers was as follows: total fat 25 ± 0.6%, crude protein 24 ± 0.5%, crude cellulose 0.1 ± 0.0%, ash 7 ± 0.6%, Ca 0.8 ± 0.1%, Na 0.5 ± 0.2%, P 0.7 ± 0.0%, Fe 36 ± 4.0 mg/kg, Cu 3 ± 1.7 mg/kg, Zn 52 ± 18.8 mg/kg, Mn 42 ± 14.4 mg/kg, I 0.22 ± 0.06 mg/kg, Se 0.1 ± 0.06 mg/kg, and BHT 65 ± 30 ppm. Kids of the NM rearing system suckled directly from dams with no additional feedstuff. Dams grazed in farm facilities for eight h per day and the rest of the time, they were housed with their kids in a stable. Kids from both rearing systems had free access to water 24 h a day.

The numbers of kids evaluated are shown in Table 1. The 246 kids were slaughtered at an estimated body weight of 8 kg to achieve a target hot carcass weight of 5.0 kg. The night before slaughtering, all animals were transported to the abattoir. Kids had access to water overnight but not to feed and were stalled without access to feed but with access to water. The animals were weighed just prior to slaughter (slaughter weight, SW). Standard commercial procedures according to the European normative of protection of animals at the time of killing (E.U., 2009) were followed. A head-only electrical stunning was applied (1.00 A) to kids, which were then exsanguinated and dressed. Since the traditional method of presenting carcasses in Spain is with the head and kidneys, thoracic viscera were removed and weighed [16], and thereafter, hot carcasses, including the head and kidneys, were weighed. Afterwards, the head was separated and weighed, and then, carcasses were hung by the calcaneus tendon and chilled for 24 h at 4 °C. Dressing percentage (DP) was calculated as HCW × 100/SW.

Table 1. Slaughter weight, dressing percentage, and weights of the head, viscera, and kidney fat of kids reared with milk replacer (MR) or natural milk from their dams (NM).

Breed (B)	RS	n	SW (kg)	DP (%)	Head (g)	Viscera (g)	KF (g)
Florida	MR	15	8.0 efg	62.2 c	494.8 bcd	521.6 a	106.7 cd
	NM	15	7.9 gh	62.5 bc	497.3 bcd	447.1 b	196.6 a
Guadarrama	MR	15	7.5 h	66.9 a	504.7 bc	393.0 cd	52.5 fg
	NM	16	7.9 gh	62.4 c	491.3 bcd	376.2 dc	148.1 b
Majorera	MR	16	9.5 a	51.2 g	654.3 a	528.9 a	83.1 ef
	NM	16	8.9 bc	54.7 ef	486.6 cd	519.0 a	93.4 cde
Palmera	MR	15	9.2 ab	52.1 fg	505.2 bc	521.2 a	50.7 fg
	NM	16	8.9 bc	55.0 e	451.4 ef	516.9 a	80.3 def
Payoya	MR	16	8.4 de	59.1 d	515.6 b	505.1 a	37.5 g
	NM	14	8.4 de	59.3 cd	474.1 de	419.1 bc	52.6 fg
Retinta	MR	15	7.9 fgh	62.4 c	501.2 bc	348.7 e	63.0 efg
	NM	15	8.3 def	59.2 d	494.7 bcd	271.6 f	156.5 b
Tinerfeña	MR	16	9.2 ab	52.7 efg	501.6 bc	532.5 a	76.8 def
	NM	16	9.3 a	52.4 efg	510.4 b	531.6 a	94.2 cde
Verata	MR	15	8.4 dc	59.7 cd	444.8 f	355.7 e	123.3 bc
	NM	15	7.6 gh	65.4 ab	449.1 f	356.3 e	113.3 cd
	s.e.		0.16	1.09	8.86	12.85	11.65
	B		0.0001	0.0001	0.0001	0.0001	0.0001
	RS		0.1552	0.2817	0.0001	0.0001	0.0001
	B*RS		0.0028	0.0002	0.0001	0.0001	0.0001

B, breed; RS, rearing system; SW, slaughter weight; DP, dressing percentage (HCW*100/SW); KF, kidney fat. s.e., standard error. Least square means were adjusted for an HCW of 4.965 kg. Different superscripts (a,b,c,d,e,f,g,h) indicate significant differences ($p < 0.05$).

At 24 h postmortem, kidney fat from the left carcass side was removed and weighed (KF). Then, it was cut with a knife, and the internal colour of the kidney fat was measured using a Minolta CM-2006d Spectrophotometer (Konica Minolta Holdings, Inc., Osaka, Japan) in CIEL*a*b* space (CIE, 1986) with the specular component including 0% UV, observer angles of 10° and 0°, and white calibration. The integrating sphere had a 52 mm diameter, and the measurement area (with a diameter of 8 mm) was covered with a CMA149 dust cover (Konica Minolta Holdings, Inc.). The illuminant used was D65. The spectrophotometer was rotated 90° on the horizontal plane before each reading, and the mean of three readings was used for analysis. The lightness (L*), redness (a*), and yellowness (b*) indices were recorded using SpectraMagic NX software (Minolta Co. Ltd., Osaka, Japan), and the hue angle [$h_{ab} = \tan^{-1}\left(\frac{b*}{a*}\right) \cdot \frac{180°}{\pi}$] and chroma [$C^*_{ab} = \sqrt{(a*)^2 + (b*)^2}$] were calculated.

Thereafter, the following carcass measurements were recorded: rump circumference (maximum circumference measurement in a horizontal plane on the hanging carcass), rump width (maximum distance in a horizontal plane at the femur trochanter level), hind limb length (length from the perineum to the distal edge of the tarsus), and internal carcass length (L, length from the cranial edge of the symphysis pelvis to the cranial edge of the first rib) [17]. From these values, carcass compactness (HCW/L) was calculated.

Then, the left forelimb was separated from the carcass in a standardized manner [18,19], vacuum packed and stored at $-20\,°C$ until sampling. Once thawed at $4\,°C$ overnight, the forelimb was weighed and dissected into muscle, intermuscular fat, subcutaneous fat, and bone (major blood vessels, ligaments, tendons, and thick connective tissue sheets associated with some muscle), according to Panea, Ripoll, Albertí, Joy, and Teixeira [18]. The tissue composition of the forelimb was expressed as percentages of muscle, subcutaneous fat, intermuscular fat, total fat (intermuscular plus subcutaneous), and bone plus others (tendons, vessels, etc.).

2.2. Statistical Analysis

All statistics were calculated using XLSTAT statistical package v.3.05 (Addinsoft, New York, NY, USA). Studied variables were analysed using the ANCOVA procedure with the breed (B) and the rearing system (RS) as fixed effects and the hot carcass weight (HCW) as a covariate, while farm was considered as nested effect. Least square means were estimated, and differences were tested with the Bonferroni test at a 0.05 level of significance. Principal component analysis (PCA) was performed with the tissue composition variables. Only factors accounting for more variation than any individual type trait (eigenvalue P 1) were retained. A Varimax rotation was applied to the retained components to redistribute the variance among factors to obtain factor pattern coefficients. The resulting rotated factors are considerably less correlated than the original ones, making it easier to interpret the components without changing their explanatory power.

3. Results and Discussion

3.1. Carcass, Head, Viscera, and Kidney Fat Weights

Table 1 shows the slaughter weight (SW), dressing percentage (DP), head weight, visceral weight, and kidney fat weight (KF) as a function of the breed and rearing system.

There were significant interactions between breed and the rearing system for all variables. Hence, to achieve a hot carcass weight (HCW) of 5 kg, Majorera and Palmera kids fed MR and all Tinerfeña kids were slaughtered with the greatest SW. Conversely, Florida and Verata kids fed NM and Guadarrama kids were slaughtered with the lowest SW. The least square in the table has been adjusted for an HCW of 4.965 kg. Because the HCW was the same for all kids, breeds with greater SW had a lower dressing percentage (DP). Within breeds, Guadarrama and Retinta fed MR had greater DP than their counterparts fed NM ($p < 0.05$), while Majorera, Palmera, and Verata fed NM had greater DP than their counterparts fed MR ($p < 0.05$). Independent of the rearing system, the three Canarian breeds (Palmera, Tinerfeña, and Majorera) presented the lowest DP. In the MR rearing system, Guadarrama presented the highest DP values, whereas in the NM rearing system, Verata presented the highest values, although they were no different from Florida. Current DP agreed with those reported by several authors in similar breeds [10,16,20].

The use of MR increased the head weights ($p < 0.05$) of Majorera, Palmera, and Payoya, while the other breeds were not affected by the rearing system ($p > 0.05$). Verata from the two rearing systems had the lightest head, while Majorera fed MR had the heaviest head.

The use of MR increased the viscera weights ($p < 0.05$) of Florida, Payoya, and Retinta, while the other breeds were not affected by the rearing system ($p > 0.05$). Verata reared on both systems had the lightest viscera.

According to the results of this study, the influence of the rearing system is clearly conditioned by the breed but there is no pattern associated with the dairy or meat-production aptitude of the breeds used, that is, the differences are due to the breed and not to its usefulness.

Differences in dressing percentage and head and visceral weights between breeds of suckling kids have been reported previously [15,16], although some inconsistencies can be found in the literature. Since dressing percentage is mainly affected by the weight of the digestive tract [21], some authors have reported that natural milk increases the ruminal and intestinal weights in lambs because these lambs have a lower rumen functionality [22], whereas other authors reported opposite findings, supporting our results [23]. Panea, Ripoll, Horcada, Sañudo, Teixeira, and Alcalde [16] and Perez, et al. [24] found that the use of milk replacer did not affect the DP and head weights of Creole, Malagueña, and Murciano-Granadina suckling kids, but other studies reported that the visceral weight of Murciano-Granadina changed with the rearing system [16].

The KF weight increased when animals were fed NM in Florida, Guadarrama, and Retinta breeds, without changes in the other breeds. This increase is especially noticeable in the Guadarrama breed. Florida from NM system presented the highest KF weights whereas Payoya from MR presented the lowest values, although without differences with Payoya fed NM, Palmera fed MR, and Guadarrama fed MR. The effect of rearing system on fat carcass content agreed with the results from other authors [10,25] as well as the fact that in natural milk feeding regimes, dairy breeds presented higher fat amounts than meat-specialized breeds [10,16].

3.2. Kidney Fat Colour

The colour of kidney fat is not a quality cue per se because it is not an eaten fat, but since goats have very little subcutaneous fat, kidney fat is a good option to measure the influence of the rearing system on fat colour. Therefore, there were statistical interactions between the rearing system and the breed for all colour variables ($p < 0.001$). Colour L* and H_{ab} parameters are depicted in Figure 1. In general, L* values were higher for the NM rearing system, although no differences between rearing systems were detected in FL and GU. The h_{ab} was affected only in the Florida and Majorera breeds, with MR values higher than those of the NM breeds.

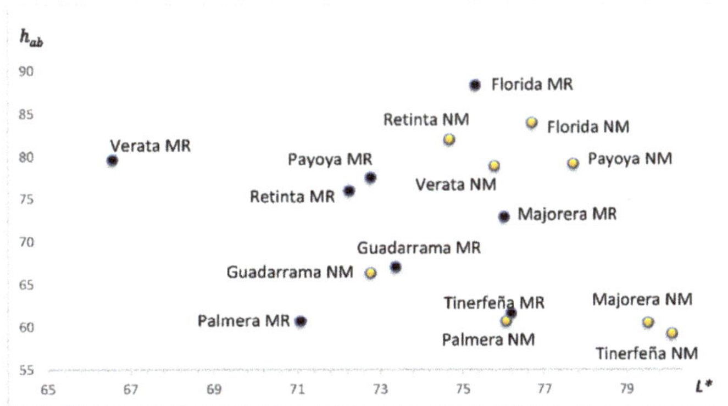

Figure 1. L* versus H_{ab} colour parameters of the kidney fat of kids reared with milk replacer (MR) or natural milk from their dams (NM).

3.3. Carcass Measurements

Means for carcass measurements are presented in Table 2. There were significant interactions between effects for all studied variables ($p < 0.005$). Both the round perimeter (RP) and the hind limb width (LWI) were affected by the rearing system only in the Majorera breed, with NM presenting higher values than MR. Florida presented the lowest values for both variables, independent of the rearing system, whereas Verata presented the highest values. The use of MR resulted in longer carcasses in the three Canarian breeds (Majorera, Palmera, and Tinerfeña), without influence on the other breeds. Majorera, Payoya, and Palmera presented lengthier carcasses than the other breeds, especially in animals from the Payoya breed fed NM. The rearing system affected forelimb length only in Majorera and Tinerfeña breeds, with NM values lower than MR values. The Payoya breed presented a longer forelimb, whereas Palmera and Tinerfeña presented shorter forelimbs. Finally, the rearing system affected the carcass compacity index in all breeds except in Majorera, Payoya, and Verata, with NM presenting higher values than MR.

Table 2. Carcass measurements of kids reared with milk replacer (MR) or natural milk from their dams (NM).

Breed (B)	RS	n	RP (cm)	LWI (cm)	CL (cm)	LL (cm)	CI (g/cm)
Florida	MR	15	29.8 [h]	8.3 [f]	40.6 [cde]	28.8 [bc]	122.3 [cd]
	NM	15	30.3 [gh]	8.2 [f]	39.0 [e]	28.4 [bc]	126.7 [a]
Guadarrama	MR	15	34.0 [def]	9.4 [de]	41.6 [bcd]	28.7 [bc]	119.0 [e]
	NM	16	37.0 [abcd]	10.2 [cde]	39.7 [de]	28.0 [bc]	124.2 [ab]
Majorera	MR	16	35.4 [cde]	9.2 [ef]	44.7 [a]	28.6 [bc]	111.1 [gh]
	NM	16	39.7 [a]	10.3 [cd]	41.7 [bcd]	20.4 [de]	117.8 [g]
Palmera	MR	15	37.4 [abc]	10.5 [cd]	43.1 [ab]	20.4 [de]	113.9 [h]
	NM	16	38.7 [ab]	10.0 [cde]	40.9 [cd]	20.5 [de]	120.2 [fg]
Payoya	MR	16	34.9 [cdef]	10.6 [c]	43.1 [ab]	30.5 [a]	114.8 [ef]
	NM	14	35.9 [cbde]	10.2 [cde]	43.1 [ab]	29.7 [ab]	115.0 [ef]
Retinta	MR	15	32.1 [fgh]	10.9 [c]	40.6 [cde]	27.6 [c]	121.1 [ef]
	NM	15	33.2 [efg]	11.1 [bc]	40.1 [de]	27.1 [c]	122.8 [c]
Tinerfeña	MR	16	37.7 [abc]	10.1 [cde]	42.4 [bc]	20.9 [d]	116.1 [gh]
	NM	16	37.8 [abc]	10.2 [cde]	39.1 [e]	19.2 [e]	125.7 [ef]
Verata	MR	15	33.9 [def]	12.3 [a]	40.2 [de]	27.6 [c]	123.1 [bc]
	NM	15	36.8 [abcd]	11.9 [ab]	40.2 [de]	27.6 [c]	123.5 [c]
	s.e.		0.23	0.10	0.15	0.30	1.44
	B		0.0001	0.0001	0.0001	0.0001	0.0001
	RS		0.0001	0.473	0.0001	0.0001	0.0001
	B*RS		0.006	0.001	0.0001	0.0001	0.0001

RS, rearing system; RP, round perimeter; LWI, forelimb width, CL, carcass length, LL, forelimb length, CI, carcass compactness index. s.e., standard error. Least square means were adjusted for an HCW of 4.965 kg. Different superscripts (a,b,c,d,e,f,g,h) indicate significant differences ($p < 0.05$).

In general, the MR rearing system increased the length measurements, whereas the NM rearing system increased compactness. Rodríguez, et al. [26] reported greater quality carcasses from kids reared with milk replacers, which agreed with the current results for carcass length and forelimb length.

3.4. Tissue Composition

The tissue composition of the forelimb is shown in Table 3. In general, animals fed NM had greater percentages of fat (subcutaneous, intermuscular, and total), while animals fed MR had greater percentages of muscle. Palmera fed MR had the greatest muscle percentage, and Verata fed NM had the lowest ($p < 0.05$). Majorera, Palmera, and Tinerfeña fed NM had the greatest percentages of subcutaneous fat, and Retinta fed MR had the lowest percentage ($p < 0.05$). Florida, Guadarrama, Majorera, Palmera, and Tinerfeña fed MR had the lowest percentages of intermuscular fat, and Tinerfeña and Verata fed NM had the greatest percentages ($p < 0.05$). Florida, Guadarrama, and Payoya had the greatest percentages of bone, and Palmera and Verata had the lowest ($p < 0.05$).

Table 3. Means and standard errors for forelimb tissue composition from kids reared with milk replacer (MR) or natural milk from their dams (NM).

Breed (B)	RS †	%M	%SF	%IF	%TF	%B
Florida	MR	66.2 [bcd]	0.8 [cdef]	7.7 [f]	8.5 [h]	25.2 [bc]
	NM	64.4 [de]	1.1 [bcde]	10.3 [de]	11.4 [defg]	24.2 [bcde]
Guadarrama	MR	63.8 [fg]	0.8 [cde]	7.2 [ef]	8.0 [efgh]	28.2 [a]
	NM	64.5 [de]	2.1 [a]	9.5 [def]	11.6 [defg]	23.9 [cdef]
Majorera	MR	67.7 [bc]	0.5 [ef]	6.0 [ef]	6.4 [gh]	25.9 [cde]
	NM	66.9 [cde]	1.0 [bc]	6.8 [def]	7.8 [cde]	25.3 [defg]
Palmera	MR	70.3 [a]	0.8 [bcd]	5.0 [ef]	5.8 [fgh]	23.9 [i]
	NM	67.3 [bcd]	1.4 [b]	7.8 [cd]	9.1 [bcd]	23.7 [hi]
Payoya	MR	65.1 [ef]	0.6 [def]	7.6 [def]	8.2 [efgh]	26.7 [b]
	NM	65.6 [de]	0.6 [def]	8.7 [cd]	9.2 [cdef]	25.2 [cdef]
Retinta	MR	67.5 [bc]	0.4 [f]	8.0 [cd]	8.3 [defg]	24.2 [fghi]
	NM	64.2 [ef]	0.7 [def]	12.7 [ab]	13.4 [b]	22.4 [ghi]
Tinerfeña	MR	68.2 [b]	0.9 [bc]	5.9 [ef]	6.8 [efgh]	25.0 [efgh]
	NM	65.9 [de]	1.3 [b]	6.9 [def]	8.1 [defg]	26.0 [bcd]
Verata	MR	65.6 [bcd]	0.9 [cdef]	11.7 [bc]	12.5 [bc]	21.8 [i]
	NM	62.8 [g]	1.1 [bc]	12.8 [a]	13.9 [a]	23.3 [ghi]
	s.e.	0.479	0.146	0.513	0.547	0.466
	B	0.0001	0.0001	0.0001	0.0001	0.0001
	RS	0.0001	0.0001	0.0001	0.0001	0.0372
	B*RS	0.0001	0.0055	0.1734	0.3895	0.0001

† RS, Rearing system; s.e., standard error; % M, percentage of muscle; % SF, percentage of subcutaneous fat; % IF, percentage of intermuscular fat; % TF, percentage of total fat; % B, percentage of bone and other tissues. Least square means were adjusted for an HCW of 4.965 kg. Different superscripts(a,b,c,d,e,f,g,h,i) indicate significant differences ($p < 0.05$).

The forelimb is often dissected into different tissues because it is easily disjointed, and it is said to be well related to the tissue composition of the carcasses of suckling kids [27,28]. The influence of the rearing system on tissue composition is not conclusive because it is conditioned by breed. De Palo, et al. [29] and Todaro, et al. [30] reported no effect of feed on limb weight or tissue composition, whereas Napolitano, et al. [31] did not find differences in muscle or bone percentages, but forelimb fat was greater when natural milk was used. Similarly, other authors [32,33] reported greater intramuscular amounts when animals were fed natural milk. Finally, Panea, Ripoll, Horcada, Sañudo, Teixeira, and Alcalde [16] reported that natural milk increased the subcutaneous and intramuscular fat percentages of Malagueña, while Murciano-Granadina was not affected by the rearing system. According to our results, the percentages of subcutaneous, intermuscular, and total fat of forelimbs are highly correlated among themselves, but as expected, the higher the bone percentage is, the lower the muscle percentage due to a well-known process called repartition [28,29]. These relationships can be clearly observed in Figure 2.

The usefulness of multivariate analysis to study variable relationships has been demonstrated by several authors [11,34]. The two axes of the biplot explained 82.27% of the variability. The first dimension separates the muscle percentage, on the left, to the fat percentages, on the right. Dimension 2 was explained by the bone percentage. Thus, the three Canary breeds were more muscled than the others, although Verata presented the heaviest carcasses. NM was related to fatness percentages, whereas MR was related to muscle percentage. As expected, muscle percentage was inversely related to fatness degree.

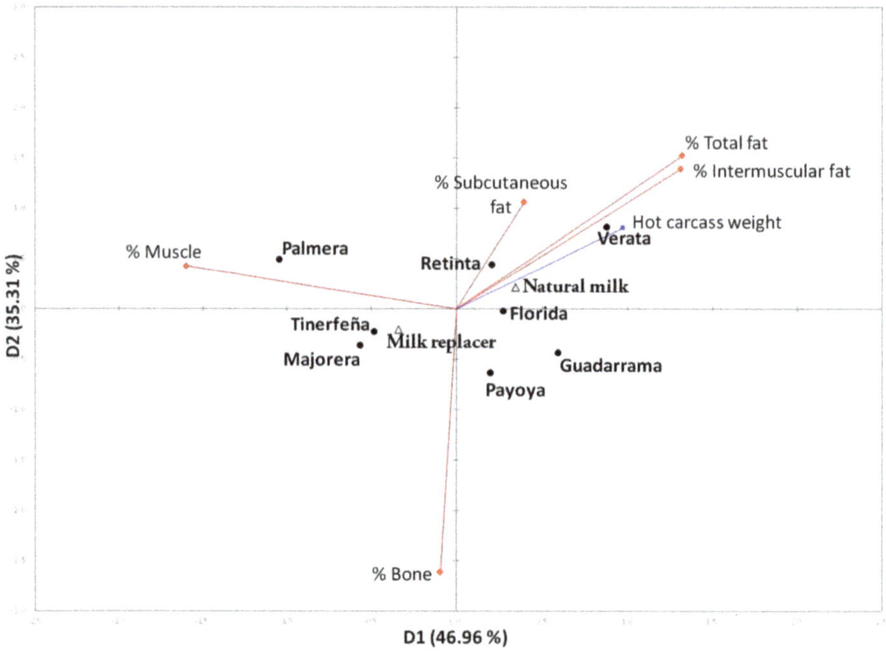

Figure 2. Biplot of the principal component analysis of tissue composition of forelimbs from kids reared with milk replacer (MR) or natural milk from their dams (NM).

4. Conclusions

For all studied variables, interactions were found between rearing system and breed. Hence, farmers should consider the selection of the breed and rearing system together to produce carcasses that the market demands. In general, the MR rearing system increased the head and visceral weights, as well as the length measurements and muscle percentages. Conversely, the NM rearing system increased carcass compactness and resulted in higher fat contents, independent of the deposit.

Author Contributions: Conceptualization, B.P. and G.R.; methodology, B.P., M.J.A., A.A., M.G.C. and G.R.; formal analysis, G.R. and B.P.; investigation, B.P., A.A., M.G.C., M.J.A. and G.R.; data curation, G.R.; writing—original draft preparation, A.A., M.G.C., G.R. and B.P.; writing—review and editing, G.R. and B.P.; funding acquisition, B.P. All authors have read and agreed to the published version of the manuscript.

Funding: This study was supported by the Ministry of Economy and Competitiveness of Spain and the European Union Regional Development Funds (RTA2012-0023-C03) and the Government of Aragón (Grant Research Group Funds, Group A14_20R).

Institutional Review Board Statement: The study was conducted according to the guidelines of the Declaration of Helsinki, and approved by the Institutional Review Board (or Ethics Committee) of CITA de Aragon (date of approval of exemption: 05 November 2021).

Data Availability Statement: Data are available under request.

Acknowledgments: Appreciation is expressed to the breeders and the staff of CITA de Aragón for their help in data collection.

Conflicts of Interest: The authors declare no conflict of interest.

References

1. Dubeuf, J.-P.; Morand-Fehr, P.; Rubino, R. Situation, changes and future of goat industry around the world. *Small Rumin. Res.* **2004**, *51*, 165–173. [CrossRef]
2. Bernués, A.; Rodríguez-Ortega, T.; Ripoll-Bosch, R.; Alfnes, F. Socio-Cultural and Economic Valuation of Ecosystem Services Provided by Mediterranean Mountain Agroecosystems. *PLoS ONE* **2014**, *9*, e102479. [CrossRef]
3. MERCASA. *Alimentación en España 2015. Producción, Industria, Distribución y Consumo*; Consumo, M.-D.Y., Ed.; Mercasa–Distribución Y Consumo: Madrid, Spain, 2016; p. 552.
4. Castel, J.M.; Mena, Y.; Ruiz, F.A.; Gutiérrez, R. Situación y evolución de los sistemas de producción caprina en España. *Tierras Caprino* **2012**, *1*, 24–37.
5. MAPAMA. Encuesta de Sacrificio de Ganado del Ministerio de Agricultura y Pesca, Alimentación y Medio Ambiente. Available online: www.mapama.gob.es/es/estadistica/temas/estadisticas-agrarias/ganaderia/encuestas-sacrificio-ganado/ (accessed on 1 January 2021).
6. Ripoll, G.; Alcalde, M.J.; Panea, B. Calidad instrumental de la carne de cabrito lechal. Revisión bibliográfica. *Inf. Tec. Econ. Agrar.* **2021**, *117*, 145–161. [CrossRef]
7. Shrestha, J.; Fahmy, M. Breeding goats for meat production: 2. Crossbreeding and formation of composite population. *Small Rumin. Res.* **2007**, *67*, 93–112. [CrossRef]
8. Caravaca, F.P.; Guzmán, J.L.; Delgado-Pertiñez, M.; Baena, J.A.; López, R.; Romero, L.; Alcalde, M.J.; Gonzalez-Redondo, P. Influencia del tipo de lactancia sobre la composición quimica de la leche de cabra de las razas Payoya y Florida Sevillana. In Proceedings of the XXIX Jornadas Científicas y VIII Internacionales de la SEOC, Lérida, España, 22–25 September 2004; pp. 312–313.
9. Argüello, A.; Castro, N.; Capote, J.; Solomon, M. Effects of diet and live weight at slaughter on kid meat quality. *Meat Sci.* **2005**, *70*, 173–179. [CrossRef]
10. Panea, B.; Ripoll, G.; Sañudo, C.; Horcada, A.; Alcalde, M.J. Influencia del sistema de lactancia sobre la calidad de la canal de cabrito de las razas Murciano-Granadina y Malagueña. In Proceedings of the XXXIX Jornadas de Estudio: XIII Jornadas Sobre Producción Animal, Zaragoza, Spain, 12–13 May 2009; pp. 493–495.
11. Herrera, P.Z.; Delgado, J.; Arguello, A.; Camacho, M. Multivariate analysis of meat production traits in Murciano-Granadina goat kids. *Meat Sci.* **2011**, *88*, 447–453. [CrossRef] [PubMed]
12. Ripoll, G.; Alcalde, M.J.; Argüello, A.; Córdoba, M.D.G.; Panea, B. Effect of the rearing system on the color of four muscles of suckling kids. *Food Sci. Nutr.* **2019**, *7*, 1502–1511. [CrossRef]
13. Ripoll, G.; Alcalde, M.J.; Argüello, A.; Cordoba, M.D.G.; Panea, B. Effect of Rearing System on the Straight and Branched Fatty Acids of Goat Milk and Meat of Suckling Kids. *Foods* **2020**, *9*, 471. [CrossRef]
14. Ripoll, G.; Alcalde, M.J.; Córdoba, M.G.; Casquete, R.; Argüello, A.; Ruiz-Moyano, S.; Panea, B. Influence of the Use of Milk Replacers and pH on the Texture Profiles of Raw and Cooked Meat of Suckling Kids. *Foods* **2019**, *8*, 589. [CrossRef]
15. Ripoll, G.; Alcalde, M.; Horcada, A.; Campo, M.; Sañudo, C.; Teixeira, A.; Panea, B. Effect of slaughter weight and breed on instrumental and sensory meat quality of suckling kids. *Meat Sci.* **2012**, *92*, 62–70. [CrossRef] [PubMed]
16. Panea, B.; Ripoll, G.; Horcada, A.; Sanudo, C.; Teixeira, A.; Alcalde, M.J. Influence of breed, milk diet and slaughter weight on carcass traits of suckling kids from seven Spanish breeds. *Span. J. Agric. Res.* **2012**, *10*, 1025–1036. [CrossRef]
17. Fisher, A.; de Boer, H. The EAAP standard method of sheep carcass assessment. Carcass measurements and dissection procedures report of the EAAP working group on carcass evaluation, in cooperation with the CIHEAM instituto agronomico Mediterraneo of Zaragoza and the CEC directorate general for agriculture in Brussels. *Livest. Prod. Sci.* **1994**, *38*, 149–159.
18. Panea, B.; Ripoll, G.; Albertí, P.; Joy, M.; Teixeira, A. Atlas of dissection of ruminant's carcass. (Atlas de disección de las canales de los rumiantes). *ITEA-Inf. Técnica Económica Agrar.* **2012**, *108*, 1–105.
19. Panea, B.; Ripoll, G. Substituting fat with soy in low-salt dry fermented sausages. *NFS J.* **2021**, *22*, 1–5. [CrossRef]
20. Santos, V.; Silva, A.; Cardoso, J.; Silvestre, A.; Silva, S.; Martins, C.; Azevedo, J. Genotype and sex effects on carcass and meat quality of suckling kids protected by the PGI "Cabrito de Barroso". *Meat Sci.* **2007**, *75*, 725–736. [CrossRef]
21. Joy, M.; Ripoll, G.; Delfa, R. Effects of feeding system on carcass and non-carcass composition of Churra Tensina light lambs. *Small Rumin. Res.* **2008**, *78*, 123–133. [CrossRef]
22. Fuente, J.; Tejon, D.; Rey, A.; Thos, J.; López-Bote, C.J. Effect of rearing system on growth, body composition and development of digestive system in young lambs. *J. Anim. Physiol. Anim. Nutr.* **1997**, *78*, 75–83. [CrossRef]
23. Otzule, L.; Ilgaža, A. Goat kids stomach morphological development depending on the milk type. In *Research for Rural Development. International Scientific Conference Proceedings*; Treija, S., Skujeniece, S., Eds.; Research for Rural Development, Latvia University of Agriculture: Jelgava, Latvia, 2014; Volume 1, pp. 185–189.
24. Pérez, P.; Maino, M.; Morales, M.; Soto, A. Effect of goat milk and milk substitutes and sex on productive parameters and carcass composition of Creole kids. *Small Rumin. Res.* **2001**, *42*, 87–94. [CrossRef]
25. Horcada, A.; Ripoll, G.; Alcalde, M.; Sañudo, C.; Teixeira, A.; Panea, B. Fatty acid profile of three adipose depots in seven Spanish breeds of suckling kids. *Meat Sci.* **2012**, *92*, 89–96. [CrossRef]
26. Rodríguez, A.; Landa, R.; Bodas, R.; Prieto, N.; Mantecon, A.R.; Giráldez, F. Carcass and meat quality of Assaf milk fed lambs: Effect of rearing system and sex. *Meat Sci.* **2008**, *80*, 225–230. [CrossRef] [PubMed]

27. Arguello, A.; Capote, J.; Ginés, R.; López, J. Prediction of kid carcass composition by use of joint dissection. *Livest. Prod. Sci.* **2001**, *67*, 293–295. [CrossRef]
28. Sanudo, C.; Campo, M.M.; Muela, E.; Olleta, J.L.; Delfa, R.; Jiménez-Badillo, R.; Alcalde, M.J.; Horcada, A.; Oliveira, I.; Cilla, I. Carcass characteristics and instrumental meat quality of suckling kids and lambs. *Span. J. Agric. Res.* **2012**, *10*, 690. [CrossRef]
29. De Palo, P.; Maggiolino, A.; Centoducati, N.; Tateo, A. Effects of different milk replacers on carcass traits, meat quality, meat color and fatty acids profile of dairy goat kids. *Small Rumin. Res.* **2015**, *131*, 6–11. [CrossRef]
30. Todaro, M.; Corrao, A.; Barone, C.; Alicata, M.; Schinelli, R.; Giaccone, P. Use of weaning concentrate in the feeding of suckling kids: Effects on meat quality. *Small Rumin. Res.* **2006**, *66*, 44–50. [CrossRef]
31. Napolitano, F.; Cifuni, G.; Pacelli, C.; Riviezzi, A.; Girolami, A. Effect of artificial rearing on lamb welfare and meat quality. *Meat Sci.* **2002**, *60*, 307–315. [CrossRef]
32. Arguello, A.; Castro, N.; Capote, J.; Solomon, M.B. The Influence of Artificial Rearing and Live Weight at Slaughter on Kid Carcass Characteristics. *J. Anim. Vet. Adv.* **2007**, *6*, 20–25.
33. Peña, F.; Domenech, V.; Acero, R.; Perea, J.; Garcia, A. Effect of sex and feed (maternal milk vs. milk substitute) on the growth and carcass characteristics in Florida goat kids. *Rev. Cient.* **2009**, *19*, 619–629.
34. Cañeque, V.; Pérez, C.; Velasco, S.; Diaz, M.; Lauzurica, S.; Álvarez, I.; de Huidobro, F.R.; Onega, E.; De La Fuente, J. Carcass and meat quality of light lambs using principal component analysis. *Meat Sci.* **2004**, *67*, 595–605. [CrossRef]

Article

Differences in Meat Quality of Six Muscles Obtained from Southern African Large-Frame Indigenous Veld Goat and Boer Goat Wethers and Bucks

Gertruida L. van Wyk [1], Louwrens C. Hoffman [1,2], Phillip E. Strydom [1] and Lorinda Frylinck [3,*]

[1] Department of Animal Sciences, University of Stellenbosch, Private Bag X1, Matieland, Stellenbosch 7602, South Africa; louisa.vanwyk@virbac.com (G.L.v.W.); louwrens.hoffman@uq.edu.au (L.C.H.); pestrydom@sun.ac.za (P.E.S.)

[2] Centre for Nutrition and Food Sciences, Queensland Alliance for Agriculture and Food Innovation (QAAFI), The University of Queensland, Coopers Plains, QLD 4108, Australia

[3] Agricultural Research Council—Animal Production, Private Bag X2, Irene 0062, South Africa

* Correspondence: lorinda@arc.agric.za; Tel.: +27-12-672-9385

Simple Summary: This study describes the meat tenderness and colour attributes of six muscles (*Longissimus thoracis et lumborum* (LTL), *Semimembranosus* (SM), *Biceps femoris* (BF), *Supraspinatus* (SS), *Infraspinatus* (IS), *Semitendinosus* (ST)) from same-aged young Boer Goat (BG) and Indigenous Veld Goat (IVG: Cape Speckled and the Cape Lob Ear) wethers and bucks. Muscle tenderness and colour characteristics differed more between wethers and bucks than between IVG and BG. Large-frame IVG bucks and wethers produced very similar meat tenderness, juiciness and colour characteristics to the BG bucks and wethers, indicating them to be just as suited for meat production. The wethers' meat, with its increased intramuscular fat in all six muscles tested, would satisfy the consumer segment that prefer juicier and more flavoursome meat. Knowledge of the muscle characteristics of goat carcasses will help the development of the formal commercial market for goat meat, which would benefit smallholder farmers, who typically produce most of the goats in the world.

Citation: van Wyk, G.L.; Hoffman, L.C.; Strydom, P.E.; Frylinck, L. Differences in Meat Quality of Six Muscles Obtained from Southern African Large-Frame Indigenous Veld Goat and Boer Goat Wethers and Bucks. *Animals* 2022, 12, 382. https://doi.org/10.3390/ani12030382

Academic Editor: Gianluca Neglia

Received: 23 November 2021
Accepted: 22 December 2021
Published: 4 February 2022

Publisher's Note: MDPI stays neutral with regard to jurisdictional claims in published maps and institutional affiliations.

Abstract: Various meat quality characteristics of six muscles (*Longissimus thoracis et lumborum* (LTL), *Semimembranosus* (SM), *Biceps femoris* (BF), *Supraspinatus* (SS), *Infraspinatus* (IS), *Semitendinosus* (ST)) from large-frame Boer Goats (BG) and Indigenous Veld Goats (IVG: Cape Speckled and the Cape Lob Ear) were studied. Weaner male BG (n = 18; 10 bucks and 8 wethers) and IVG (n = 19; 9 bucks and 10 wethers) were raised on hay and natural grass, and on a commercial pelleted diet to a live weight of 30–35 kg. All goats were slaughtered at a commercial abattoir and the dressed carcasses were chilled at 4 °C within 1 h post mortem. The muscles were dissected from both sides 24 h post mortem and aged for 1 d and 4 d. Variations in meat characteristics such as ultimate pH, water holding capacity (WHC), % purge, myofibril fragment length (MFL), intramuscular fat (IMF), connective tissue characteristics, and Warner-Bratzler shear force (WBSF) were recorded across muscles. Bucks had higher lightness (L*) and hue-angle values, whereas wethers had increased redness (a*) and chroma values. The muscle baseline data will allow informed decisions to support muscle-specific marketing strategies, which may be used to improve consumer acceptability of chevon.

Keywords: Cape Lob Ear; Cape Speckled; Boer Goat; meat goat breeds; meat tenderness; meat colour; collagen; chevon

1. Introduction

IVG are a group of specific pure-bred indigenous eco-types represented by the IVG-Association, which defines specific standards that a goat must adhere to before it can be classified as one of the eco-types such as the Cape Lob Ear and the Cape Speckled [1]. Both of these eco-types have large frames and can compete with the Boer Goat (BG) in terms of meat yield [2], whilst also having additional advantages such as adaptability to harsh climates and disease resistance [3]. The increasing global human population and the threat of global warming makes it important to promote the production of goat meat (chevon) from adapted eco-types such as the IVG. Although chevon is popular amongst the greater population of southern Africa, chevon is not available on commercial shelves in South Africa, mainly because there are insufficient commercial slaughter numbers to ensure a constant supply to the commercial retail market. Although southern Africa has relatively large numbers of meat goats (703,892 head) [4], most are produced in the informal sector and traded within this sector, thereby making it challenging to obtain official statistics of the volumes of goat meat produced and traded. Available goats are either sold alive for local traditional slaughtering practices or exported to Middle Eastern and Asian countries. Small and emerging southern African farmers are interested in IVGs as they do not require intensive management to be productive. For chevon, quality fresh meat is the most economically profitable; however, scientific knowledge on the meat quality of these breed types is scarce, compared to that of the well-known "improved" BG breed and the undefined "indigenous" goats that are usually used in comparative studies [5–10].

The term "meat quality" includes many attributes; of these, texture, juiciness, flavour and visual appeal are important to consumers. Tenderness and the mechanical properties of meat are influenced by the connective tissue, myofibrils and their interactions, which differ between muscles [11,12]. Compared to sheep and cattle, knowledge of the meat quality of BG and large-frame IVG of South Africa is limited due to a previous lack of interest. The goat carcass consists of over a hundred different muscles with different properties, which affect processing characteristics and could influence consumer acceptability [13]. There has been a continued trend in the retail sector to separate muscles, based on perceived connective tissue characteristics, to better market them and apply the knowledge in terms of the users' requirements. Notable studies on the physical and compositional traits of BG muscles have been conducted over the years [7]. These include carcass measurements and commercial yields [14], as well as cooking and juiciness related quality characteristics [15], including studies to understand the impact of carcass handling on the texture, mainly determined by the Warner-Bratzler shear force (WBSF) on different muscles [8,9,16,17]. Most studies evaluating chevon are conducted on the LTL and SM muscles in terms of tenderness and sensory quality attributes [5,6,9,10]. To establish a baseline for IVG eco-types, this paper focuses on the effect of breed (IVG vs. BG) and castration (Sex: bucks and wethers) on: ultimate muscle pH (pH_u), % purge, water holding capacity (WHC), WBSF, myofibril fragment length (MFL), intramuscular fat representing marbling, collagen characteristics, and meat colour in six different muscles (i.e., LTL, SM, BF, SS, IS, and ST) to establish baselines for these eco-types.

2. Materials and Methods

2.1. Animal and Experimental Design

This research was approved by the Agricultural Research Council-Animal Production (ARC-AP) Ethics Committee (ref no. APIEC16/021). Weaner BG (n = 18; 10 bucks and 8 wethers) and large-frame IVG (n = 19; 9 bucks and 10 wethers) were purchased from several commercial breeders at three months of age (17 kg on average for IVG and 20 kg on average for BG). The sourcing of animals from different producers provided sufficient representation of genetic variation for each breed type. When bought, the commercial breeders had already castrated the male animals on the farm. The animals were reared at the Small Stock Section of the ARC-AP situated in Irene in the Gauteng province of South Africa, where they grazed a natural grass diet supplemented with *Eragrostis curvula* hay (estimated crude protein 48.9 g/kg dry matter (DM); neutral detergent fibre 746 g/kg DM) ad libitum and an average of 250 g commercial "Ram, lamb and ewe-13" pellets (protein 130 g/kg, fat 25–70 g/kg, fibre 150 g/kg, moisture 120 g/kg, calcium 15 g/kg, phosphorus 3 g/kg, urea 10 g/kg; Meadow Feeds, Lanseria, South Africa) per day per animal. The goats were fed for, on average, 6 to 8 months until they attained a live weight (LW) of between 30 and 35 kg. After weighing

(LW), the goats were transported for 3 km to the abattoir of the ARC-AP on the day of slaughter. The experimental design is presented in Figure 1 and has been described in more detail in an earlier paper [2]. The carcasses were subjected to electrical stimulation (ES 20 s, 400 Volts peak, 5 ms pulses at 15 pulses/s), 10 min after stunning and exsanguination, after which all the carcasses were placed in the chiller at 4 °C within 60 min of post mortem. After chilling (24 h, <4 °C), the carcasses were removed from the chiller and the specific muscles removed from both sides of the carcass and cut into various slices for the different meat quality analyses (Figure 2). Temperature and pH values were measured 24 h post mortem (pH_u) on the same chilled muscles used for colour measurement with a calibrated CyberScan PC 300 (Eutech Instruments Pte Ltd., Queenstown, Singapore).

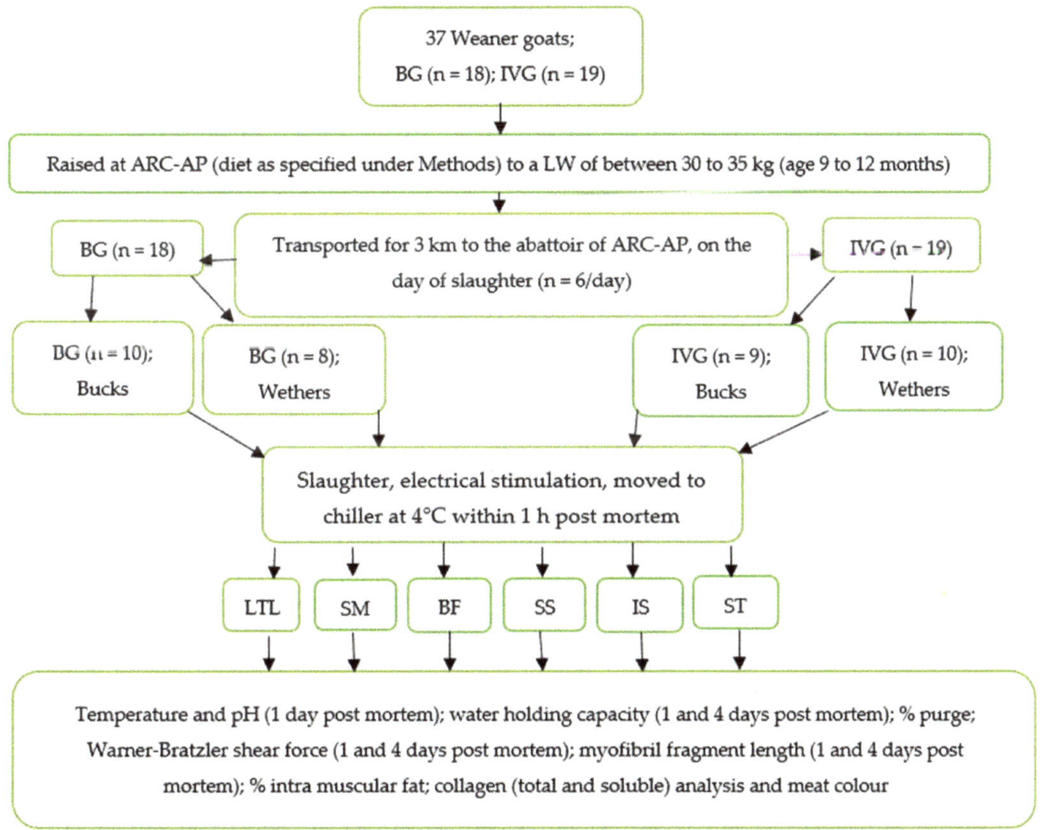

Figure 1. Experimental design to evaluate the effect of the breeds large-frame Indigenous Veld Goat (IVG, Cape Speckled and Cape Lob Ear) and Boer Goat (BG) of southern Africa, on tenderness factors, colour attributes and connective tissue characteristics of *Longissimus thoracis et lumborum* (LTL), *Semimembranosus* (SM), *Biceps femoris* (BF), *Supraspinatus* (SS), *Infraspinatus* (IS), and *Semitendinosus* (ST). ARC-AP = Agricultural Research Council-Animal Production, Irene, South Africa.

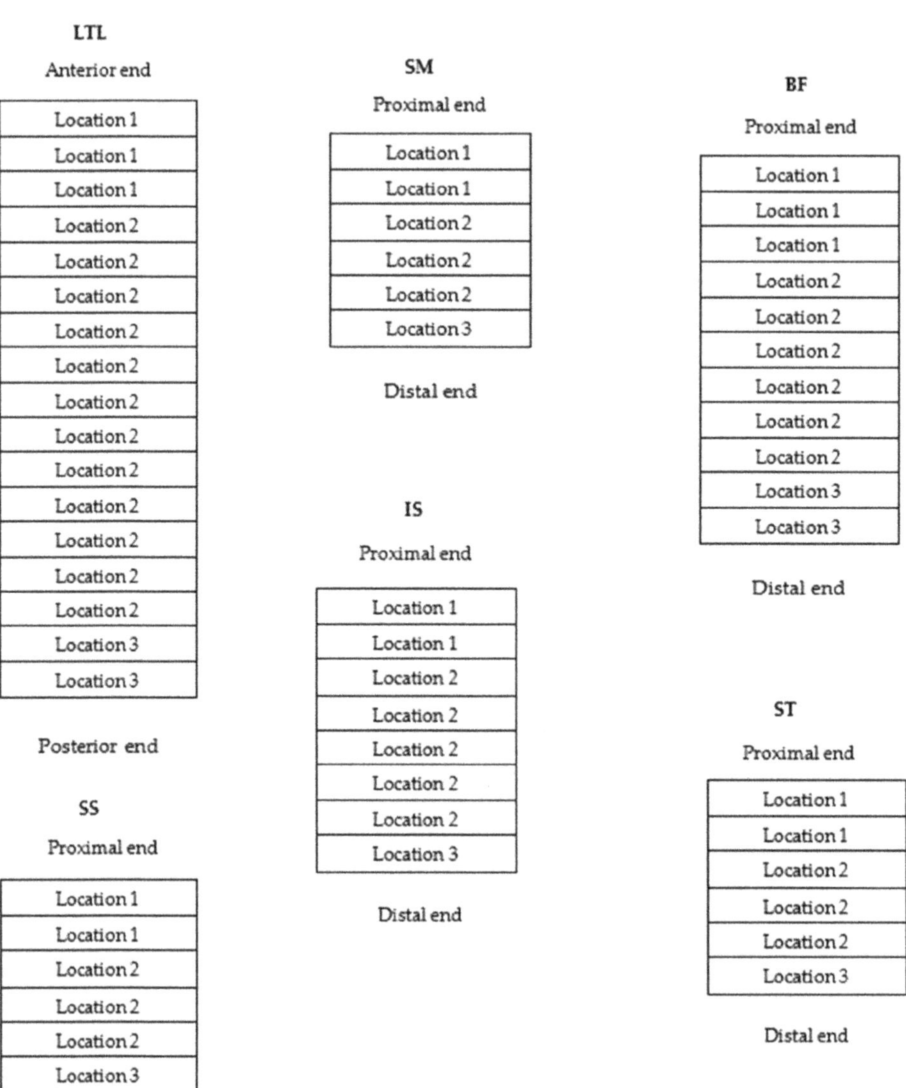

Figure 2. Sampling locations of the six different muscles (i.e., *Longissimus thoracis et lumborum* (LTL), *Semimembranosus* (SM), *Biceps femoris* (BF), *Supraspinatus* (SS), *Infraspinatus* (IS), and *Semitendinosus* (ST)). Left side of carcass for day 1 samples for location 1 (meat colour, water-holding capacity, myofibril fragment length), location 2 (Warner-Bratzler shear force) and location 3 (collagen analysis total and soluble); Right side of carcass for day 4 samples for location 1 (meat colour, water holding capacity, myofibril fragment length), location 2 (Warner-Bratzler shear force) and location 3 (collagen analysis-total and soluble, proximate analysis). Proximal = nearest to the vertebral column. Each horizontal section represents a 2.0 cm-thick slice.

2.2. Laboratory Analysis

For the chemical and physical analyses, samples were taken from the various locations of the six muscles, LTL, SM, BF, SS, IS, and ST, as described in Figure 2. Analyses were either conducted on the fresh samples (% purge, WHC, chemical and meat colour analyses) or on vacuum-packed frozen (-20 °C) and then defrosted (4 °C, 24 h) samples such as WBSF.

2.2.1. Purge and Water Holding Capacity

Purge percentage was measured using a 10 mm-thick slice of the six different muscles (LTL, SM, BF, SS, IS, and ST), vacuumed and aged for 4 d at 4 °C. The specific slices were weighed before and after storage and the weight difference indicated as purge loss percentage. The WHCs of the six fresh muscles were determined using the filter paper press method [18]. Briefly, 400 to 500 mg of meat sample was placed on Whatman 4 filter paper, (Camlab Ltd, Cambridge, UK), contained between two Perspex plates. Constant pressure was applied using a hand-operated screw for 5 min. The borders of the meat and fluid were marked out and their areas measured using a video image analyser (Soft Imaging System, Olympus, Tokyo, Japan), according to [19]. WHC was expressed as a ratio of meat area to fluid area.

2.2.2. Warner-Bratzler Shear Force

The frozen vacuum-packed muscle samples (LTL, SM, BF, SS, IS, and ST) were placed in a cold room at 4 °C to thaw for 24 h before cooking. Whole cuts were prepared according to an oven-broiling method using direct radiant heat [20]. Calibrated electric ovens (Mielé ovens, model H217, Miele & Cie. KG, Gütersloh, Germany) were set to "broil" 10 min prior to cooking at 160 °C. The LL samples were placed on an open casserole pan on a rack with no added water (dry cooking). The SM, BF, SS, IS, and ST, were placed in a casserole pan, adding 100 ml water and close with lid (moisture cooking). The cuts were broiled for approximately 20 min until they reached an internal core temperature of 70 °C. The internal temperature was monitored by placing an iron-constant thermocouple (T-type) (Hand-model Kane-Mane thermometer, Kane International Ltd., Hertfordshire, UK) in the approximate geometric centre of each sample. The cooked meat was weighed together with the pan and drip. The cooked samples were cooled for 2 h at room temperature (20 °C) before shear force measurement. Six cylindrical samples (12.5 mm core diameter) were bored parallel to the direction of the muscle fibres. Each core was sheared perpendicular to the myofibrils using a Warner-Bratzler device fitted to an Instron Universal Testing Machine (Model 4301, Instron Ltd., Buckinghamshire, UK) at a crosshead speed of 200 mm/min with one shear in the centre of each core [21]. The toughness of the meat was the average maximum force measured in Newton (N) required to shear through the cores.

2.2.3. Myofibril Fragment Length

Samples used for MFL were aged for 1 d and 4 d post mortem. Sub-samples of approximately 3 g were taken, blended with a blunt blade in cold potassium phosphate extraction buffer at 4 °C to arrest any further proteolysis [22], and determined according to [23]. The droplets of extracted MFL solution were mounted on slides, covered with a cover slip, and viewed under a microscope attached to a video image analysis (VIA). One hundred myofibril fragments per sample were examined and measured at a magnification of $40\times$.

2.2.4. Chemical Composition and Collagen Characteristics

The protein and IMF (associated with marbling) were analysed using the procedures of the Association of Official Analytical Chemists [24] at the ARC-AP Analytical Laboratories. Samples (25 g of homogenized meat) were freeze-dried according to method 934.01 [24]. The percentage fat content was determined on 5 g of freeze-dried sample using a 1:2 chloroform/methanol solution for fat extraction (SOXTEC method) as described in [25]. The total nitrogen content in the defatted muscle samples was determined after samples had been digested in a micro Kjeldahl system (Analytical Laboratory ARC-AP). The nitrogen content was multiplied by a factor of 6.25 in order to obtain the protein content of the sample, which was subsequently converted to a value per gram of wet meat (method 922.15) [24]. Soluble, insoluble and total collagen were determined in the same fresh samples.

Total collagen content in the six muscles (LTL, SM, BF, SS, IS, and ST) was determined by measuring the total hydroxyl-proline nitrogen content in the hydrolysed samples according to a modified method of [26]. Approximately 1 g of fresh sample was weighed into a hydrolysed tube and mixed with 15 mL of 6 N HCl. The samples were hydrolysed at 120 °C for 16 h, then 0.5 g active carbon was added to each tube, stirred, and filtered through Whatman 4 filter paper. The aliquots were collected in a 100 mL volumetric flask and filled up with distilled water. An aliquot of 50 mL was used for the determination of total collagen, described below.

The solubility of the muscle collagen (hydroxy-proline nitrogen content of soluble collagen) was determined according to the method of [27], with some modifications. About 2 g of fresh sample was stirred in 10 mL of 1% NaCl solution. The samples were heated in a shaking water bath at 78 °C for 60 min. The cooled samples were centrifuged at 10,000 rpm for 15 min. The supernatants were poured into hydrolysing tubes, marked as soluble. The pellet was poured into another hydrolysing tube and marked insoluble. To each tube, 7.5 mL of 6 N HCl (19.2%) was added and hydrolysed overnight at 120 °C. The following day, 0.5 g of active carbon was added to the cooled tubes, stirred, and the homogenates filtered into 50 mL volumetric flasks and filled to the mark with distilled water. Aliquots of 50 mL were used for determination of both soluble and insoluble collagen.

Hydroxy-proline concentrations were determined calorimetrically according to a modified method of [28]. About 1 mL of the final sample was added into the test tubes, to which 1 mL of 10% KOH solution was added (to neutralise the acid in the sample). A blank consisting of 2 mL distilled water was prepared. Standard solutions were prepared containing 0 to 7.5 µg/mL and 2 mL hydroxy-proline to create a new standard curve for each analysis session.

To each test tube (including standards and blanks), 1 mL of the oxidant solution (1.41 g Chloramine-T in a 100 mL, pH 6.8 buffer solution consisting of: 26 g citric acid monohydrate, 14 g sodium hydroxide, 78 g Anhydrous sodium acetate and 250 mL propan-1-ol) was added. The tubes were vortexed for 5 s and left for 20 min at room temperature. After 20 min, 1 mL of the colour reagent (10 g para-dimethylaminobenzaldehyde, 35 mL perchloric acid solution (60%), 65 mL propan-2-ol, prepared fresh) was added and the tubes vortexed. The tubes were heated to 62 ± 5 °C for 30 min, then vortexed. Thereafter, they were cooled to room temperature (a strong, aromatic, pink liquid with a white salt residue formed in the tubes). The top transparent pink liquid was pipetted into disposable micro cuvettes and the absorbance was read on a spectrophotometer at 558 nm (± 2 nm). Hydroxy-proline content was determined from the standard addition curve.

Total collagen content was determined by calculating hydroxy-proline nitrogen from hydroxy-proline (molecular mass 131.13 and nitrogen atom number 14.0067). Collagen values were expressed as mg collagen/g of muscle sample by using the hydroxy-proline conversion of 7.25 and 7.53 for insoluble and soluble collagen respectively [29].

2.2.5. Measurement of Colour and pH

The colours of muscle samples (ca. 15 mm thick) were measured fresh at 1 d and 4 d post mortem. The meat samples were allowed to bloom for 60 min at ± 4 °C before the meat colour values were recorded. A Konica-Minolta 600d spectrophotometer (Konica-Minolta Inc. Osaka, Japan) with the software package Spectra Magic NX Pro was used to measure surface D65 at three different positions on the meat samples. Three components were recorded according to the CIELAB colour space ($L^*a^*b^*$) defined by the International Commission on Illumination in 1976; lightness, L^* (dark (0) to light (100)) and the two chromatic components; a^* (green (-60, 180°) to red ($+60$, 0°)) and b^* (blue (-60, 270°) to yellow ($+60$, 90°)) which represented the myoglobin levels in the meat [30]. The spectrophotometer configuration consisted of illuminate D65 with an observer angle of 10°, and the spectral component excluded (SCE) mode after calibration using a white reference [31]. Chroma (saturation index (S) = $(a^{*2} + b^{*2})^{1/2}$ [32] and hue-angle (discolouration) = $\tan^{-1(b^*/a^*)}$ [33] were calculated from a^* and b^* values; chroma measures colour intensity, where higher values indicate a more intense red colour in meat. An increase in hue-angle between $0°$ and $90°$ corresponds to a blending of yellowness or less redness, probably due to metmyoglobin formation in fresh meat. Ultimate pH (pH_u) was measured with a portable pH meter (Eutech Instruments, Cyber Scan pH 11, Keppel Logistic, Singapore) in the same location in the LTL, SM, BF, SS, IS, and ST muscles at 24 h post mortem.

2.2.6. Statistical Analysis

The data were subjected to analysis of variance [34] to test the effect of breed (BG and IVG), and sex type (bucks and wethers) on six muscles for the following characteristics; pH and temperature (24 h post mortem, pH_u and T_u), WHC (1 and 4 d post mortem), % purge, WBSF (1 and 4 d post mortem), MFL (1 and 4 d post mortem), connective tissue characteristics, and meat colour (L^*, a^*, b^*, chroma and hue-angle, 1 and 4 d post mortem) [35]. Statistical significance (Fisher's t-test, least significant difference) was calculated at a 5% level to compare means. A value of $p \leq 0.05$ was considered statistically significant, although in some instances, data with a $p \leq 0.1$, (10% level) was considered as a trend worth discussing.

Prior to analyses, a Shapiro-Wilk test for normality was performed on the data [36] and, where applicable, outliers (classified as such when the standardized residual for an observation deviated by more than three standard deviations (SD) from the model value) were removed. Where applicable, the closeness of the linear relationships between the measured variables was determined using Pearson's correlation coefficient (r).

3. Results

The results for the carcass characteristics of the experimental animals have been described previously [2] and summarised in Table 1

Table 1. Least square means and standard error (SE) of means for carcass characteristics of Boer- (BG) and large frame Indigenous Veld (IVG) buck and wether goats (adapted from [2]).

| | Breed | | | | Significance (*p*-Values) | | |
| | BG | | IVG | | | | |
Carcass Characteristics	Bucks $n = 10$	Wethers $n = 8$	Bucks $n = 9$	Wethers $n = 10$	Breed	Sex	Breed × Sex
Live weight (kg)	35.40 [a,b] ± 4.01	36.13 [a] ± 3.02	36 67 [a] ± 2.68	32.8 [b] ± 2.39	0.293	0.118	0.032
Cold carcass weight (kg)	15.26 ± 2.31	16.25 ± 1.66	15.88 ± 1.83	14.86 ± 0.97	0.541	0.938	0.094
Dressing (%)	42.99 [a] ± 2.44	44.95 [b] ± 1.08	43.28 [a] ± 3.23	45.42 [b] ± 2.49	0.508	0.017	0.912

[a,b] Means in the same row per main effect bearing different letters differ ($p \leq 0.05$).

The choice of the six muscles studied was intended to obtain a set of muscles representing a variation in tenderness and other quality parameters due to their different anatomical positions, functions and commercial value. Means and standard errors of breed and sex on pH_u, T_u, muscle WHC, % purge, WBSF, MFL, IMF, collagen characteristics, and meat colour (L^*, a^*, b^*, chroma and hue-angle) for each of these six muscles are presented in Tables 2–7, respectively.

Table 2. Least square means and standard error of means for meat tenderness, meat colour and related physiological characteristics of buck and wether Boer Goats (BG) and Indigenous Veld Goats (IVG) of the *Longissimus thoracis et lumborum* (LTL) muscle.

Meat Quality Characteristics	Breed				Significance (*p*-Values)		
	BG		IVG				
	Bucks	Wethers	Bucks	Wethers	Breed	Sex	Breed × Sex
pHu	5.54 [a] ± 0.18	5.60 [a] ± 0.05	5.67 [b] ± 0.11	5.72 [b] ± 0.18	0.011	0.241	0.944
Water holding capacity							
1-dpm [1]	0.41 ± 0.03	0.39 ± 0.06	0.38 ± 0.04	0.37 ± 0.05	0.101	0.384	0.642
4-dpm	0.38 [a] ± 0.04	0.45 [b] ± 0.08	0.39 [a] ± 0.08	0.43 [b] ± 0.07	0.979	0.018	0.515
Purge (%)	1.71 ± 0.84	1.86 ± 0.78	2.00 ± 1.02	1.96 ± 0.79	0.495	0.836	0.721
Warner Bratzler Shear force							
1-dpm (N [2])	58.5 ± 1.10	59.0 ± 1.17	57.4 ± 1.15	59.5 ± 1.05	0.958	0.752	0.834
4-dpm (N)	46.5 ± 1.14	40.5 ± 1.12	43.3 ± 0.88	42.9 ± 1.22	0.842	0.395	0.499
Myofibril fragment length							
1-dpm (μm)	37.16 ± 5.46	35.55 ± 4.83	35.26 ± 5.05	37.42 ± 5.04	0.351	0.220	0.319
4-dpm (μm)	33.62 ± 6.21	29.63 ± 2.01	30.32 ± 5.07	29.85 ± 6.14	0.471	0.332	0.426
Marbling [3]							
IMF (%)	1.97 [a] ± 1.11	2.58 [b] ± 1.35	1.49 [a] ± 0.94	2.59 [b] ± 0.70	0.620	0.017	0.473
Collagen characteristics							
Collagen solubility (%)	36.68 ± 10.69	37.55 ± 11.25	38.63 ± 9.83	35.49 ± 11.13	0.973	0.722	0.707
Soluble collagen (mg/g [4])	1.37 [x] ± 0.58	1.40 [x] ± 0.42	1.66 [y] ± 0.48	1.27 [x] ± 0.38	0.958	0.501	0.080
Insoluble collagen (mg/g)	2.40 ± 0.54	2.50 ± 0.91	2.71 ± 0.42	2.40 ± 0.71	0.549	0.232	0.229
Total collagen (mg/g)	3.68 ± 0.85	3.80 ± 0.85	4.24 ± 0.39	3.59 ± 0.78	0.566	0.222	0.160
Meat colour characteristics							
*L** 1-dpm	35.61 [a] ± 2.12	33.50 [b] ± 1.20	35.11 [a] ± 2.60	33.20 [b] ± 2.47	0.877	0.010	0.545
*L** 4-dpm	36.65 ± 3.18	34.75 ± 2.67	35.28 ± 1.35	34.84 ± 2.79	0.755	0.471	0.238
*a** 1-dpm	9.45 [a] ± 0.84	11.25 [b] ± 0.76	9.90 [a] ± 1.60	10.53 [b] ± 1.27	0.966	0.004	0.139
*a** 4-dpm	9.75 ± 1.25	10.91 ± 1.12	10.09 ± 0.96	10.43 ± 1.44	0.736	0.168	0.208
*b** 1-dpm	11.16 ± 1.41	11.26 ± 1.18	11.10 ± 1.81	12.14 ± 1.41	0.371	0.236	0.354
*b** 4-dpm	13.04 ± 0.94	12.64 ± 0.65	12.52 ± 0.85	12.48 ± 0.91	0.209	0.413	0.499
Chroma 1-dpm	14.66 [a] ± 1.30	15.95 [b] ± 1.02	14.93 [a] ± 1.96	16.13 [b] ± 1.39	0.486	0.015	0.898
Chroma 4-dpm	16.34 ± 1.13	16.74 ± 0.06	16.11 ± 1.10	16.18 ± 1.27	0.340	0.577	0.680
Hue angle 1-dpm	49.58 [x] ± 4.02	44.96 [y] ± 3.51	48.76 [x] ± 6.09	47.74 [y] ± 2.73	0.388	0.059	0.139
Hue angle 4-dpm	53.36 [a] ± 3.86	49.36 [b] ± 2.62	51.16 [a] ± 2.39	50.16 [b] ± 3.49	0.724	0.026	0.116

[a,b] Means in the same row per main effect bearing different letters differ ($p \leq 0.05$). [x,y] Means in the same row per main effect bearing different letters was considered a tendency to differ ($p \leq 0.1$). [1] dpm = day post mortem; [2] N = Newton; [3] Marbling = chemically determined intramuscular fat (IMF); [4] mg/g = milligram per gram fresh sample; *L** = lightness; *a** = redness; *b** = yellowness; Chroma = saturation index; Hue angle = discolouration.

Table 3. Least square means and standard error of means for meat tenderness, meat colour and related physiological characteristics of buck and wether Boer Goat (BG) and Indigenous Veld Goats (IVG) of *Semimembranosus* (SM) muscle.

Meat Quality Characteristics	Breed				Significance (*p*-Values)		
	BG		IVG				
	Bucks	Wethers	Bucks	Wethers	Breed	Sex	Breed × Sex
pH_u	5.89 [a] ± 0.27	5.98 [a,b] ± 0.11	5.91 [a] ± 0.12	6.17 [b] ± 0.25	0.092	0.017	0.267
Water holding capacity							
1-dpm [1]	0.35 [x] ± 0.03	0.35 [x] ± 0.03	0.35 [x] ± 0.06	0.31 [y] ± 0.04	0.205	0.078	0.165
4-dpm	0.35 [a,b] ± 0.03	0.35 [a,b] ± 0.04	0.36 [a] ± 0.06	0.41 [b] ± 0.03	0.019	0.026	0.185
Purge (%)	1.89 ± 0.48	2.21 ± 1.12	1.60 ± 1.03	1.92 ± 1.00	0.384	0.306	0.999
Warner Bratzler Shear force							
1-dpm (N [2])	37.6 ± 0.44	37.4 ± 0.60	39.7 ± 0.50	35.8 ± 0.71	0.908	0.415	0.230
4-dpm (N)	33.1 ± 0.43	31.9 ± 0.84	34.7 ± 0.49	30.0 ± 0.69	0.968	0.177	0.420
Myofibril fragment length							
1-dpm (μm)	41.06 ± 5.85	45.03 ± 5.03	44.08 ± 4.74	42.13 ± 2.73	0.883	0.560	0.066
4-dpm (μm)	38.64 ± 6.78	37.85 ± 5.78	40.22 ± 3.62	35.46 ± 4.60	0.803	0.130	0.276
Marbling [3]							
IMF (%)	1.94 [a] ± 1.09	3.05 [b] ± 1.53	1.76 [a] ± 1.05	2.76 [b] ± 0.80	0.689	0.008	0.888
Collagen characteristics							
Collagen solubility (%)	35.19 ± 11.59	27.58 ± 9.62	32.91 ± 5.68	33.03 ± 12.27	0.935	0.236	0.572
Soluble collagen (mg/g [4])	2.55 ± 1.30	1.76 ± 0.76	2.09 ± 0.53	2.04 ± 1.01	0.602	0.624	0.388
Insoluble collagen (mg/g)	4.43 ± 0.45	4.60 ± 0.67	4.39 ± 0.56	4.11 ± 0.78	0.647	0.207	0.384
Total collagen (mg/g)	6.82 ± 1.60	6.21 ± 1.03	6.32 ± 0.81	5.99 ± 0.97	0.705	0.175	0.467
Meat colour characteristics							
L^* 1-dpm	35.74 [a] ± 3.03	33.78 [b] ± 1.84	37.24 [a] ± 2.36	33.01 [b] ± 1.47	0.894	0.0003	0.199
L^* 4-dpm	36.94 [a] ± 3.22	34.06 [b] ± 2.99	36.33 [a] ± 2.08	34.14 [b] ± 2.72	0.501	0.012	0.270
a^* 1-dpm	10.55 [a] ± 1.40	12.36 [b] ± 1.66	10.30 [a] ± 1.32	11.74 [b] ± 1.72	0.388	0.003	0.060
a^* 4-dpm	9.85 [a] ± 2.03	12.30 [b] ± 1.84	11.17 [b] ± 1.63	10.37 [a] ± 2.21	0.066	0.111	0.018
b^* 1-dpm	11.91 ± 1.31	12.06 ± 1.37	12.31 ± 0.67	12.07 ± 1.31	0.318	0.474	0.580
b^* 4-dpm	12.71 ± 1.21	12.68 ± 0.63	13.26 ± 0.67	12.23 ± 1.38	0.828	0.353	0.512
Chroma 1-dpm	15.99 [a] ± 1.49	17.33 [b] ± 1.91	16.12 [a] ± 0.90	16.89 [b] ± 1.84	0.754	0.018	0.375
Chroma 4-dpm	16.14 [a] ± 2.06	17.71 [b] ± 1.61	17.41 [b] ± 1.43	16.16 [a] ± 1.99	0.078	0.185	0.024
Hue angle 1-dpm	48.71 [a] ± 4.36	44.49 [b] ± 3.34	50.39 [a] ± 4.10	44.9 [b] ± 2.28	0.395	0.001	0.011
Hue angle 4-dpm	52.71 [a] ± 4.11	46.21 ± 3.61 [b]	50.34 [a] ± 3.46	48.29 [b] ± 4.23	0.215	0.003	0.236

[a,b] Means in the same row per main effect bearing different letters differ ($p \leq 0.05$). [x,y] Means in the same row per main effect bearing different letters was considered a tendency to differ ($p \leq 0.1$). [1] dpm = day post mortem; [2] N = Newton; [3] Marbling = chemically determined intramuscular fat (IMF); [4] mg/g = milligram per gram fresh sample; L^* = lightness; a^* = redness; b^* = yellowness; Chroma = saturation index; Hue angle = discolouration.

Table 4. Least square means and standard error of means for meat tenderness, meat colour and related physiological characteristics of buck and wether Boer Goats (BG) and Indigenous Veld Goats (IVG) of *Biceps femoris* (BF) muscle.

Meat Quality Characteristics	Breed				Significance (*p*-Values)		
	BG		IVG		Breed	Sex	Breed × Sex
	Bucks	Wethers	Bucks	Wethers			
pH$_u$	5.74 [a] ± 0.11	5.71 [a] ± 0.14	5.82 [b] ± 0.13	5.91 [b] ± 0.16	0.003	0.477	0.204
Water holding capacity							
1-dpm [1]	0.38 [y] ± 0.04	0.38 [y] ± 0.05	0.36 [x] ± 0.04	0.35 [x] ± 0.05	0.096	0.550	0.686
4-dpm	0.35 ± 0.04	0.41 ± 0.06	0.37 ± 0.04	0.37 ± 0.06	0.647	0.167	0.074
Purge (%)	0.96 ± 0.34	1.00 ± 0.40	0.97 ± 0.27	0.70 ± 0.35	0.182	0.282	0.188
Warner Bratzler Shear force							
1-dpm (N [2])	55.8 ± 1.06	47.1 ± 1.52	49.9 ± 1.09	47.6 ± 1.43	0.444	0.211	0.455
4-dpm (N)	44.5 ± 0.82	34.4 ± 0.78	40.9 ± 0.96	42.1 ± 1.36	0.652	0.213	0.102
Myofibril fragment length							
1-dpm (µm)	43.57 [a] ± 9.93	35.01 [b] ± 5.51	40.81 [a] ± 6.80	38.89 [b] ± 6.50	0.989	0.046	0.188
4-dpm (µm)	35.11 [a] ± 5.76	28.26 [b] ± 3.54	33.29 [a] ± 7.04	32.21 [b] ± 5.27	0.724	0.044	0.128
Marbling [3]							
IMF (% Fat)	2.75 [a] ± 1.85	4.18 [b] ± 2.46	1.88 [a] ± 1.29	3.74 [b] ± 0.74	0.345	0.005	0.694
Collagen characteristics							
Collagen solubility (%)	37.88 ± 14.34	34.50 ± 7.73	27.93 ± 9.14	37.33 ± 16.13	0.450	0.418	0.143
Soluble collagen (mg/g [4])	2.80 ± 1.67	2.46 ± 1.44	1.82 ± 0.78	2.43 ± 1.21	0.218	0.286	0.646
Insoluble collagen (mg/g)	4.27 ± 0.97	4.49 ± 0.87	4.67 ± 0.43	4.09 ± 1.11	0.519	0.505	0.974
Total collagen (mg/g)	6.92 ± 2.22	6.81 ± 2.19	6.33 ± 0.91	6.36 ± 1.25	0.466	0.467	0.938
Meat colour characteristics							
*L** 1-dpm	37.60 [a] ± 3.05	33.29 [b] ± 2.18	37.11 [a] ± 2.38	34.06 [b] ± 1.50	0.744	<0.0001	0.246
*L** 4-dpm	38.00 ± 2.56	35.83 ± 1.76	36.68 ± 2.09	36.24 ± 2.93	0.965	0.432	0.160
*a** 1-dpm	9.95 [a] ± 1.16	12.29 [b] ± 0.99	10.33 [a,b] ± 1.62	10.64 [a,b] ± 1.41	0.267	0.006	0.027
*a** 4-dpm	8.76 [a] ± 1.17	10.84 [b] ± 1.36	9.78 [a,b] ± 1.33	9.25 [a,b] ± 1.19	0.648	0.085	0.004
*b** 1-dpm	11.81 ± 1.33	11.98 ± 1.10	11.89 ± 1.14	12.02 ± 1.52	0.860	0.729	0.997
*b** 4-dpm	11.71 ± 1.31	12.19 ± 1.15	11.84 ± 1.10	11.99 ± 1.23	0.985	0.445	0.671
Chroma 1-dpm	15.49 [x] ± 1.40	17.16 [y] ± 1.24	15.79 [x] ± 1.57	16.11 [y] ± 1.67	0.574	0.056	0.179
Chroma 4-dpm	14.66 [x] ± 1.62	16.39 [z] ± 1.59	15.38 [y] ± 1.64	15.23 [y] ± 1.29	0.809	0.143	0.072
Hue angle 1-dpm	49.84 [a] ± 3.94	44.25 [b] ± 2.98	49.21 [a] ± 4.49	47.54 [b] ± 2.43	0.243	0.005	0.064
Hue angle 4-dpm	53.22 [b] ± 2.66	48.95 [a] ± 3.26	50.56 [a,b] ± 2.33	51.84 [a,b] ± 3.73	0.723	0.398	0.010

[a,b] Means in the same row per main effect bearing different letters differ ($p \leq 0.05$). [x,y] Means in the same row per main effect bearing different letters was considered a tendency to differ ($p \leq 0.1$). [1] dpm = day post mortem; [2] N = Newton; [3] Marbling = chemically determined intramuscular fat (IMF); [4] mg/g = milligram per gram fresh sample; *L** = lightness; *a** = redness; *b** = yellowness; Chroma = saturation index; Hue angle = discolouration.

Table 5. Least square means and standard error of means for meat tenderness, meat colour and related physiological characteristics of buck and wether Boer Goats (BG) and Indigenous Veld Goats (IVG) of *Supraspinatus* (SS) muscle.

Meat Quality Characteristics	Breed				Significance (*p*-Values)		
	BG		IVG				
	Bucks	Wethers	Bucks	Wethers	Breed	Sex	Breed × Sex
pH_u	5.89 [a] ± 0.27	5.98 [b] ± 0.11	5.91 [a] ± 0.12	6.17 [b] ± 0.25	0.092	0.017	0.267
Water holding capacity							
1-dpm [1]	0.35 [x] ± 0.03	0.35 [x] ± 0.03	0.35 [x] ± 0.06	0.31 [y] ± 0.04	0.205	0.078	0.165
4-dpm	0.35 [a,b] ± 0.03	0.35 [a,b] ± 0.04	0.36 [a] ± 0.06	0.41 [b] ± 0.03	0.019	0.026	0.185
Purge (%)	1.89 ± 0.48	2.21 ± 1.12	1.60 ± 1.03	1.92 ± 1.00	0.384	0.306	0.999
Warner Bratzler Shear force							
1-dpm (N [2])	37.6 ± 0.44	37.4 ± 0.60	39.7 ± 0.50	35.8 ± 0.71	0.908	0.415	0.230
4-dpm (N)	33.1 ± 0.43	31.9 ± 0.84	34.7 ± 0.49	30.0 ± 0.69	0.968	0.177	0.420
Myofibril fragment length							
1-dpm (μm)	41.06 ± 5.85	45.03 ± 5.03	44.08 ± 4.74	42.13 ± 2.73	0.883	0.560	0.066
4-dpm (μm)	38.64 ± 6.78	37.85 ± 5.78	40.22 ± 3.62	35.46 ± 4.60	0.803	0.130	0.276
Marbling [3]							
IMF (%)	1.94 [a] ± 1.09	3.05 [b] ± 1.53	1.76 [a] ± 1.05	2.76 [b] ± 0.80	0.089	0.000	0.888
Collagen characteristics							
Collagen solubility (%)	35.19 ± 11.59	27.58 ± 9.62	32.91 ± 5.68	33.03 ± 12.27	0.741	0.297	0.202
Soluble collagen (mg/g [4])	2.55 ± 1.30	1.76 ± 0.76	2.09 ± 0.53	2.04 ± 1.01	0.697	0.575	0.179
Insoluble collagen (mg/g)	4.43 ± 0.45	4.60 ± 0.67	4.39 ± 0.56	4.11 ± 0.78	0.498	0.359	0.838
Total collagen (mg/g)	6.82 ± 1.60	6.21 ± 1.03	6.32 ± 0.81	5.99 ± 0.97	0.987	0.946	0.128
Meat colour characteristics							
*L** 1-dpm	35.74 [a] ± 3.03	33.78 [b] ± 1.84	37.24 [a] ± 2.36	33.01 [b] ± 1.47	0.649	0.0003	0.222
*L** 4-dpm	36.94 [a] ± 3.22	34.06 [b] ± 2.99	36.33 [a] ± 2.08	34.14 [b] ± 2.72	0.991	0.012	0.450
*a** 1-dpm	10.55 [a] ± 1.40	12.36 [b] ± 1.66	10.30 [a] ± 1.32	11.74 [b] ± 1.72	0.558	0.003	0.720
*a** 4-dpm	9.85 [a] ± 2.03	12.30 [b] ± 1.84	11.17 [a,b] ± 1.63	10.37 [a,b] ± 2.21	0.788	0.224	0.018
*b** 1-dpm	11.91 ± 1.31	12.06 ± 1.37	12.31 ± 0.67	12.07 ± 1.31	0.623	0.885	0.597
*b** 4-dpm	12.71 ± 1.21	12.68 ± 0.63	13.26 ± 0.67	12.23 ± 1.38	0.853	0.131	0.153
Chroma 1-dpm	15.99 [x] ± 1.49	17.33 [y] ± 1.91	16.12 [x] ± 0.90	16.89 [y] ± 1.84	0.934	0.054	0.591
Chroma 4-dpm	16.14 [a] ± 2.06	17.71 [a,b] ± 1.61	17.41 [a,b] ± 1.43	16.16 [b] ± 1.99	0.911	0.811	0.024
Hue angle 1-dpm	48.71 [a] ± 4.36	44.49 [b] ± 3.34	50.39 [a] ± 4.10	44.9 [b] ± 2.28	0.351	0.001	0.934
Hue angle 4-dpm	52.71 [a] ± 4.11	46.21 [b] ± 3.61	50.34 [a] ± 3.46	48.29 [b] ± 4.23	0.800	0.003	0.054

[a,b] Means in the same row per main effect bearing different letters differ ($p \leq 0.05$). [x,y] Means in the same row per main effect bearing different letters was considered a tendency to differ ($p \leq 0.1$). [1] dpm = day post mortem; [2] N = Newton; [3] Marbling = chemically determined intramuscular fat (IMF); [4] mg/g = milligram per gram fresh sample; *L** = lightness; *a** = redness; *b** = yellowness; Chroma = saturation index; Hue angle = discolouration.

Table 6. Least square means and standard error of means for meat tenderness, meat colour and related physiological characteristics of buck and wether Boer Goats (BG) and Indigenous Veld Goats (IVG) of *Infraspinatus* (IS) muscle.

Meat Quality Characteristics	Breed				Significance (*p*-Values)		
	BG		IVG		Breed	Sex	Breed × Sex
	Bucks	Wethers	Bucks	Wethers			
pH_u	5.97 ± 0.26	6.11 ± 0.10	6.09 ± 0.24	6.12 ± 0.21	0.324	0.247	0.446
Water holding capacity							
1-dpm [1]	0.36 ± 0.05	0.38 ± 0.07	0.34 ± 0.05	0.34 ± 0.05	0.195	0.791	0.606
4-dpm	0.35 ± 0.05	0.39 ± 0.06	0.38 ± 0.04	0.37 ± 0.05	0.686	0.419	0.199
Purge (%)	$0.97^a \pm 0.35$	$1.20^a \pm 0.57$	$0.82^b \pm 0.49$	$0.62^b \pm 0.23$	0.015	0.960	0.129
Warner Bratzler Shear force							
1-dpm (N [2])	33.8 ± 0.63	31.9 ± 0.45	29.9 ± 0.40	30.0 ± 0.68	0.155	0.641	0.588
4-dpm (N)	$26.9^x \pm 0.37$	$28.9^x \pm 0.42$	$25.7^y \pm 0.39$	$24.8^y \pm 0.54$	0.083	0.726	0.331
Myofibril fragment length							
1-dpm (μm)	46.53 ± 6.51	42.70 ± 4.59	44.63 ± 5.51	44.43 ± 8.29	0.886	0.367	0.403
4-dpm (μm)	$41..41 \pm 7.32$	39.36 ± 6.25	38.78 ± 4.06	37.46 ± 5.89	0.232	0.407	0.856
Marbling [3]							
IMF (%)	$1.49^a \pm 0.59$	$2.70^b \pm 1.10$	$1.10^a \pm 0.66$	$2.09^b \pm 0.41$	0.092	<0.0001	0.641
Collagen characteristics							
Collagen solubility (%)	37.05 ± 10.26	39.39 ± 9.81	38.31 ± 11.58	34.79 ± 9.69	0.513	0.873	0.538
Soluble collagen (mg/g [4])	2.83 ± 1.14	2.76 ± 1.22	2.89 ± 1.11	2.33 ± 0.74	0.793	0.396	0.200
Insoluble collagen (mg/g)	4.89 ± 1.06	4.18 ± 1.15	4.67 ± 0.81	4.47 ± 0.81	0.848	0.133	0.733
Total collagen (mg/g)	7.55 ± 1.55	6.79 ± 2.04	7.39 ± 1.21	6.61 ± 0.89	0.891	0.131	0.598
Meat colour characteristics							
L^* 1-dpm	$36.96^a \pm 3.39$	$34.64^b \pm 2.57$	$38.36^a \pm 2.32$	$37.0^b \pm 2.15$	0.057	0.048	0.537
L^* 4-dpm	37.61 ± 3.34	36.03 ± 2.64	38.21 ± 2.43	37.19 ± 3.88	0.461	0.221	0.785
a^* 1-dpm	$8.22^a \pm 1.92$	$10.28^b \pm 1.45$	$8.43^a \pm 1.64$	$9.03^b \pm 2.18$	0.519	0.040	0.244
a^* 4-dpm	$8.86^a \pm 1.70$	$10.84^b \pm 2.08$	$8.69^a \pm 1.74$	$9.60^b \pm 2.42$	0.402	0.039	0.447
b^* 1-dpm	$10.60^a \pm 1.52$	$10.89^a \pm 1.47$	$12.13^b \pm 0.71$	$11.17^b \pm 0.97$	0.042	0.815	0.411
b^* 4-dpm	12.41 ± 1.28	12.20 ± 1.34	12.46 ± 1.22	11.98 ± 1.02	0.831	0.364	0.712
Chroma 1-dpm	13.52 ± 2.18	15.03 ± 1.98	14.53 ± 1.98	14.48 ± 1.84	0.642	0.289	0.254
Chroma 4-dpm	15.36 ± 1.67	16.36 ± 2.20	15.32 ± 1.72	15.54 ± 1.86	0.545	0.332	0.508
Hue angle 1-dpm	$52.73^a \pm 4.83$	$47.01^b \pm 2.23$	$55.90^a \pm 4.28$	$50.57^b \pm 4.40$	0.034	0.001	0.586
Hue angle 4-dpm	54.86 ± 5.07	49.16 ± 4.10	55.66 ± 4.87	52.18 ± 7.07	0.409	0.017	0.544

[a,b] Means in the same row per main effect bearing different letters differ ($p \leq 0.05$). [x,y] Means in the same row per main effect bearing different letters was considered a tendency to differ ($p \leq 0.1$). [1] dpm = day post mortem; [2] N = Newton; [3] Marbling = chemically determined intramuscular fat (IMF); [4] mg/g = milligram per gram fresh sample; L^* = lightness; a^* = redness; b^* = yellowness; Chroma = saturation index; Hue angle = discolouration.

Table 7. Least square means and standard error of means for meat tenderness, meat colour and related physiological characteristics of buck and wether Boer Goats (BG) and Indigenous Veld Goats (IVG) of *Semitendinosus* (ST) muscle.

Meat Quality Characteristics	Breed				Significance (p-Values)		
	BG		IVG				
	Bucks	Wethers	Bucks	Wethers	Breed	Sex	Breed × Sex
pH_u	5.66 [a] ± 0.11	5.69 [a] ± 0.06	5.71 [b] ± 0.13	5.89 [b] ± 0.18	0.004	0.021	0.091
Water holding capacity							
1-dpm [1]	0.37 ± 0.04	0.35 ± 0.05	0.38 ± 0.03	0.37 ± 0.04	0.432	0.394	0.705
4-dpm	0.38 ± 0.07	0.39 ± 0.06	0.39 ± 0.04	0.41 ± 0.05	0.265	0.421	0.750
Purge (%)	1.49 ± 0.97	1.62 ± 0.83	1.93 ± 1.53	1.54 ± 0.92	0.624	0.708	0.479
Warner Bratzler Shear force							
1-dpm (N [2])	50.8 [a] ± 0.51	44.8 [b] ± 0.48	44.8 [b] ± 0.48	44.1 [b] ± 1.19	0.440	0.047	0.736
4-dpm (N)	47.3 ± 0.61	41.4 ± 0.32	43.0 ± 0.64	40.8 ± 1.23	0.288	0.137	0.483
Myofibril fragment length							
1-dpm (μm)	46.48 ± 4.56	45.63 ± 3.40	44.06 ± 5.03	46.66 ± 5.38	0.662	0.553	0.274
4-dpm (μm)	40.58 ± 5.24	38.44 ± 4.41	40.12 ± 6.19	38.51 ± 8.17	0.864	0.371	0.899
Marbling [3]							
IMF (%)	2.12 [a] ± 1.53	2.76 [b] ± 1.50	1.84 [a] ± 1.07	2.93 [b] ± 0.68	0.980	0.040	0.590
Collagen characteristics							
Collagen solubility (%)	37.09 ± 11.22	33.60 ± 9.82	35.31 ± 7.75	32.94 ± 8.66	0.821	0.404	0.690
Soluble collagen (mg/g [4])	1.85 [x] ± 0.52	1.41 [y] ± 0.47	1.74 [x] ± 0.75	1.57 [y] ± 0.55	0.058	0.059	0.757
Insoluble collagen (mg/g)	3.36 ± 1.07	2.89 ± 0.52	3.10 ± 0.30	3.21 ± 0.47	0.688	0.128	0.136
Total collagen (mg/g)	5.08 ± 1.05	4.20 ± 0.57	4.72 ± 0.10	4.70 ± 0.71	0.823	0.104	0.160
Meat colour characteristics							
L^* 1-dpm	40.11 [x] ± 2.05	38.73 [y] ± 1.68	39.36 [y] ± 0.98	39.46 [y] ± 2.62	0.963	0.882	0.090
L^* 4-dpm	39.89 ± 2.21	39.58 ± 2.99	39.52 ± 1.68	38.28 ± 3.03	0.781	0.849	0.899
a^* 1-dpm	7.58 [b] ± 1.22	9.25 [b] ± 0.94	8.17 [b] ± 0.85	7.63 [a] ± 1.27	0.342	0.891	0.005
a^* 4-dpm	7.21 [a] ± 1.28	8.50 [b] ± 1.63	8.09 [a] ± 1.06	8.96 [b] ± 1.61	0.347	0.029	0.392
b^* 1-dpm	12.40 ± 0.78	12.79 ± 1.09	12.84 ± 0.89	12.76 ± 0.73	0.428	0.618	0.408
b^* 4-dpm	12.47 ± 0.91	12.73 ± 0.93	12.80 ± 1.23	13.23 ± 0.80	0.178	0.285	0.785
Chroma 1-dpm	14.64 [a] ± 0.93	15.89 [b] ± 1.11	15.60 [b] ± 0.82	14.79 [a] ± 1.08	0.959	0.594	0.004
Chroma 4-dpm	14.47 [x] ± 1.23	15.41 [y] ± 1.33	15.19 [x] ± 1.49	15.86 [y] ± 0.90	0.110	0.059	0.744
Hue angle 1-dpm	59.04 [a,b] ± 4.34	54.45 [a] ± 3.42	58.40 [a,b] ± 3.14	59.51 [b] ± 3.76	0.236	0.936	0.029
Hue angle 4-dpm	60.12 ± 4.16	56.94 ± 5.15	58.16 ± 2.41	55.96 ± 5.36	0.671	0.335	0.421

[a,b] Means in the same row per main effect bearing different letters differ ($p \leq 0.05$). [x,y] Means in the same row per main effect bearing different letters was considered a tendency to differ ($p \leq 0.1$). [1] dpm = day post mortem; [2] N = Newton; [3] Marbling = chemically determined intramuscular fat (IMF); [4] mg/g = milligram per gram fresh sample; L^* = lightness; a^* = redness; b^* = yellowness; Chroma = saturation index; Hue angle = discolouration.

IVG presented higher pH_u values ($p \leq 0.05$) compared to those of BG for LTL, BF, and ST muscles, with SM and SS having a tendency ($p \leq 0.10$) to show breed differences. Sex differences for pH_u were more prominent ($p \leq 0.05$) for SM, SS, with ST showing both breed and sex differences, and therefore a tendency ($p \leq 0.10$) to have breed × sex interactions. The IS muscle (~6.1) showed, on average, the highest pH_u, but no differences between breed and sex. In the muscles where pH_u differences were found, the IVG seemed to have the higher pH_u compared to BG. When sex differences arose, the wethers always tended to have higher pH_u than the bucks. On average, the SS had a pH_u of ~5.9, followed by BF and ST with pH_u between 5.7 and 5.9.

Although there are some tendencies towards breed and sex differences at 1 d post mortem for some muscles, it is only after 4 d post mortem that significant differences were observed in pressed-out water (WHC). WHC varied between 0.35 to 0.40, measured at

4 d post mortem, but LTL measured 0.43 to 0.45, respectively for BG and IVG wethers, compared to 0.38 and 0.39, respectively for BG and IVG bucks. Significant breed and sex effects for WHC at 4 d post mortem were recorded for SM and SS muscles, although the ratio was not as high as for the LTL. Only IS presented a breed difference for % purge, with that of IVG (0.62–0.82%) significantly lower than that of BG (0.97–1.20%). It was observed that, overall, IS and BF seemed to have lower % purge than that of the other muscles (>1.5%) (Results not shown).

Tenderness-related sex differences were recorded for the BF (MFL 1 and 4 d post mortem) and ST (WBSF 1 d post mortem) muscles, while a tendency ($p \leq 0.1$) for an interaction between sex and breed was recorded for MFL at 1 d post mortem for the SM muscles (Table 3). The BF wether muscle measured shorter MFL than that of the buck muscle (Table 4). Differences were found between the different muscles (results not shown). Some numerical tenderisation from 1 to 4 d post mortem can be observed in each of the Tables 2–7, with SM, SS and IS being the most tender after 4 d post mortem.

All the muscles showed differences by sex ($p \leq 0.05$) for IMF (Tables 2–7). Wether muscles overall recorded a higher percentage IMF than that of bucks in LTL, SM, BF, SS, IS and ST. IVG bucks recorded the lowest values (1.1%) in the IS muscles (Table 6) and BG wethers recorded the highest values of 4.18% in the BF muscle (Table 4). In most muscles, the bucks had about 1% less IMF than that of the wethers, whilst buck BF muscle had up to 2% less IMF than that of its equivalent wether muscle.

There were no significant differences in breed and sex in any of the collagen characteristics among the six muscles studied (Tables 2–7). However, there were tendencies ($p \leq 0.1$) observed for IVG buck LTL as well as BG and IVG buck ST to have higher collagen solubility levels.

Meat colour differences related to sex were noted; L^* (lightness) differences were observed in LTL (1 d post mortem), SM (1 and 4 d post mortem), BF (1 d post mortem), SS (1 and 4 d post mortem), and IS (1 d post mortem), with a trend in the ST for a breed $\times$ sex interaction. For these muscles, wethers recorded lower L^* values (darker meat) than the bucks. Sex differences for a^* and chroma (saturation index) were recorded in LTL (1 d post mortem), SM (1 d post mortem), BF (1-d post mortem), and SS (1 d post mortem). These muscles from wethers seem darker and a brighter red than those of bucks, especially at 1 d post mortem. At 4 d post mortem, the hue-angles (discolouration) of wether LTL, SM, SS and IS were lower than that of the corresponding buck muscles. Significant breed * sex interactions were observed for the chroma of the SM and ST, along with a trend in BF at 4 d post mortem indicating towards a higher saturation index for BG wethers and IVG bucks. No breed or sex differences were detected for b^* for any of the muscles.

4. Discussion

Compared to extensive studies on the influence of muscle source on meat quality indicators such as pH_u, chemical composition, tenderness, juiciness and colour attributes in other livestock, only limited studies examined these phenomena in chevon (goat meat); with the focus mainly being on the LTL and SM muscles [8,37,38]. The present study investigated the meat quality of six different muscles: LTL, SM, BF, SS, IS, and ST. It is expected that different muscles will show different results for the various meat quality characteristics [39]. However, when considering differences between breeds and sexes in muscles, few and often negligible differences were found.

Except for the LTL muscle (which is usually the standard position for monitoring pH), pH_u for all muscles were in general above 5.8 or even 6.0, suggesting an effect of long-term stress, irrespective of breed or sex. This effect was displayed despite relatively high energy supplementation of the animals during growth and limited pre-slaughter stress due to the short transport distance from the grower facility to abattoir, combined with a short lairage period. IVG animals recorded higher pH_u for BF and ST muscles, while the SM, SS and ST muscles of wethers, often of the IVG, were higher than those of bucks, suggesting a slightly higher susceptibility to stress by this breed type when muscles other than the LTL were considered. Similar final pH_u values for LTL were reported when high-energy supplements were fed to animals [8,40]. Pophiwa [8] found no differences in LTL pH_u between breeds. When animals were kept on natural pasture, it seems that higher final pH values (>5.8) for LTL accompanied by a dark, firm, and dry final meat condition (DFD) prevailed [41–43]. A possible reason could be that goats reared on pastures (extensively) have limited, if any, interaction with humans compared to goats reared in feedlot systems, resulting in the former being more stressed because of exposure to humans during the transport, lairage and slaughter processes.

Goats tend to deposit most of their fat in the visceral, rather than carcass depots and produce leaner carcasses [44–46] whilst the "indigenous" goat groups usually give inferior results compared to that of the BG [8–10]. However, improved feeding conditions in our study benefited castrated animals of both breeds concerning intramuscular fat deposition in all muscles. A faster rate of deposition for carcass and non-carcass fat, as well as total fat, has been reported for does and wethers raised under intensive management compared to bucks [42,43]. Higher marbling levels could contribute to eating quality, although it did not seem to have any effect on WBSF in our or other studies. Shahrai [47] could find no effect of % IMF and WBSF for beef with IMF values between 6.8% and 20.9%. Only when % IMF levels reached 33.9% did the effect become significant [48]. The IMF levels in our study were much lower, between 1.1% and 7.7%. According to Corbin [49], marbling levels varying between 1.96% and 3.8% had no effect on consumers' scores for beef tenderness (not WBSF), but scores increased significantly at 5.6% and higher.

Variation in connective tissue characteristics across muscles were consistent with the muscle type, which agrees with the findings for cattle [29]. The LTL recorded low total collagen levels while the other muscles recorded much higher levels. Collagen solubility varied between 27% and 39%. MFL, measuring the amount of ageing, varied between 27 and 47 µm [50]. However, the differences in WBSF within muscle were only recorded for 1 d post mortem ST muscle in favour of BG wethers, but were not complimented by any expected differences in MFL or connective tissue characteristics. MFLs of BF muscle were shorter in wethers than rams and only showed numerically lower WBSF values for BG wethers. Under the conditions of this trial, considering the use of electrical stimulation and normal commercial chilling conditions, it can therefore be concluded that neither goat breed nor sex had any effect on WBSF, irrespective of muscle. It is, however, important to note that on average, the LTL muscles had the most advantageous post mortem proteolytic activity (as indicated by the MFL) and lowest total collagen [51], and even though ES was applied during slaughter, the LTL was still tough, as indicated by the high WBSF. No plausible explanation for the tougher LTL muscles could be found, since the choice of a dry cooking method (broiling) also corresponded with the low connective tissue of the cut.

The differences in muscle physiology between species could also explain some of the colour differences noted. Neethling [52] reported muscle-specificity in fresh meat from a medium-sized wild ungulate, the blesbok (*Damaliscus pygargus phillipsi*) and observed that the blesbok *Infraspinatus* muscle was more colour-stable than the LTL and BF. This observation is different from that previously reported for fresh beef [51], and suggests that game species have unique biology and that the influence of muscle source on colour stability is species-dependent [51]. These observations may support the idea that the goat is a unique species, and that chevon should be approached differently from the other better-known red meats such as beef and mutton. In general, the rate and extent of post mortem glycolysis and ultimate pH of the muscle are critical factors that determine goat meat quality, in particular WHC and meat colour [53]. Contrary to Simela [54], no breed differences in meat colour characteristics for the various BG and IVG muscles were observed in this study. For the SM and SS muscles, wethers recorded higher pH_u values, which also coincided with slightly darker muscles, i.e., lower L^* values and higher values for a^* and consequently chroma. Incidentally, there were no differences in purge between breed nor sex. In general, this might have been due to the high pH_u, as all values were above 5.8, suggesting higher stress susceptibility in these specific animals [55].

Meat from intact male animals (bulls and rams) is generally darker compared to that of females and castrated males [56]. This is in contrast to the present study, where the wethers had darker meat ($L^* < 35.0$) compared to bucks ($L^* > 36.9$). Small and sometimes significant differences were found for other colour parameters, where muscles of wethers in most cases tend to show more vivid colours (higher chroma) and lower discolouration (lower hue angle values). It is known that energy status immediately after slaughter has an influence on meat colour (lightness) and tenderness [57,58].

5. Conclusions

Knowledge about meat quality of specific indigenous eco-types is limited as studies usually compare nonspecific "indigenous" goats with BG (well described). This study alleviates some misconceptions that exist about the potential quality of "indigenous" goat meat. More muscle meat quality differences were found between bucks and wethers than between BGs and large-frame IVGs consisting of a mixture of the different goat eco-types. This study showed that the muscles of IVG large-frame goats differed minimally from the same muscles derived from BG when finished off in the same feedlot. This study further showed that goat muscles have different characteristics from those of other red meat animals; further research is warranted to better understand these species' meat quality characteristics and the factors that influence it. More studies should also focus on understanding how to adapt/manage pre- and post-slaughter procedures to produce the best goat meat (chevon) eating experience.

Author Contributions: L.F., P.E.S. and L.C.H. were responsible for Conceptualization. G.L.v.W. was responsible for methodology, formal analysis, investigation and writing the original draft preparation. L.F., L.C.H. and P.E.S. were responsible for reviewing and editing. L.F. was responsible for resources, supervision, project administration and funding acquisition. All authors have read and agreed to the published version of the manuscript.

Funding: This work was supported by the Red Meat Research and Development of South Africa (RMRDSA) and Technology and Human Resources for Industry Programme (THRIP) of the Department of Trade and Industry, Pretoria, Government of South Africa, Grant number TP14080787990. The authors thank the Agricultural Research Council (ARC) for facilities and human resources.

Institutional Review Board Statement: This research was approved by the Agricultural Research Council-Animal Production (ARC-AP) Ethics Committee (ref no. APIEC16/021).

Data Availability Statement: Except for the carcass quality, supportive data provided in Table 1 that was published in Van Wyk et al. (2020), none of the project related data was published yet. Van Wyk, G.L. did complete a thesis on the full project from which additional publications will follow. Van Wyk, G.L.; Hoffman, L.C.; Strydom, P.E.; Frylinck, L. Effect of Breed Types and Castration on Carcass Characteristics of Boer and Large Frame Indigenous Veld Goats of Southern Africa. Animals 2020, 10, 1884. https://doi.org/10.3390/ani10101884.

Acknowledgments: The authors acknowledge the support given by the personnel of the ARC-Animal Production, i.e., Small Stock, Abattoir, Biochemistry and Technology sections for technical support.

Conflicts of Interest: The authors declare no conflict of interest.

References

1. Indigenous Goat Breeds of South Africa. Available online: https://southafrica.co.za/indigenous-goat-breeds-of-south-africa.html (accessed on 21 October 2021).
2. Van Wyk, G.L.; Hoffman, L.C.; Strydom, P.E.; Frylinck, L. Effect of Breed Types and Castration on Carcass Characteristics of Boer and Large Frame Indigenous Veld Goats of Southern Africa. *Animals* **2020**, *10*, 1884. [CrossRef]
3. Visser, C. A review on goats in southern Africa: An untapped genetic resource. *Small Rumin. Res.* **2019**, *176*, 11–16. [CrossRef]
4. Food and Agriculture Organization of the United Nations, Statistics Division (FAOSTAT). Available online: https://www.fao.org (accessed on 28 December 2020).
5. Tshabalala, P.A.; Strydom, P.E.; Webb, E.C.; de Kock, H.L. Meat quality of designated South African indigenous goat and sheep breeds. *Meat Sci.* **2003**, *65*, 563–570. [CrossRef]

6.   Simela, L. Meat Characteristics and the Acceptability of Chevon from South African Indigenous Goats. Ph.D. Thesis, University of Pretoria, Pretoria, South Africa, 2005. Available online: http://hdl.handle.net/2263/29932 (accessed on 28 January 2021).

7.   Webb, E.C.; Casey, N.H.; Simela, L. Goat meat quality. *Small Rumin. Res.* **2005**, *60*, 153–166. [CrossRef]

8.   Pophiwa, P.; Webb, E.C.; Frylinck, L. Meat quality characteristics of two South African goat breeds after applying electrical stimulation or delayed chilling of carcasses. *Small Rumin. Res.* **2016**, *145*, 107–114. [CrossRef]

9.   Pophiwa, P.; Webb, E.C.; Frylinck, L. Carcass and meat quality of Boer and indigenous goats of South Africa under delayed chilling conditions. *S. Afr. J. Anim. Sci.* **2017**, *47*, 794–803. [CrossRef]

10.   Pophiwa, P.; Webb, E.C.; Frylinck, L. A review of factors affecting goat meat quality and mitigating strategies. *Small Rumin. Res.* **2020**, *183*, 106035. [CrossRef]

11.   Sacks, M.S.; Kronick, P.L.; Buechler, P.R. Contribution of intramuscular connective tissue to the viscoelastic properties of *post-rigor* bovine muscle. *J. Food Sci.* **1988**, *53*, 19–24. [CrossRef]

12.   Listrat, A.; Lebret, B.; Louveau, I.; Astruc, T.; Bonnet, M.; Lefaucheur, L.; Picard, B.; Bugeon, J. How Muscle Structure and Composition Influence Meat and Flesh Quality. *Sci. World J.* **2016**, *2016*, 3182746. [CrossRef]

13.   Font-i-Furnols, M.; Guerrero, L. Consumer preference, behaviour and perception about meat and meat products: An overview. *Meat Sci.* **2014**, *98*, 361–371. [CrossRef]

14.   Sheridan, R.; Hoffman, L.C.; Ferreira, A.V. Meat quality of Boer Goat kids and Mutton merino lambs. 1. Commercial yields and chemical composition. *Anim. Sci.* **2003**, *76*, 63–71. [CrossRef]

15.   Schönfeldt, H.C.; Naude, R.T.; Bok, W.; van Heerden, S.M.; Smit, R.; Boshoff, E. Flavour and tenderness related quality characteristics of goat and sheep meat. *Meat Sci.* **1993**, *34*, 363–379. [CrossRef]

16.   Schönfeldt, H.C.; Naude, R.T.; Bok, W.; van Heerden, S.M.; Swoden, L.; Boshoff, E. Cooking and juiciness related quality characteristics of goat and sheep meat. *Meat Sci.* **1993**, *34*, 381–394. [CrossRef]

17   Schönfeldt, H.C.; Strydom, P.E. Effect of age and cut on tenderness of South African beef. *Meat Sci.* **2011**, *87*, 206–208. [CrossRef]

18.   Strydom, P.E.; Frylinck, L.; Smith, M.F. Should electrical stimulation be applied when cold shortening is not a risk? *Meat Sci.* **2005**, *70*, 733–742. [CrossRef] [PubMed]

19.   Irie, M.; Izumo, A.; Mohri, S. Rapid method for determining water-holding capacity in meat using video image analysis and simple formulae. *Meat Sci.* **1996**, *42*, 95–102. [CrossRef]

20.   *Research Guidelines for Cookery and Evaluation and Instrumental Tenderness Measurements of Meat*, 2nd ed.; Wheeler, T.L., Ed.; Version, 1.02; American Meat Science Association (AMSA): Ithaca, NY, USA, 2016.

21.   Honikel, J.L. Reference methods for the assessment of physical characteristics of meat. *Meat Sci.* **1998**, *49*, 447–457. [CrossRef]

22.   Culler, R.D.; Parrish, J.R.; Smith, G.C.; Cross, H.R. Relationship of myofibril fragmentation index to certain chemical, physical and sensory characteristics of bovine Longissimus muscle. *J. Food Sci.* **1978**, *43*, 1177–1180. [CrossRef]

23.   Heinze, P.H.; Bruggemann, D. Ageing of beef: Influence of two ageing methods on sensory properties and myofibrillar proteins. *Sci. Des. Aliment.* **1994**, *14*, 387–399.

24.   Association of Official Analytical Chemist (AOAC). *Official Methods of Analyses*, 15th ed.; Helrich, K., Ed.; AOAC: Arlington, VA, USA, 1990. Available online: https://law.resource.org/pub/us/cfr/ibr/002/aoac.methods.1.1990.pdf (accessed on 15 December 2021).

25.   Foster, M.L.; Gonzales, S.E. Soxtec Fat Analyzer for Determination of Total Fat in Meat: Collaborative Study. *J. AOAC Int.* **1992**, *75*, 288–292. [CrossRef]

26.   Bergman, I.; Loxley, R. Two improved and simplified methods for the spectrophotometric determinations of hydroxy-proline. *Anal. Chem.* **1963**, *35*, 1961–1965. [CrossRef]

27.   Hill, F. The solubility of intermuscular collagen in meat animals of various ages. *J. Food Sci.* **1966**, *31*, 161–166. [CrossRef]

28.   Boccard, R.L.; Naudé, R.T.; Cronje, D.E.; Smit, M.; Venter, H.J.; Rossouw, E. The influence of age, sex and breed of cattle on their muscle characteristics. *Meat Sci.* **1979**, *3*, 261–280. [CrossRef]

29.   Cross, H.R.; Carpenter, Z.L.; Smith, G.C. Effects of intramuscular collagen and elastin on bovine muscle tenderness. *J. Food Sci.* **1973**, *38*, 998–1003. [CrossRef]

30.   *Commission Internationale de l'Eclairage (CIE) (1986) Colorimetry*, 2nd ed.; Publication CIE No. 15.2; Commission International de l'Eclairage: Vienna, Austria, 1986.

31.   Krzywicki, K. Assessment of relative content of myoglobin oxymyoglobin and metmyoglobin at the surface of beef. *Meat Sci.* **1978**, *3*, 1–10. [CrossRef]

32.   MacDougall, D.B. Colour in meat. In *Sensory Properties of Foods*; Birch, G.G., Brennan., J.G., Parker, K., Eds.; Applied Science Publishers: London, UK, 1977; p. 59.

33.   Young, O.A.; Priolo, A.; Simmons, N.J.; West, J. Effects of rigor attainment temperature on meat blooming and colour on display. *Meat Sci.* **1999**, *52*, 47–56. [CrossRef]

34.   *SAS/STAT User's Guide*; SAS Institute Inc.: Cary, NC, USA, 2004.

35.   Snedecor, G.W.; Cochran, W.G. *Statistical Methods*, 7th ed.; Times; Iowa State University Press: Ames, IA, USA, 1980.

36.   Shapiro, S.S.; Wilk, M.B. An analysis of variance test for normality (complete samples). *Biometrika* **1965**, *52*, 591–611. [CrossRef]

37.   Babiker, S.A.; El Khider, I.A.; Shafie, S.A. Chemical composition and quality attributes of goat meat and lamb. *Meat Sci.* **1990**, *28*, 273–277. [CrossRef]

38.   Dhanda, J.S.; Taylor, D.G.; Murray, P.J.; McCosker, J.E. The influence of goat genotype on the production of Capretto and Chevon carcasses. 2. Meat quality. *Meat Sci.* **1999**, *52*, 363–367. [CrossRef]

39. Adeyemi, K.D.; Sazili, A.Q. Efficacy of carcass electrical stimulation in meat quality enhancement: A review. *Asian-Aust. J. Anim. Sci.* **2014**, *27*, 447–456. [CrossRef]
40. Safari, J.; Mushi, D.E.; Mtenga, L.A.; Kifaro, G.C.; Eik, L.O. Effects of concentrate supplementation on carcass and meat quality attributes of feedlot finished Small East African goats. *Livest. Sci.* **2009**, *125*, 266–274. [CrossRef]
41. Hogg, B.W.; Mercer, G.J.K.; Mortimer, B.J.; Kirton, A.H.; Duganzick, D.M. Carcass and meat quality attributes of commercial goats in New Zealand. *Small Rumin. Res.* **1992**, *8*, 243–256. [CrossRef]
42. Swan, J.E.; Esguerra, C.M.; Farouk, M.M. Some physical, chemical and sensory properties of chevon products from three New Zealand breeds. *Small Rumin. Res.* **1998**, *28*, 273–280. [CrossRef]
43. Kannan, G.; Kouakou, B.; Gelaye, S. Colour changes reflecting myoglobin and lipid oxidation in chevon cuts during refrigerated display. *Small Rumin. Res.* **2001**, *42*, 67–75. [CrossRef]
44. Mahgoub, O.; Khan, A.J.; Al-Maqbaly, R.S.; Al-Sabahi, J.N.; Anna-Malai, K.; Al-Sakry, N.M. Fatty acid composition of muscle and fat tissues of Omani Jebel Akhdar goats of different sexes and weights. *Meat Sci.* **2002**, *61*, 381–387. [CrossRef]
45. Mahgoub, O.; Kadim, I.T.; Al-Saqry, N.M.; Al-Busaidi, R.M. Effect of body weight and sex on carcass tissue distribution in goats. *Meat Sci.* **2004**, *67*, 577–585. [CrossRef] [PubMed]
46. Devendra, C.; Owen, J.E. Quantitative and qualitative aspects of meat production from goats. *World Anim. Rev.* **1983**, *47*, 19–29.
47. Shahrai, N.N.; Babji, A.S.; Maskat, M.Y.; Razali, A.F.; Yusop, S.M. Effects of marbling on physical and sensory characteristics of ribeye steaks from four different cattle breeds. *Anim. Biosci.* **2021**, *34*, 904–913. [CrossRef] [PubMed]
48. Carpenter, Z.L.; King, G.T. Tenderness of lamb rib chops. *Food Technol.* **1965**, *19*, 1706.
49. Corbin, C.H.; O'Quinn, T.G.; Garmyn, A.J.; Legako, J.F.; Hunt, M.R.; Dinh, T.T.N.; Rathmann, R.J.; Brooks, J.C.; Miller, M.F. Sensory evaluation of tender beef strip loin steaks of varying marbling levels and quality treatments. *Meat Sci.* **2015**, *100*, 24–31. [CrossRef]
50. Purslow, P.P. Contribution of collagen and connective tissue to cooked meat toughness; some paradigms reviewed. *Meat Sci.* **2018**, *144*, 127–134. [CrossRef]
51. Berry, B.W. Tenderness of beef loin steaks as influenced by marbling level, removal of subcutaneous fat, and cooking method. *J. Anim. Sci.* **1993**, *71*, 2412–2419. [CrossRef]
52. Neethling, N.E.; Suman, S.P.; Sigge, G.O.; Hoffman, L.C. Muscle-specific colour stability of blesbok (Damaliscus pygargus phillipsi) meat. *Meat Sci.* **2016**, *119*, 69–79. [CrossRef]
53. Casey, N.H.; Webb, E.C. Managing goat production for meat quality. *Small Rumin. Res.* **2010**, *89*, 218–224. [CrossRef]
54. Simela, L.; Webb, E.C.; Frylinck, L. Effect of sex, age and pre-slaughter conditioning on pH, temperature, tenderness and colour of indigenous South African goats. *S. Afr. J. Anim. Sci.* **2004**, *34*, 208–211.
55. Gardener, G.E.; Kenny, L.; Milton, J.T.B.; Pethick, D.W. Glycogen metabolism and ultimate pH in Merino, first cross and second cross wether lambs as affected by stress before slaughter. *Aust. J. Agric. Res.* **1999**, *50*, 175–181. [CrossRef]
56. Seideman, S.C.; Cross, H.R.; Oltjen, R.R.; Schanbacher, B.D. Utilization of the Intact Male for Red Meat Production: A Review. *J. Anim. Sci.* **1982**, *55*, 826–840. [CrossRef]
57. Monin, G.; Seller, P. Pork of low technological quality with normal rate of muscle pH fall in the immediate *post-mortem* period: The case of the Hampshire breed. *Meat Sci.* **1985**, *13*, 49–63. [CrossRef]
58. Scheffler, T.L.; Park, S.; Gerrard, D.E. Lessons to learn about post-mortem metabolism using AMPKγ3R200Q mutation in the pig. *Meat Sci.* **2011**, *89*, 244–250. [CrossRef] [PubMed]

Article

The Inclusion of Pea in Concentrates Had Minor Effects on the Meat Quality of Light Lambs

Mireia Blanco *, Guillermo Ripoll , Sandra Lobón , Juan Ramón Bertolín, Isabel Casasús and Margalida Joy

Centro de Investigación y Tecnología Agroalimentaria de Aragón (CITA), Instituto Agroalimentario de Aragón–IA2 (CITA-Universidad de Zaragoza), Avda. Montañana 930, 50059 Zaragoza, Spain; gripoll@cita-aragon.es (G.R.); slobon@cita-aragon.es (S.L.); jrbertolin@cita-aragon.es (J.R.B.); icasasus@cita-aragon.es (I.C.); mjoy@cita-aragon.es (M.J.)
* Correspondence: mblanco@cita-aragon.es

Simple Summary: The use of local protein sources, such as pea (*Pisum sativum*), has been encouraged to reduce the dependency of Europe on soybean meal imports. Changes in the ingredients of iso-energetic concentrates may affect the fatty acid profiles of the concentrates and other secondary compounds and, therefore, affect meat quality parameters. The objective of the study was to compare the carcass colour, and the meat quality parameters (texture, chemical composition, lipid oxidation, fatty acids) of lambs fed concentrates with different proportions of pea for replacing soybean meal. The inclusion of pea had no effects on carcass colour and minor effects on the fatty acid profile. Therefore, the inclusion of pea can be recommended to increase the self-sufficiency of lamb production systems.

Abstract: The use of pea (*Pisum sativum*) has been recommended to replace soybean meal in the diet of ruminants, but it may affect meat quality. The aim of this study was to evaluate the effect of the proportion of pea (0%, 10%, 20% and 30%) in fattening concentrates fed to light lambs for 41 days on carcass colour and on the meat quality. Pea inclusion affected neither the colour of the lamb carcasses nor affected most of the parameters of the meat quality. However, the inclusion of pea affected the cholesterol content, and the 20%pea concentrate yielded meat with greater cholesterol contents than the 30%pea concentrate did ($p < 0.05$). The inclusion of pea had minor effects on individual FAs but affected the total saturated fatty acids ($p < 0.01$) and the thrombogenicity index ($p < 0.05$). A greater total saturated fatty acid content was recorded for the 20%pea concentrate than for the rest of the concentrates, and a greater thrombogenicity index was recorded for the 20% concentrate than for the 10%pea concentrate. The results indicated the viability of the inclusion of pea in the fattening concentrate of light lambs without impairing meat quality, with the 30%pea concentrate being the most suitable to reduce the soya-dependency.

Keywords: *Pisum sativum*; fatty acids; colour; texture; soybean

Citation: Blanco, M.; Ripoll, G.; Lobón, S.; Bertolín, J.R.; Casasús, I.; Joy, M. The Inclusion of Pea in Concentrates Had Minor Effects on the Meat Quality of Light Lambs. *Animals* **2021**, *11*, 2385. https://doi.org/10.3390/ani11082385

Academic Editor: Gianluca Neglia

Received: 1 July 2021
Accepted: 8 August 2021
Published: 12 August 2021

Publisher's Note: MDPI stays neutral with regard to jurisdictional claims in published maps and institutional affiliations.

1. Introduction

There is a worldwide interest in enhancing the use of pea (*Pisum sativum* L.) in the diets of ruminants. Using pea in place of soybean meal (SBM) has been encouraged to reduce dependency on imports in Europe [1]. Moreover, European consumers reject soybean, as it is mainly a genetically modified organism [2] and is therefore banned by organic production regulations [3] and because of sustainability issues linked to deforestation in soybean production areas [4].

Pea has a high crude protein (CP) content [5,6] and highly soluble and degradable starch [7], and the net energy for the weight gain of pea is at least equal to that of corn and is greater than that of SBM [6]. Up to 15% pea inclusion in the fattening diets of light lambs is recommended [8], but the use of pea has been studied scarcely. Nevertheless, minor effects were observed when lambs were fattening with greater rates of pea inclusion [9–12]. Purroy et al. [13] observed greater internal fat depositions in lambs that were fed concentrate including pea instead of SBM, and they related these results to the differences in the net energies for the weight gain of these protein sources. Therefore, the intramuscular fat (IMF) deposition in the studied lambs might also have been altered, although this effect was not studied. The inclusion of pea can also affect the fatty acid (FA) composition of lambs [6,12,14] and improve the health properties of meat because of the deposition of n-3 polyunsaturated fatty acids (PUFAs) is increased when pea is used compared to SBM [12,15]. Moreover, pea can be a source of carotenoid [3], which affects fat colour and lipid oxidation during storage and have a minor effect on meat colour [9]. Therefore, the aim of this study was to evaluate the effects of increasing the proportion of pea in the fattening concentrate of light lambs on the fat colour and meat quality.

2. Materials and Methods

The experiment was conducted in the facilities of the CITA Research Centre (41°3′ N, 0°47′ W, 216 m above sea level) in Zaragoza (Spain).

2.1. Animal Management and Experimental Design

At weaning, 54 male Rasa Aragonesa lambs (13.4 ± 0.16 kg in BW; 31 ± 0.6 d in age) were randomly selected among 98 single raised lambs of the experimental flock. The lambs were randomly assigned to 1 of 4 groups, balanced by BW and age at weaning and by weight gain during suckling. Each group received a pelleted concentrate with a different proportion of pea (0%, 10%, 20% or 30%) and barley straw on an ad libitum basis for 41 (±1.4) days (d), at which point they reached 23.1 (±0.11) kg in BW. The inclusion of pea mainly replaced SBM, and the pea concentrates were formulated to be iso-energetic (13.3 MJ metabolic energy kg^{-1} DM) and iso-proteic (198 g CP kg^{-1} DM) [16] considering a barley straw intake of 10%. The main ingredients were barley, corn, soybean meal, wheat, pea, wheat bran, sugarcane molasses and palm oil. Further information on the ingredients and chemical composition of concentrates, the management of the lambs during the fattening period, and information on the slaughtering procedures are reported elsewhere [11]. Each week, samples of the concentrates were collected for chemical analyses.

2.2. Slaughtering Conditions and Sampling

After slaughter, the carcasses were chilled at 4 °C for 24 h in total darkness. The carcasses were split along the dorsal line. The *Rectus abdominis* (RA), *Longissimus thoracis et lumborum* (LTL), *Semimembranosus* (SM) and *Semitendinosus* (ST) muscles were removed. The LTL muscle from the 4th to the 6th lumbar *Vertebrae* of the left half of the carcass was sliced, freeze-dried and minced to determine its chemical composition and retinol, cholesterol and tocopherol contents. The same portion from the right half of the carcass was identically processed to analyse the fatty acid composition. The LTL muscles from the 6th to the 13th thoracic *Vertebrae* were sliced into 2.5-cm-thick samples, which were randomly placed in trays wrapped with oxygen-permeable PVC film and kept in darkness at 4 °C until the colour was measured (0, 2, 5, 7 and 9 d of air exposure). The 0-d samples were also allowed to bloom in darkness at 4 °C for 1 h before being measured. Immediately after the colour measurements were conducted, the samples were vacuum-packed and frozen (at −20 °C) until lipid oxidation analysis. The ST and SM muscles were immediately vacuum-packed and stored at −20 °C until texture determinations.

2.3. Determinations

The colours of the RA and LTL muscles, subcutaneous caudal fat and perirenal fat were measured using a Minolta CM-2006d spectrophotometer (Konica Minolta Holdings, Inc., Osaka, Japan) in CIEL*a*b* space [17]. The lightness (L*), redness (a*) and yellowness (b*) were recorded and used to calculate the hue angle ($(h_{ab}) = \tan^{-1}\left(\frac{b^*}{a^*}\right) \times 57.29$, expressed in degrees), and the chroma ($(C^*_{ab}) = \sqrt{(a^{*2} + b^{*2})}$). The relative contents of metmyoglobin (MMb), oxymyoglobin (MbO$_2$) and deoxymyoglobin (DMb) in the meat were estimated [18].

The absolute value of the integral of the translated spectrum (SUM) was calculated in the fat deposits [19].

The intramuscular lipid oxidation of the LTL muscle was determined following the procedure reported by Ripoll, González-Calvo, Molino, Calvo and Joy [17]. The ST muscle was used to study the shear force in cooked meat using an Instron machine model 5543 (Instron Limited, Cerdanyola, Spain). The muscles were thawed in tap water until they reached an internal temperature of 16–19 °C and then cooked in a water bath at 75 °C (internal temperature of 70 °C). The temperatures were controlled with a Testo 108-2 waterproof food thermometer with a Type T thermocouple (Instrumentos Testo S.A., Cabrils, Spain). The steaks were cooled overnight at room temperature. Meat blocks of $1 \times 1 \times 3$ cm were sheared perpendicularly to the long axis of the block using a Warner–Bratzler device with a cross-head speed of 2.5 mm s^{-1}. The shear force and toughness of each block were determined. The SM muscle was used to determine the texture in raw meat using a modified compression device that avoids the transversal elongation of the sample [20]. Cores of $30 \times 10 \times 10$ mm were cut. The stress was measured when the probe compressed the core by 20% and 80%, with a cross-head speed of 0.83 mm s^{-1}.

2.4. Chemical Analyses

The samples were weighed before and after freeze-drying to obtain the DM content. The CP content was determined following the Dumas procedure [21] using a nitrogen analyser (Model NA 2100, CE Instruments, Thermoquest SA, Barcelona, Spain). The IMF was determined following the Ankom procedure [22] with an XT10 Ankom extractor (Ankom Technology Corporation, New York, NY, USA). The extraction of carotenoids and tocopherols in the concentrates was performed following the procedure described in Blanco et al. [23], whereas the extraction of retinol, tocopherols and cholesterol in the meat was performed following the methodology of Bertolín et al. [24]. In both analyses, an Acquity UPLC H-Class liquid chromatograph (Waters, Mildford, KA, USA) equipped with a silica-based bonded phase column (Acquity UPLC HSS T3, 1.8-μm $\times$ 2.1-mm $\times$ 150-mm column, Waters, Mildford, KA, USA), an absorbance detector (Acquity UPLC Photodiode Array PDA eλ Detector; Waters, Mildford, KA, USA) and a fluorescence detector (2475 Multi λ Fluorescence Detector, Waters, Mildford, KA, USA) was used. The carotenoids, retinol and cholesterol were detected by measuring the absorbance at 450, 325 and 220 nm, respectively, and the tocopherols were detected by measuring the fluorescent emissions at λexc = 295 and λemi = 330 nm.

Fatty acids were extracted and derivatised from the concentrates (extracted with heptane) [25] and from the meat samples [26]. Fatty acid methyl esters (FAMEs) were determined using a GC (Bruker 436 Scion gas, Billerica, MA, USA) equipped with a cyanopropyl capillary column (BR-2560, 100 m $\times$ 0.25 mm ID $\times$ 0.20 μm thick, Bruker, Billerica, MA, USA) with a flame ionisation detector and Compass CDS software. FA identification was performed using the GLC-532, GLC-401, GLC-643, GLC-642, GLC-463, C18:1 t11, C19:0 and C23:0 standard references (Nu-Chek-Prep Inc., Elysian MN, USA) and the relative retention times observed in the literature [26,27]. FA quantification was performed following the UNE-EN 12966-4 Official Method (2015). After individual FA determination, the sum of the saturated fatty acids (SFA), monounsaturated FAs (MUFA), polyunsaturated FAs (PUFA), PUFA n-6, PUFA n-3 and n-6:n-3 ratio were calculated. The indexes of atherogenicity (IA) and thrombogenicity (TI) were calculated by the following formulas according to the method used by Ulbricht and Southgate [28]:

atherogenicity index

$$= \frac{(\text{C12:0} + 4 \times \text{C14:0} + \text{C16:0})}{(\sum \text{MUFA} + \sum \text{n} - 6 + \sum \text{n} - 3)} \tag{1}$$

and

thrombogenicity index

$$= \frac{(C14{:}0 + C16{:}0 + C18{:}0)}{\left(0.5 \times \sum MUFA + 0.5 \times \sum n-6 + 3 \times \frac{\sum n-3}{\sum n-6}\right)} \tag{2}$$

2.5. Statistical Analyses

The data were analysed with SAS 9.4 statistical software (SAS Inst. Inc., Cary, NC, USA). The chemical composition, tocopherol content, cholesterol content and FA composition of the LTL muscle were analysed using the GLM procedure with the pea proportion as the fixed effect. The colour and lipid oxidation of the LTL muscle were analysed with a mixed model (MIXED procedure) using repeated measurements. The pea proportion, meat display time and the interaction between the two factors were included as fixed effects, and the lamb was included as a random effect. The degrees of freedom were adjusted with the Kenward–Roger correction to account for unequal observations or missing values. To model the error, different variance-covariance matrices were tested, and the matrix with the lowest Aikake and Bayesian information criteria was chosen. Multiple comparisons among treatments were performed using Tukey's method. The least-square smeans and standard errors were obtained, and differences were considered significant when $p < 0.05$. The trends were discussed when $0.10 < p \leq 0.05$.

3. Results

3.1. Feedstuffs

The main FAs in the concentrates were C16:0, C18:0, C18:1 c9 and C18:2 n-6, all of which were affected by the pea proportion ($p < 0.001$; Table 1). The percentage of C16:0 was greatest in the 20%pea concentrate, intermediate in the 10%pea and 30%pea concentrates and lowest in the 0%pea concentrate ($p < 0.001$). The percentage of C18:0 was greater in the 0%pea and 30%pea concentrates than in the other two concentrates ($p < 0.001$). The percentage of C18:1 c9 was lower in the 0%pea concentrate than in the other concentrates, whereas the percentage of C18:2 n-6 was higher in the 0%pea concentrate than in the rest of the concentrates ($p < 0.001$).

Table 1. Fatty acid (FA) profile, carotenoids and tocopherols of the concentrates with different proportions of pea.

	0%pea	10%pea	20%pea	30%pea	s.e.	*p*-Value
n	4	4	4	4		
			FA, g/100 g			
C12:0	0.08 c	0.13 ab	0.11 b	0.15 a	0.004	<0.001
C14:0	0.52 b	0.68 ab	0.74 ab	0.77 a	0.029	0.04
C16:0	34.01 d	37.18 b	38.15 a	35.49 c	0.118	<0.001
C17:0	0.19	0.15	0.17	0.21	0.014	0.52
C18:0	10.61 a	8.79 b	8.54 b	10.61 a	0.185	0.002
C18:1 c9	19.18 b	23.28 a	24.86 a	23.19 a	0.211	<0.001
C18:2 n-6	32.18 a	26.95 b	24.66 b	26.56 b	0.294	<0.001
C20:0	0.26	0.23	0.27	0.25	0.011	0.70
C18:3 n-3	2.44	2.20	2.15	2.38	0.041	0.09
C23:0	0.44	0.27	0.21	0.19	0.071	0.61
			Carotenoids, µg/g DM			
Lutein	0.9 b	1.2 b	1.3 b	1.7 a	0.04	<0.001
Zeaxanthin	0.69 a	0.51 b	0.32 c	0.47 bc	0.02	<0.001
13 Z-β-carotene	0.8 c	1.7 b	2.2 a	1.4 b	0.04	<0.001
9 Z-β-carotene	0.5 c	1.0 b	1.4 a	0.9 b	0.04	<0.001
All E-β-carotene	1.1 b	1.8 a	2.3 a	1.9 a	0.06	<0.001
			Tocopherols, µg/g DM			
α-tocopherol	4.4 a	4.8 a	4.9 a	3.3 b	0.07	<0.001
γ-tocopherol	10.0 b	11.0 b	11.4 b	13.6 a	0.21	<0.001
δ-tocopherol	3.4 c	6.2 b	7.8 a	6.3 b	0.06	<0.001

Within a parameter, means with different letters differ at $p < 0.05$.

Regarding the carotenoids, lutein, zeaxanthin, all-E-, 13Z- and 9Z-β-carotene were detected in low quantities in all concentrates, with differences obtained among the concentrates ($p < 0.001$; Table 1) with no clear pattern. The lutein content increased with the pea proportion, whereas the concentrate with 0%pea had the greatest zeaxanthin content, and the concentrate with 20%pea had the greatest β-carotene content ($p < 0.001$). Contents of γ-, α- and δ-tocopherols were detected in all concentrates and were affected by the inclusion of pea. The content of γ-tocopherol increased with the proportion of pea, whereas the 20%pea concentrate presented the greatest δ-tocopherol content ($p < 0.001$) and α-tocopherol content that was only greater than that of the 30%pea concentrate.

3.2. Carcass Colour

The colour parameters of the studied RA muscle and fat deposits were similar among the concentrates ($p > 0.05$; Table 2).

Table 2. Effect of the proportion of pea in the concentrate on the colour and the estimator of carotenoids (SUM) of *Rectus abdominis* muscle, subcutaneous fat and perirenal fat.

	0%pea	10%pea	20%pea	30%pea	s.e.m.	*p*-Value
n	13	13	14	14		
		Rectus abdominis muscle				
Lightness (L*)	47.8	48.0	47.3	48.1	0.3	0.81
Redness (a*)	9.1	8.5	9.0	9.0	0.2	0.71
Yellowness (b*)	9.8	10.1	9.5	10.3	0.2	0.53
Chroma (C^*_{ab})	13.4	13.3	13.3	13.8	0.2	0.75
Hue angle (h_{ab})	47.1	49.6	45.9	49.0	1.0	0.55
		Perirenal fat				
Lightness (L*)	71.4	72.3	71.6	72.4	0.3	0.66
Redness (a*)	4.8	4.2	4.6	4.3	0.2	0.56
Yellowness (b*)	11.6	11.0	11.8	11.4	0.2	0.57
Chroma (C^*_{ab})	12.5	11.8	12.8	12.2	0.2	0.52
Hue angle (h_{ab})	67.6	69.2	68.4	69.7	0.6	0.68
SUM	120	110	123	104	5.7	0.63
		Subcutaneous caudal fat				
Lightness (L*)	69.7	69.5	70.1	70.1	0.3	0.86
Redness (a*)	2.8	2.9	2.7	2.7	0.2	0.98
Yellowness (b*)	11.3	12.1	11.5	12.3	0.3	0.59
Chroma (C^*_{ab})	11.6	12.5	11.9	12.7	0.3	0.62
Hue angle (h_{ab})	76.4	76.8	76.5	78.4	0.6	0.67
SUM	93	102	109	121	5.1	0.28

3.3. Meat Quality

The colours, pigments and lipid oxidation levels of the LTL muscle were only affected by the time of air exposure ($p < 0.001$), nor by pea proportion, neither by the interaction of time and pea proportion ($p > 0.05$). All colour parameters increased between 0 and 2 d ($p < 0.001$), with slight changes observed thereafter (Figure 1). Similarly, MMb and MbO2 increased with concomitant decreases in DMb between 0 and 2 d, but no changes were registered thereafter (Figure 1). Lipid oxidation increased linearly during the time of air exposure ($p < 0.001$), irrespective of the proportion of pea in the concentrate (Figure 2).

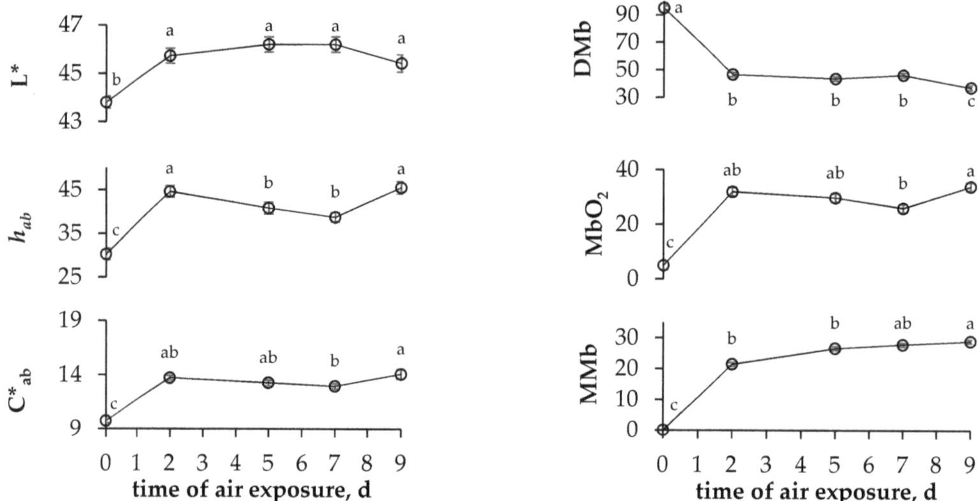

Figure 1. Evolution of colour parameters (lightness (**L***), hue angle (h_{ab}) and chroma (**C***$_{ab}$)) and haem pigments (deoxy-myoglobin (**DMb**), oxymyoglobin (**MbO₂**) and metmyoglobin (**MMb**)) of LTL muscle throughout display. Vertical bars indicate the standard error of the mean; different letters indicate differences between times ($p < 0.05$).

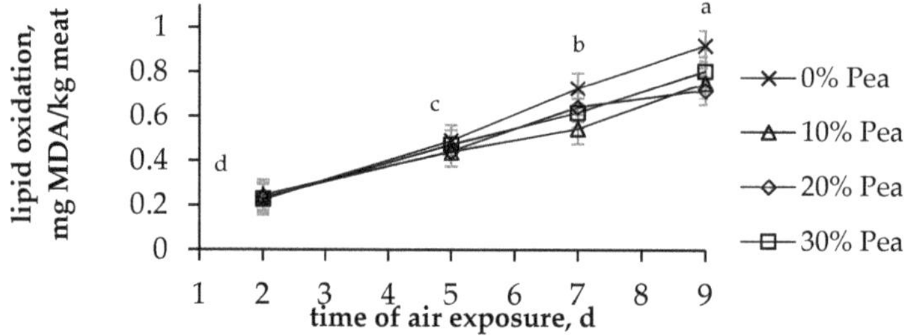

Figure 2. Effect of the proportion of pea in the concentrate on lipid oxidation during air exposure. Vertical bars indicate the standard error of the mean; different letters indicate differences between times ($p < 0.001$); MDA: malondialdehyde.

The texture of the meat was not affected by the inclusion of pea either in the ST muscle or in the SM muscle ($p > 0.05$, Table 3). Regarding the chemical composition of the meat, the proportion of pea only affected the cholesterol and retinol contents ($p < 0.05$, Table 3). The meat of lambs fed 20%pea concentrate presented greater cholesterol content than the meat of lambs fed 30%pea concentrate ($p < 0.05$), but this content was similar to those observed under the rest of the treatments.

Regarding retinol, the meat of the lambs fed 10%pea concentrate presented greater retinol content than the meat of lambs fed 0%pea concentrate ($p < 0.05$), but this value was similar to those observed for the rest of the concentrates.

The proportion of pea in the concentrate had minor effects on the percentages of individual FAs in the meat (Tables 4 and 5), affecting only the percentages of C13:0, C15:0, C17:0 and C16:1 t9 ($p < 0.05$) and the percentages of C16:0, C17:1 c9, iC18:0 and C18:2 n-6 t9, t12 ($p < 0.10$).

Table 3. Effect of the proportion of pea in the concentrate on texture, chemical composition, cholesterol, retinol and tocopherols in meat.

	0%pea	10%pea	20%pea	30%pea	s.e.m.	*p*-Value
	Raw *Semitendinosus* muscle					
Shear force, N cm^{-2}	41.9	43.5	42.2	41.7	0.88	0.89
Toughness, N cm^{-2}	12.6	13.0	12.1	13.0	0.31	0.68
	Cooked *Semimembranosus* muscle					
Stress-compression 20%, N	14.8	17.1	15.5	16.3	0.41	0.24
Stress-compression 80%, N	52.1	49.6	52.2	53.5	1.62	0.86
	LTL muscle					
Dry matter, %	22.00	21.79	22.23	21.53	0.112	0.13
Crude protein, %FM	20.30	20.27	20.41	19.86	0.097	0.25
Intramuscular fat, %FM	1.65	1.67	1.93	1.72	0.051	0.26
Cholesterol, mg/g FM	0.51 [ab]	0.51 [ab]	0.53 [a]	0.49 [b]	0.005	0.02
Retinol, µg/g FM	0.023 [b]	0.028 [a]	0.026 [ab]	0.024 [ab]	0.001	0.02
α-tocopherol, µg/g FM	0.63	0.69	0.66	0.57	0.026	0.35
γ-tocopherol, µg/g FM	0.17	0.21	0.21	0.24	0.011	0.11

Within a parameter, means with different letters differ at $p < 0.05$.

Table 4. Effect of the proportion of pea in the individual saturated and monounsaturated fatty acids (FA) in LTL muscle.

	0%pea	10%pea	20%pea	30%pea	s.e.m.	Pr > F
	Saturated FA, g/100 g					
C10:0	0.12	0.13	0.13	0.11	0.01	0.78
C12:0	0.24	0.20	0.28	0.24	0.02	0.37
aC13:0	0.94	1.14	0.94	1.06	0.04	0.20
C13:0	0.06 [ab]	0.09 [a]	0.05 [b]	0.07 [ab]	0.01	0.04
iC14:0	0.68	0.82	0.61	0.67	0.03	0.11
C14:0	3.38	3.12	3.46	3.05	0.09	0.32
iC15:0	0.11	0.11	0.11	0.09	0.01	0.50
aC15:0	0.66	0.68	0.63	0.67	0.02	0.83
C15:0	0.42 [b]	0.41 [b]	0.49 [a]	0.47 [ab]	0.01	0.01
DMA C16:0	0.87	0.84	0.82	0.86	0.03	0.94
iC16:0	0.13	0.11	0.12	0.12	0.00	0.36
aC16:0	0.48	0.52	0.48	0.50	0.02	0.86
C16:0	22.12	22.00	22.81	21.74	0.15	0.07
C17:0	0.99 [b]	0.96 [b]	1.20 [ab]	1.27 [a]	0.04	0.01
DMA C18:0	0.18	0.14	0.15	0.18	0.01	0.56
iC18:0	0.12	0.11	0.09	0.10	0.00	0.07
C18:0	13.29	13.40	14.12	13.78	0.17	0.28
C20:0	0.10	0.09	0.10	0.10	0.00	0.70
C22:0	0.08	0.08	0.07	0.08	0.00	0.56
C24:0	0.02	0.02	0.02	0.02	0.00	0.79
	Monounsaturated FA g/100 g					
C14:1 c9	0.15	0.15	0.14	0.13	0.00	0.39
C16:1 t9	0.44 [a]	0.41 [ab]	0.38 [b]	0.42 [ab]	0.01	0.04
C16:1 c7	0.21	0.23	0.21	0.21	0.00	0.50
C16:1 c9	2.28	2.20	2.15	2.11	0.03	0.14
C17:1 c9	0.79	0.80	0.85	0.98	0.03	0.08
C18:1 t11	2.43	2.57	2.50	2.20	0.14	0.79
C18:1 c9	33.48	33.29	33.45	34.01	0.31	0.85
C18:1 t15	0.18	0.20	0.15	0.16	0.01	0.24
C18:1 c11	0.16	0.18	0.14	0.14	0.01	0.15
C18:1 c12	0.17	0.16	0.18	0.18	0.00	0.72
C18:1 c13	0.10	0.10	0.09	0.08	0.01	0.83
C18:1 t16	0.21	0.19	0.20	0.18	0.00	0.37
C18:1 c15	0.08	0.09	0.09	0.08	0.00	0.97
C24:1 c9	0.15	0.21	0.16	0.18	0.01	0.10

Within a parameter, means with different letters differ at $p < 0.05$. DMA: dimethylacetals.

Table 5. Effect of the proportion of pea in the individual polyunsaturated fatty acids (PUFA) in LTL muscle.

	0%pea	10%pea	20%pea	30%pea	s.e.m.	Pr > F
		PUFA, g/100 g				
C18:2 n-6 t9, t12	0.17	0.15	0.15	0.13	0.00	0.06
C18:2 c9, t11	0.31	0.28	0.28	0.28	0.01	0.74
C18:2 t10, c12	0.10	0.10	0.10	0.11	0.00	0.63
C20:3 n-9	0.44	0.47	0.43	0.46	0.01	0.61
C18:2 n-6	7.84	7.72	6.89	7.38	0.18	0.25
C20:2 n-6	0.16	0.12	0.12	0.15	0.01	0.17
C20:3 n-6	0.24	0.25	0.22	0.25	0.01	0.16
C20:4 n-6	2.99	3.20	2.68	3.07	0.08	0.13
C22:4 n-6	0.28	0.29	0.25	0.28	0.01	0.32
C18:3 n-3	0.42	0.41	0.39	0.41	0.01	0.76
C20:5 n-3	0.29	0.30	0.27	0.27	0.01	0.76
C22:5 n-3	0.59	0.61	0.53	0.56	0.02	0.40
C22:6 n-3	0.26	0.28	0.24	0.27	0.01	0.71

Regarding the sums of FAs in the meat, only the total SFAs were affected by the pea proportion ($p < 0.01$), and the concentrate with 20%pea yielded greater SFA contents than did the other concentrates ($p < 0.05$; Table 6). The proportion of pea affected the thrombogenicity index ($p < 0.05$), which was higher in the 20%pea concentrate than in the 10%pea concentrate ($p < 0.05$) and tended to affect the atherogenicity index ($p < 0.10$).

Table 6. Effect of the proportion of pea in the sums and ratios of fatty acids in LTL muscle.

	0%pea	10%pea	20%pea	30%pea	s.e.m.	Pr > F
Total Saturated FA	44.98 [b]	44.97 [b]	46.67 [a]	45.16 [b]	0.20	0.01
Total Monounsaturated FA	40.19	40.15	40.11	40.47	0.33	0.98
Total Polyunsaturated FA	14.16	14.25	12.62	13.72	0.30	0.20
n-3	1.57	1.60	1.44	1.53	0.04	0.61
n-6	8.75	8.59	7.69	8.26	0.19	0.22
n-6:n-3	5.62	5.43	5.41	5.58	0.11	0.86
Atherogenicity index	0.71	0.69	0.75	0.68	0.01	0.09
Thrombogenicity index	1.32 [ab]	1.31 [b]	1.43 [a]	1.33 [ab]	0.01	0.02

Within a parameter, means with different letters differ at $p < 0.05$.

4. Discussion

The inclusion of pea to reduce the proportion of SBM in iso-energetic and iso-proteic concentrates involved a modification of the proportions of other ingredients, but the resulting concentrates had similar chemical compositions [16]. Numerous studies that have included pea in concentrates have used concentrates that varied by more than two ingredients [9,12,15,29], and thus, the concentrates varied in some chemical components, such as the FA profile [5]. In the present study, the FA profiles differed among the concentrates. The differences observed in the content of C16:0 are in line with the proportion of palm oil in the concentrate, which was 1.0%, 2.4%, 2.6% and 1.4% in 0%pea, 10%pea, 20%pea and 30%pea, respectively [11]. In concordance with the present results, the literature showed that when pea replaced SBM in the diets of lambs, the resulting differences in C16:0, C18:0, C18:1 c9, C18:2 n-6 and C18:3 n-3 ranged from 12% to 32%, 12% to 30%, 3% to 16%, 1% to 24% and 15% to 57%, respectively [12,15,30]. Similarly, the ingredients also contain different contents of carotenoids, especially tocopherols, causing differences in the overall contents of the concentrates. The contents in the concentrates, however, were low, and the differences in the contents of carotenoids and tocopherols among concentrates, although significant, were narrow.

The effects of pea inclusion on the colours of the RA muscle and fat deposits have seldom been evaluated in fattening lambs. The proportion of pea in the concentrate did not affect the RA colour parameters, with similar values to those previously reported in light lambs fed concentrates [17,31]. Regarding the fat colour, Lanza, Fabro, Scerra, Bella, Pagano, Brogna and Pennisi [15] found no differences when lambs were fed 40% pea or 38% fava bean compared to 18% soybean meal for 79 d [15]. Similarly, Bonanno, Tornambè, DiGrigoli, Genna, Bellina, DiMiceli and Giambalvo [3] studied the effects of four different protein sources on fat and meat colour and concluded that only fat redness and chroma were affected by the source of protein, and overall, the carcass and meat characteristics were similar to those obtained with conventional SBM. In the current experiment, the LTL muscle did not show any effect from the substitution of SBM with pea on the meat colour; this result was in line with the similar colours reported in lambs that were fed concentrates with SBM or 40%pea for 79 d [15], concentrates with 18%pea and 39%pea for 43 d [29], and concentrates with 25%pea for 49 d [30]. This lack of effect of the proportion of pea on the fat and meat colour may be due to the narrow differences in carotenoid contents among concentrates, regardless of statistical significance, in addition to the short experimental period (41 d). The time of air exposure affected all colour parameters, and haem pigments studied in the LTL, as reported in light lambs of the same breed that were fed commercial concentrates [17,32]. Usually, colour variables increase due to blooming and increased MMb contents, a plateau of approximately 5 days follows, and then discolouration occurs in the meat of light lambs [33]. Accordingly, in the current experiment, the colour variables increased at the beginning, plateaued until day 9 of air exposure when there was an increase of the Hue angle, an abrupt change is a good indicator of discolouration in the meat of light lambs regardless of the absolute value, and the greater MMb content, which indicates discolouration [34] and the end of the shelf life of meat. The lack of effects observed on either the fat or the meat colour can be considered positive because the meat colour is the main trait influencing consumer choice of light lamb [35].

The absence of any effect on the texture and shear force was in agreement with the similar values reported for lambs fed 86%pea for 48 d [3] or for lambs fed concentrates with 40%pea for 42–72 d [9,15] when compared to those fed SBM. This result indicates that the ingredients in the fattening diets of lambs have scarce effects on the texture parameters of lamb meat. The oxidative stability of meat depends on the balance between the pro-oxidant compound (i.e., total unsaturated FA, cholesterol and DMb) and antioxidant compound (tocopherols, carotenoids, . . . [36]) contents. In the current experiment, the mild differences observed in the unsaturated FAs and antioxidant compound contents among concentrates were not enough to elicit an effect on the lipid oxidation of the meat.

The absence of any effect on the chemical composition of the meat was expected because all diets were iso-proteic and iso-energetic; this result is in agreement with previous experiments that studied the effects of the inclusion of pea in concentrates in several doses [29,37]. The greater cholesterol content observed in the 20%pea treatment than in the 30%pea treatment agrees with the differences observed in the concentration of C16:0 in the LTL muscle (see below). This FA increases the plasma total and LDL cholesterol content [38]. However, this difference did not reach statistical significance in the plasma of these lambs at the time of slaughter [11].

The differences observed in the FA content of the concentrates were not exactly mirrored in the FA profile of the lamb meat due to the process of biohydrogenation of PUFA conducted by ruminal microorganisms [39]. The inclusion of pea had a minor effect on individual FAs in the muscle, as reported in light lambs fed pea instead of SBM in concentrates for 42–48 d [3,30]. However, when lambs were fed peas or SBM in concentrates for longer feeding periods, the most relevant FAs were affected, but these effects differed depending on the studies. The total replacement of SBM by pea for 72 d increased the C18:1 c9 and C18:3 n-3 contents but decreased the C18:1 t11 content [15], whereas the total replacement of SBM with pea for 98 d increased the C18:2 n-6 and C18:3 n-3 contents while decreasing the C16:0 and C18:0 contents [12]. The discrepancies among studies can be partially related to the amplitudes of the differences in the major FAs in the applied diets. The FAs in meat were affected only when the differences in the C16:0 and C18:0 contents in the diets of the lambs were above 30% [12] or when the differences in the C18:3 n-3 contents among diets were above 28% [12,15]. In the current experiment, the slight increase in certain individual SFAs in the meat of the lambs fed 20%pea concentrate concomitantly increased the total SFAs when compared to their counterparts, with no effects on the other FA sums. However, the total replacement of SBM in concentrate by 86%pea or 40%pea had no effect on the SFA, MUFA or PUFA contents in meat [3,15], whereas the inclusion of 24–25%pea decreased the total SFA content in meat and increased the MUFA and n-3 PUFA contents [12,30]. The slight differences in fatty acids led to differences in the thrombogenicity index, with the 20%pea concentrate yielding the greatest value, making this concentrate the least advisable. In this sense, the consumption of food with a low IA and IT has a better nutritional quality, which may reduce the risk of coronary heart disease, but no organisation has yet provided the recommended values for the IA and IT [40]. However, the impact of this difference on human health would be mild.

5. Conclusions

The inclusion of pea in the fattening concentrates of light lambs had no effects on fat, meat colour and lipid oxidation. Time of air exposure affected the evolution of colour and lipid oxidation, especially between 7 to 9 days. The effect on the fatty acid profile was minimal and had no effect on most FAs related to human health. However, the greater cholesterol and thrombogenicity index of the 20%pea concentrate should be considered. From the present results, the inclusion of 30%pea in concentrate would be the most advisable proportion in order to reduce the dependency on soybean meal, although the prices of the feedstuffs should be taken into account.

Author Contributions: Conceptualization, M.B., I.C. and M.J.; formal analysis, M.B.; investigation, M.B., G.R., J.R.B. and M.J.; resources, M.B.; writing—original draft preparation, M.B. and J.R.B.; writing—review and editing, M.J., S.L., G.R., I.C. and M.B. project administration, M.B.; funding acquisition, M.B. All authors have read and agreed to the published version of the manuscript.

Funding: This research was funded by the Ministry of Economy and Competitiveness of Spain and the European Union Regional Development Funds (INIA RTA2014-00038-C02-01 and RZP2017-00001) and the by the Research Group Funds of the Aragón Government (A14_17R; A14_20R).

Institutional Review Board Statement: The Animal Ethics Committee of the Research Centre approved the experimental procedures (protocol no. CEEA-03-2014-26), which were in compliance with the European Union Directive 2010/63 on the protection of animals used for experimental and other scientific purposes.

Data Availability Statement: The data that support the findings of this study are available from the corresponding author upon reasonable request.

Acknowledgments: We are grateful to the technical staff of the CITA Research Centre in Zaragoza.

Conflicts of Interest: The authors declare no conflict of interest.

References

1. Karlsson, J.O.; Parodi, A.; van Zanten, H.H.E.; Hansson, P.-A.; Röös, E. Halting European Union soybean feed imports favours ruminants over pigs and poultry. *Nat. Food* **2021**, *2*, 38–46. [CrossRef]
2. Fischer, E.; Cachon, R.; Cayot, N. Pisum sativum vs. Glycine max, a comparative review of nutritional, physicochemical, and sensory properties for food uses. *Trends Food Sci. Technol.* **2020**, *95*, 196–204. [CrossRef]
3. Bonanno, A.; Tornambè, G.; Di Grigoli, A.; Genna, V.; Bellina, V.; Di Miceli, G.; Giambalvo, D. Effect of legume grains as a source of dietary protein on the quality of organic lamb meat. *J. Sci. Food Agric.* **2012**, *92*, 2870–2875. [CrossRef] [PubMed]
4. Schreuder, R.; de Visser, C.L.M. *EIP-AGRI Focus Group; Protein CROPS: Final Report*; European Commission: Brussels, Belgium, 2014.
5. Bonanno, A.; Di Grigoli, A.; Vitale, F.; Alabiso, M.; Giosuè, C.; Mazza, F.; Todaro, M. Legume grain-based supplements in dairy sheep diet: Effects on milk yield, composition and fatty acid profile. *Anim. Prod. Sci.* **2016**, *56*, 130–140. [CrossRef]
6. Loe, E.R.; Bauer, M.L.; Lardy, G.P.; Caton, J.S.; Berg, P.T. Field pea (*Pisum sativum*) inclusion in corn-based lamb finishing diets. *Small Rumin. Res.* **2004**, *53*, 39–45. [CrossRef]

7.   Bachmann, M.; Kuhnitzsch, C.; Okon, P.; Martens, S.D.; Greef, J.M.; Steinhöfel, O.; Zeyner, A. Ruminal in vitro protein degradation and apparent digestibility of energy and nutrients in sheep fed native or ensiled + toasted pea (*Pisum sativum*) grains. *Animals* **2019**, *9*, 401. [CrossRef] [PubMed]

8.   FEDNA. *Tablas FEDNA de Composición y Valor Nutritivo de Alimentos Para la Fabricación de Piensos Compuestos*; FEDNA: Madrid, Spain, 2019.

9.   Colonna, M.A.; Giannico, F.; Marsico, G.; Vonghia, G.; Ragni, M.; Jambrenghi, A.C. Effect of pea (*Pisum sativum* L.) as alternative to soybean meal on the productive performances and meat quality traits of Merino crossbred lamb types. *Prog. Nutr.* **2014**, *16*, 39–51.

10.   Karlsson, L.; Martinsson, K. Growth performance of lambs fed different protein supplements in barley-based diets. *Livest. Sci.* **2011**, *138*, 125–131. [CrossRef]

11.   Lobón, S.; Joy, M.; Casasús, I.; Rufino-Moya, P.J.; Blanco, M. Field pea can partially replace soybean meal in fattening concentrate without deleterious effects on the digestibility and performance of light lambs. *Animals* **2020**, *10*, 243. [CrossRef] [PubMed]

12.   Scerra, M.; Caparra, P.; Foti, F.; Cilione, C.; Zappia, G.; Motta, C.; Scerra, V. Intramuscular fatty acid composition of lambs fed diets containing alternative protein sources. *Meat Sci.* **2011**, *87*, 229–233. [CrossRef]

13.   Purroy, A.; Surra, J.; Muñoz, F.; Morago, E. Empleo de leguminosas grano en el pienso para cebo de corderos: Y III. Guisantes. *Inf. Tec. Econ. Agrar.* **1992**, *88A*, 63–69.

14.   Turner, T.D.; Karlsson, L.; Mapiye, C.; Rolland, D.C.; Martinsson, K.; Dugan, M.E.R. Dietary influence on the m. longissimus dorsi fatty acid composition of lambs in relation to protein source. *Meat Sci.* **2012**, *91*, 472–477. [CrossRef] [PubMed]

15.   Lanza, M.; Fabro, C.; Scerra, M.; Bella, M.; Pagano, R.; Brogna, D.M.R.; Pennisi, P. Lamb meat quality and intramuscular fatty acid composition as affected by concentrates including different legume seeds. *Ital. J. Anim. Sci.* **2011**, *10*, 87–94. [CrossRef]

16.   Joy, M.; Rufino-Moya, P.; Lobón, S.; Blanco, M. The effect of the inclusion of pea in lamb fattening concentrate on in vitro and in situ rumen fermentation. *J. Sci. Food Agric.* **2021**, *101*, 3041–3048. [CrossRef]

17.   Ripoll, G.; González-Calvo, L.; Molino, F.; Calvo, J.H.; Joy, M. Effects of finishing period length with vitamin E supplementation and alfalfa grazing on carcass color and the evolution of meat color and the lipid oxidation of light lambs. *Meat Sci.* **2013**, *93*, 906–913. [CrossRef]

18.   Krzywicki, K. Assessment of relative content of myoglobin, oxymyoglobin and metmyoglobin at the surface of beef. *Meat Sci.* **1979**, *3*, 1–10. [CrossRef]

19.   Prache, S.; Theriez, M. Traceability of lamb production systems: Carotenoids in plasma and adipose tissue. *Anim. Sci.* **1999**, *69*, 29–36. [CrossRef]

20.   Lepetit, J. Deformation of collagenous, elastin and muscle fibres in raw meat in relation to anisotropy and length ratio. *Meat Sci.* **1989**, *26*, 47–66. [CrossRef]

21.   AOAC. *Official Methods of Analysis*, 16th ed.; AOAC International: Gaithersburg, MD, USA, 1999.

22.   AOCS. *Rapid Determination of Oil/Fat Utilizing High Temperature Solvent Extraction*; AOCS Press: Urbana, IL, USA, 2005.

23.   Blanco, M.; Ripoll, G.; Casasús, I.; Bertolín, J.R.; Joy, M. Carotenoids and tocopherol in plasma and subcutaneous fat colour to trace forage-feeding in growing steers. *Livest. Sci.* **2019**, *219*, 104–110. [CrossRef]

24.   Bertolín, J.R.; Joy, M.; Rufino-Moya, P.J.; Lobón, S.; Blanco, M. Simultaneous determination of carotenoids, tocopherols, retinol and cholesterol in ovine lyophilised samples of milk, meat, and liver and in unprocessed/raw samples of fat. *Food Chem.* **2018**, *257*, 182–188. [CrossRef]

25.   Sukhija, P.S.; Palmquist, D.L. Rapid method for determination of total fatty-acid content and composition of feedstuffs and feces. *J. Agric. Food Chem.* **1988**, *36*, 1202–1206. [CrossRef]

26.   Lee, M.R.F.; Tweed, J.K.S.; Kim, E.J.; Scollan, N.D. Beef, chicken and lamb fatty acid analysis—A simplified direct bimethylation procedure using freeze-dried material. *Meat Sci.* **2012**, *92*, 863–866. [CrossRef] [PubMed]

27.   Bravo-Lamas, L.; Barron, L.J.R.; Kramer, J.K.G.; Etaio, I.; Aldai, N. Characterization of the fatty acid composition of lamb commercially available in northern Spain: Emphasis on the trans-18:1 and CLA content and profile. *Meat Sci.* **2016**, *117*, 108–116. [CrossRef]

28.   Ulbricht, T.L.V.; Southgate, D.A.T. Coronary heart disease: Seven dietary factors. *Lancet* **1991**, *338*, 985–992. [CrossRef]

29.   Lanza, M.; Bella, M.; Priolo, A.; Fasone, V. Peas (*Pisum sativum* L.) as an alternative protein source in lamb diets: Growth performances, and carcass and meat quality. *Small Rumin. Res.* **2003**, *47*, 63–68. [CrossRef]

30.   Facciolongo, A.M.; De Marzo, D.; Ragni, M.; Lestingi, A.; Toteda, F. Use of alternative protein sources for finishing lambs. 2. Effects on chemical and physical characteristics and fatty acid composition of meat. *Prog. Nutr.* **2015**, *17*, 165–173.

31.   Lobón, S.; Blanco, M.; Sanz, A.; Ripoll, G.; Joy, M. Effects of feeding strategies during lactation and the inclusion of quebracho in the fattening on performance and carcass traits in light lambs. *J. Sci. Food Agric.* **2019**, *99*, 457–463. [CrossRef] [PubMed]

32.   Lobón, S.; Blanco, M.; Sanz, A.; Ripoll, G.; Bertolín, J.R.; Joy, M. Meat quality of light lambs is more affected by the dam's feeding system during lactation than by the inclusion of quebracho in the fattening concentrate. *J. Anim. Sci.* **2017**, *95*, 4998–5011. [CrossRef] [PubMed]

33.   Carrasco, S.; Panea, B.; Ripoll, G.; Sanz, A.; Joy, M. Influence of feeding systems on cortisol levels, fat colour and instrumental meat quality in light lambs. *Meat Sci.* **2009**, *83*, 50–56. [CrossRef]

34.   Liu, Q.; Lanari, M.C.; Schaefer, D.M. A Review of Dietary Vitamin-E Supplementation for Improvement of Beef Quality. *J. Anim. Sci.* **1995**, *73*, 3131–3140. [CrossRef]

35. Bernués, A.; Ripoll, G.; Panea, B. Consumer segmentation based on convenience orientation and attitudes towards quality attributes of lamb meat. *Food Qual. Prefer.* **2012**, *26*, 211–220. [CrossRef]
36. Domínguez, R.; Pateiro, M.; Gagaoua, M.; Barba, F.J.; Zhang, W.; Lorenzo, J.M. A comprehensive review on lipid oxidation in meat and meat products. *Antioxidants* **2019**, *8*, 429. [CrossRef] [PubMed]
37. Lestingi, A.; Facciolongo, A.M.; Jambrenghi, A.C.; Ragni, M.; Toteda, F. The use of peas and sweet lupin seeds alone or in association for fattening lambs: Effects on performance, blood parameters and meat quality. *Small Rumin. Res.* **2016**, *143*, 15–23. [CrossRef]
38. Salter, A.M. Dietary fatty acids and cardiovascular disease. *Animal* **2013**, *7*, 163–171. [CrossRef] [PubMed]
39. Woods, V.B.; Fearon, A.M. Dietary sources of unsaturated fatty acids for animals and their transfer into meat, milk and eggs: A review. *Livest. Sci.* **2009**, *126*, 1–20. [CrossRef]
40. Chen, J.; Liu, H. Nutritional indices for assessing fatty acids: A mini-review. *Int. J. Mol. Sci.* **2020**, *21*, 5695. [CrossRef]

Article

Carcass Traits and Meat Quality of Fat-Tailed Lambs Fed Rosemary Residues as a Part of Concentrate

Yathreb Yagoubi [1], Samir Smeti [1], Samia Ben Saïd [2], Houssem Srihi [1], Ilyes Mekki [1], Mokhtar Mahouachi [2] and Naziha Atti [1,*]

[1] Laboratoire de Productions Animales et Fourragères, INRA-Tunisia, University of Carthage, rue Hédi Karray, 2049 Ariana, Tunisia; yagoubiyathreb@hotmail.fr (Y.Y.); sam_fsb@live.fr (S.S.); houssemsrihi60@gmail.com (H.S.); ilyesyassinemekki@gmail.com (I.M.)

[2] Laboratoire Appui à la Durabilité des Systèmes de Production Agricole dans la Région du Nord-Ouest, ESAK, Le Kef, Tunisia, University of Jendouba, 7100 Jendouba, Tunisia; sbensaid@gmail.com (S.B.S.); taymallahmah@gmail.com (M.M.)

* Correspondence: naziha.atti@gmail.com or naziha.belhaj@iresa.agrinet.tn

Simple Summary: This study aims to investigate the carcass and meat quality from lambs fed a dietary treatment including rosemary residues obtained after distillation as cereal substitute in concentrate knowing that cereals are the main component of concentrate. Twenty-four male lambs from local fat-tailed Barbarine breed were allocated into three groups. They received individually oat hay as roughage and as complementation standard concentrate for control group (C) and two concentrate types containing rosemary residues (RR) for the other groups. The protein source was soybean (S) for RRS group while faba bean (F, *Vicia Faba*) which is a legume was the protein source for RRF group. The results suggest a positive action of rosemary by-products in improving phenolic and tocopherol compounds given their richness in these components. In addition, growth, the non-carcass and carcass traits and the meat physical properties were not altered.

Abstract: Facing climate change implications on feeds unavailability, unconventional resources are being considered with a growing interest such as aromatic plant distillation residues with a two-fold object, enhancing meat quality by increasing the antioxidant properties and reducing feed prices which are often imported though expensive. Hence, this study aims to assess the effects of rosemary distillation residues (RR) incorporation in concentrate associated to two nitrogen sources as a substitute for standard concentrate on lamb's growth, carcass traits and meat quality. For this, 24 Barbarine male lambs (3 months old, 17.83 ± 2.6 kg body weight) were divided into three groups. All lambs received individually 600 g of oat hay as roughage and 600 g of standard concentrate for control group, 600 g of concentrate based on RR and soybean meal for RRS group and 600 g of concentrate based on RR and faba bean for RRF group. After 65 days of experiment, all lambs were slaughtered. Phenolic and tocopherol intakes were significantly higher for both RR groups compared to control ($p < 0.05$). Growth, carcass weights, dressing percentages and non-carcass component weights were unaffected by the diet ($p > 0.05$). Moreover, regional and tissular compositions and meat physical properties were similar irrespective of the diet ($p > 0.05$). All color parameters were similar among groups ($p > 0.05$). However, meat produced by lambs receiving RR-based concentrate was richer on vitamin E and polyphenol contents than control lambs ($p < 0.05$). Rosemary by-products may substitute the standard concentrate resulting in similar lamb's growth and carcass traits, while improving meat quality by increasing vitamin E content, which could improve its antioxidant power.

Keywords: lambs; carcass characteristics; meat quality; vitamin E; rosemary residue

Citation: Yagoubi, Y.; Smeti, S.; Ben Saïd, S.; Srihi, H.; Mekki, I.; Mahouachi, M.; Atti, N. Carcass Traits and Meat Quality of Fat-Tailed Lambs Fed Rosemary Residues as a Part of Concentrate. *Animals* **2021**, *11*, 655. https://doi.org/10.3390/ani11030655

Received: 2 December 2020
Accepted: 27 December 2020
Published: 1 March 2021

Publisher's Note: MDPI stays neutral with regard to jurisdictional claims in published maps and institutional affiliations.

1. Introduction

Sheep farming has always been a vital sector in the economy of many countries worldwide since historical records began. However, in recent years it has been affected by climate changes, which directly disturbs livestock health, growth and reproduction, while the indirect effects are on the shortage of productive pastures, forages and feeds [1]. Hence, the scarcity of forage and some conventional feeds accentuated by the volatilizing prices of the imported concentrate is increasingly worrying problem for breeders. Consequently, a considerable interest has been currently given to the use of the unconventional feeds such as shrubs and agro-industrial by-products as a viable alternative for enhancing animal performance [2–5]. Among the agro-industrial by-products, those of aromatic plants are used specially to improve animal product quality to meet consumer demand for safe and high-quality foods [5,6]. The improvement of meat quality depends on their richness of numerous bioactive compounds such as the phenolic compounds and vitamins that provide an antioxidant activity to reduce meat oxidation, thus, extending the meat shelf life [7,8]—especially when the use of synthetic antioxidant become rejected [9] by consumers given their toxicological consequences. Aromatic plants have been used since antiquity [10] as folk medicine and as preservatives in foods and the best known aromatic plants are rosemary, thyme, oregano and sage which are widespread in the Mediterranean area. In Tunisia, the industry of rosemary essential oil extraction generates a great amount of residues (5460 Tm/year; [11]), which could be valorized as alternative feed for livestock given their free availability. Several investigations have studied the use of rosemary residues or essential oil as additive to the basal diet of lambs or ewes [8,12,13]. On the other hand, the concentrate for fattening lambs is based on cereals (>70%) in arid and semi-arid regions [14]. These regions are marked by the low hay production and the irregular availability of cereals such as barley and corn. For this, there is a growing resort to importation of these products with increasing prices. Moreover, soybean meal is the main protein source in lamb diets which is often imported; however, it could be replaced by faba bean (*Vicia Faba*) which is a legume and a local protein source that leads to similar animal performances [15]. Therefore, the use of rosemary residues could be widespread in sheep feeding. However, research on their use at high rates is scarce. To the best of our knowledge, only one study has dealt with their use as roughage substituted with the hay [4] and there are no studies on their use as concentrate. Therefore, our hypothesis resides in enhancing Barbarine lamb's growth, carcass and non-carcass traits and meat quality by substituting standard concentrate with two types of concentrate based on rosemary residues associated to soybean or local protein resources such as the faba bean.

2. Material and Methods

2.1. Experimental Design, Feeds and Animals

Rosemary residues (RR) were collected from a forest in the Northwest of Tunisia after essential oil extraction. Fresh rosemary leaves were harvested and mixed for essential oils extraction by hydrodistillation (10 kg of fresh plant in 50 L of distilled water) using a Clevenger-type apparatus for 5 h; the by-product of distillation, the rosemary residues, were air-dried. Then they were ground in a manufactory and mixed with the remaining ingredients to obtain two types of concentrate based on RR that substitute the standard concentrate. The protein source was soybean (S) in RRS concentrate (31% RR, 39% barley, 16% S and 11% wheat bran) and faba bean in RRF one (33% RR, 22% barley, 29% faba bean, 8% wheat bran and 5% molasses). The standard concentrate was composed by 30% corn, 20% barley, 7.5% S and 37.5% wheat bran. All concentrates contain 3% of mineral vitamin supplement (10.0% Ca, 3.5% P, 8.0% Na, 4.4% Mg, 0.4% S, 0.4% Zn, 0.2 Mn, 0.2% Fe). The Dry matter (DM) and the chemical composition (% DM) of the experimental feeds are shown in Table 1.

Table 1. Chemical composition of diets fed to Barbarine lambs.

Item	Oat Hay	RR	Standard Concentrate	RRF	RRS
Dry matter (%)	91.95	84.89	88.69	92.35	90.67
Crude Protein (%DM)	5.47	7.31	16.33	17.30	17.38
Organic Matter (%DM)	93.04	92.62	80.65	89.56	91.22
Neutral detergent fiber (%DM)	69.15	38.53	20.31	29.47	34.01
α-Tocopherol (μg/g DM)	4.42	217.20	0.45	52.71	62.98
γ-Tocopherol (μg/g DM)	3.97	3.78	0.78	11.85	7.38

RR: rosemary residues; RRF: rosemary residues + faba bean; RRS: rosemary residues + soybean; DM: Dry matter.

2.2. Animals and Feeding

The experiment was carried at the farm of the High School of Agriculture of Kef. Twenty-four male lambs (3-month-old, 17.83 ± 2.6 kg of body weight (BW)), from fat-tailed Barbarine breed, were divided into three groups of 8 lambs each according to BW. Animals were allocated in individual pens and had free access to fresh water during the 65 days of trial. Lambs in each group received individually 600 g of oat hay and 600 g of standard concentrate for control group (C), 600 g of concentrate based on RR and soybean meal for RRS group and 600 g of concentrate based on RR and faba bean for RRF group. Throughout this period, the amount of feed offered and what was refused the previous day was daily recorded and then the intake calculated. The lamb's BW was recorded at the beginning of the trial, and then monitored regularly once a week prior to the morning feeding. Average daily gain (ADG) was calculated.

2.3. Slaughter Procedure and Measurements

At the end of the growth trial, all lambs were transported to the abattoir of the National Institute of Agronomic Researches of Tunisia (INRAT) where they were slaughtered after 12 h fasting with only access to water. They were weighed just before slaughter (slaughter body weight (SBW)). After slaughter, internal fats (omental and mesenteric) and non-carcass components such as skin, head, feet, gastro-intestinal tract, red organs (heart, liver, lungs and trachea) were removed. All fractions of the digestive tract (reticulo-rumen + omasum (rumen), abomasum, and intestine) were weighed full then empty after hand rinsing, in order to determine the weight of digestive contents. Hot carcass weight (HCW) was recorded and then carcasses were stored at 4 °C.

2.4. Carcass Cutting and Dissection

The cold carcass weight (CCW) was recorded 24 h post-mortem after chilling at 4 °C. The kidneys, kidney fat, testis and the fat tail were removed from cold carcasses and weighed; then each carcass was split longitudinally into two halves. The left half was cut into 4 joints (Leg, shoulder, neck and a block composed by ribs, loin and breast (RLB)). From RLB, the *Longissimus thoracis et lumborum* (LTL) muscle was removed and sampled to determine meat quality. The shoulders were dissected to estimate the tissular composition given the shoulder composition is representative of the carcass composition [16]. The first operation in the dissection process was the removal of subcutaneous fat. Muscles were then removed singly from bones; finally, inter-muscular fat was trimmed from muscles and bones. Each tissue was weighed individually; the sum of weights of each tissue in shoulder was used for calculation of carcass composition. The carcass composition data were reported as percentages.

2.5. Meat Physical Properties Measurement

The pH was measured in *Longissimus thoracis et lumborum* (LTL) muscle at 1 and 24 h (ultimate pH) post-mortem using a penetrating electrode connected to a portable pH-meter (HI 99163; Hanna Instruments, Cluj-Napoca, Romania) after calibration with two buffers (7.00 and 4.01). Meat color was measured directly on the LTL muscle surface with measured area of 8 mm, standard illuminant D65 and an observer angle of 10°, 24 h post-mortem using a Minolta chroma Meter CR-400 (Konica Minolta Holdings, Japan) according to the CIE L* a* b* space (CIE, 1978) and the bloom time from carcass to carcass was the same (3 min). The lightness (L*), redness (a*) and yellowness (b*) were directly recorded while, Hue angle (H*) and Chroma (C*) were calculated as: $H^* = \tan^{-1}(b^*/a^*) \times 57.29$, expressed in degrees, and $C^* = a^{*2} + b^{*2}$. To determine water cooking loss (WCL), meat samples were weighed (Wi: initial weight) and held in plastic bags and then immersed in a water-bath at 75° and heated for 30 min until the internal temperature reached 75 °C, which was monitored with thermocouple. Then, the bags were cooled under running tap water and blotted dry with paper towels. The cooked meat was weighed again (Wf: final weight). WCL was calculated as the difference between of sample weight before and after cooking and it was expressed as a percentage: $100 \times (\text{Wi} - \text{Wf})/\text{Wi}$.

2.6. Meat Vitamin E and Total Phenolic Content (TPC) Analyses

Vitamin E analysis was performed according to the method described by [17] using high performance liquid chromatography. Vitamin E analysis in meat samples was previously described in details in [5]. To determine the total phenolic content (TPC) in meat, the method of [18] was used with some modifications. Briefly, 1 g of ground meat was mixed with 9 mL of milli-Q water (Ultramatic GR Wasserlab, Navarra, Spain), then, 10 mL of aqueous solution of methanol (50/50; v/v) was added. The obtained solution was shaken with vortex for 5 min. After 5 min of homogenization 500 μL of Carrez I solution (Scharlau, Barcelona, Spain) was added, while vortexing for 1 min. Then, 5 mL of acetonitrile was added to the mixture, while vortexing for 5 more minutes. The tubes were left to stand for 25 min, and then centrifuged at $4000 \times g$ for 15 min at 4 °C. Finally, two phases were obtained; a solid one formed by protein and lipid fraction and a liquid phase. To filter the supernatant, a 0.22 μm Polytetrafluoroethylene (PTFE) filter was used in a 15 mL tube. Then, the extract obtained was used to determine the TPC. The TPC in the liquid extract were estimated using the Folin-Ciocalteu method [18]. For that, 147 μL of water milli-Q, 13 μL of Folin-Ciocalteu reagent and 125 μL of 7% Na_2CO_3 were added to 15 μL of the extract in a micro plate. Samples were held to stand for 1.5 h in the dark. The samples' absorbance was measured with a spectrophotometer at 750 nm (Epoch 2 Microplate Spectrophotometer, Biotek, VT, USA) and the results were expressed as μg gallic acid equivalents (GAE)/g dried sample.

2.7. Calculation and Statistical Analysis

Empty body weight (EBW) was calculated as the difference between SBW and weight of digestive contents. Commercial and real dressing percentage (CDP, RDP) were calculated according to the following equations:

$$\text{CDP (\%)} = 100 \times \text{HCW}/\text{SBW} \tag{1}$$

$$\text{RDP (\%)} = 100 \times \text{CCW}/\text{EBW} \tag{2}$$

The effect of using rosemary residues in dietary treatment on carcass and non-carcass traits and meat quality was assessed by one-way ANOVA, using the General Linear Model (GLM) procedure of SAS (2004) [19] according to the following model:

$$\text{Yij} = \mu + \text{Di} + \text{eij} \tag{3}$$

(Yij = j^{th} measure of the ith diet; μ = overall mean; Di = effect of the ith diet (C, RRF and RRS); eij = error term).

The differences between means were compared using the Duncan's Multiple Range Test (DMRT) and the statistical significance was defined at $p < 0.05$.

3. Results

3.1. Feed Intake and Lamb's Growth

All concentrate types were iso-nitrogenous and all lambs consumed comparable amounts of roughages and concentrate in the proportion of 50–50%, respectively, resulting in similar final body weight (25.2 ± 2.9 kg) and similar average daily gain (114 ± 23 g). Given this result and the free availability of rosemary residues, their inclusion in concentrate reduced its cost from 850 Tunisian Dinar/Ton for control to 562 and 539 Tunisian Dinar/Ton for RRS and RRV, respectively. In addition, the α-Tocopherol intake was higher for RRS (26.74 mg/day) and RRF (21.67 mg/day) groups than control (2.04 mg/day). Additionally, the daily intake of total phenolic compounds was similar for RRF and RRS but higher than that of control lambs (1.17 and 1.24 vs. 0.67 g/day, respectively; $p < 0.05$).

3.2. Carcass Weights and Dressing Percentage

The empty body weight, hot carcass weight, cold carcass weight and both dressing percentages were similar for all groups irrespective of the lamb's diet (Table 2).

Table 2. Body weight, empty body weight, carcass weight and dressing percentage in Barbarine lambs fed RR-based concentrate.

Item	C	RRF	RRS	SEM	p
Slaughter body weight (kg)	25.75	24.25	25.12	1.91	0.67
Empty body weight(kg)	20.37	19.62	20.12	0.80	0.80
Hot carcass weight (kg)	11.28	10.61	11.01	0.54	0.71
Cold carcass weight (kg)	10.98	10.40	10.91	0.46	0.68
Commercial dressing percentage (%)	43.95	43.84	44.26	0.38	0.90
Real dressing percentage (%)	54.23	53.33	54.39	0.39	0.51

C: Control; RRF: rosemary residues + faba bean; RRS: rosemary residues + soybean; SEM: standard error mean.

3.3. Non-Carcass Components

All non-carcass components' weights and proportions in the EBW were unaffected by the dietary treatment (Table 3).

Table 3. Fresh organ weights and proportion (%) in EBW of Barbarine lambs fed RR-based concentrate.

Organs	C	RRF	RRS	SEM	p
Skin (kg)	2.90	2.66	2.77	0.14	0.51
Skin (%)	14.38	13.65	13.78	0.24	0.44
Head (kg)	1.49	1.39	1.41	0.06	0.43
Head (%)	7.39	7.15	7.07	0.10	0.45
Gut (kg)	5.48	4.71	4.85	0.51	0.52
Gut (%)	26.76	24.33	24.07	1.18	0.59
Red organs (g)	874.00	749.89	719.49	0.03	0.60
Red organs (%)	4.34	3.92	3.73	0.34	0.76
Liver (g)	384.00	369.63	403.25	0.02	0.52
Liver (%)	1.90	1.90	2.02	0.05	0.63
Testis (g)	73.06	48.63	62.21	0.02	0.54
Testis (%)	0.34	0.24	0.30	0.03	0.56

C: Control; RRF: rosemary residues + faba bean; RRS: rosemary residues + soybean; SEM: standard error mean.

3.4. Carcass Sectional and Tissular Composition

The carcass sections' weights were unaffected ($p > 0.05$) by the dietary treatments. In addition, their proportions in the tailed or untailed carcass were similar for all groups (Figure 1). The tail weight was similar for all lambs averaging 728 g and 6% of carcass weight. The substitution of standard concentrate with both types of concentrate based on RR led to the same amount and proportions of muscle, fat and bone for all groups

(Figure 2). Within fat tissue in the shoulder, regardless of the diet, all lambs deposed the same amount of subcutaneous and inter-muscular fat averaging 67 and 59 g, respectively. However, the kidney fat was higher for Control ($p < 0.05$) than both RR groups (89 vs. 53 g).

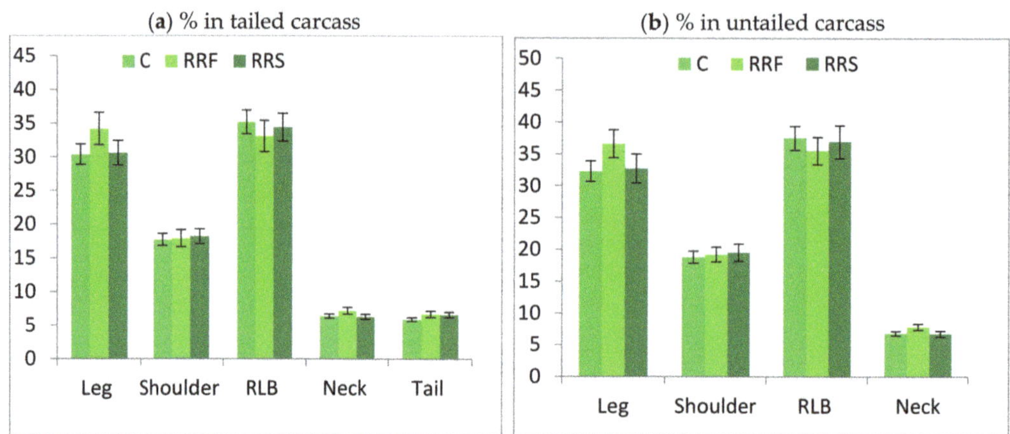

Figure 1. Carcass cut proportions in the tailed and untailed carcasses of Barbarine lambs fed RR-based concentrate.

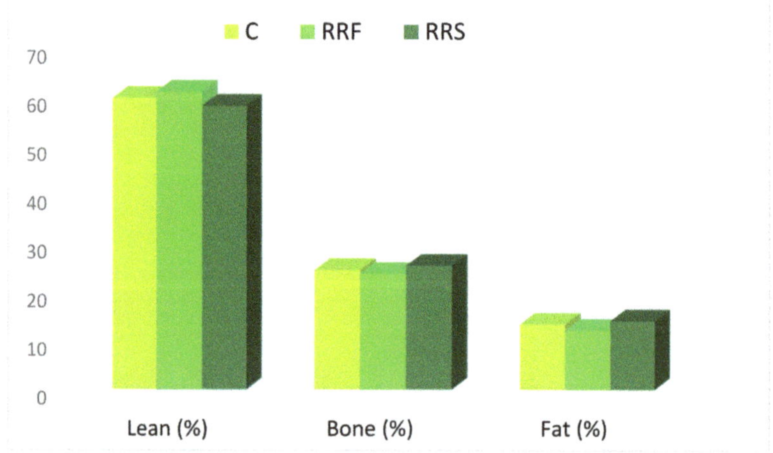

Figure 2. Carcass tissular proportion of Barbarine lambs fed RR-based concentrate.

3.5. Meat Quality

Initial pH was significantly higher for C and RRF groups compared to RRS but all the values are acceptable. After 24 h post-mortem, the ultimate pH ranged from 5.51 to 5.86 (Table 4). Although the ultimate pH of control was higher than either rosemary treatments, all values were acceptable and varied similarly given the dpH (pH24–pH1) was similar among groups ($p > 0.05$). Water cooking loss was unaffected by using RR in concentrate and was about 21.8% for all groups. All color parameters were similar among groups.

The α-Tocopherol and phenolic contents were higher ($p < 0.05$) for the meat of both experimental groups than the control one (Table 5).

Table 4. Meat physical properties in Barbarine lambs fed RR-based concentrate.

Meat Physical Parameters	C	RRF	RRS	SEM	p
Initial pH	6.33 [a]	6.18 [ab]	5.96 [b]	0.05	0.03
Ultimate pH	5.86 [a]	5.59 [b]	5.51 [b]	0.03	0.01
dpH	−0.46	−0.59	−0.44	0.04	0.39
Water cooking loss	21.96	23.21	20.29	0.91	0.43
Lightness (L*)	43.94	42.65	43.33	0.81	0.81
Redness (a*)	17.05	16.92	16.81	0.32	0.95
Yellowness (b*)	4.05	4.39	3.04	0.27	0.13
Chroma (C*)	17.64	17.50	17.11	0.33	0.83
Hue angle (H*)	13.07 [ab]	14.41 [a]	10.09 [b]	0.74	0.07

a, b: values within a row different superscript differ significantly at $p < 0.05$. C: Control; RRF: rosemary residues + faba bean; RRS: rosemary residues + soybean; SEM: standard error mean; L*: Black to White (0 to 100); a*: green to red (−60 to +60); b*: blue to yellow (−60 to +60).

Table 5. Meat vitamin E and total phenolic content in Barbarine lambs fed RR-based concentrate.

Item	C	RRF	RRS	SEM	p-Value
α-tocopherol (μg/g DM)	3.36 [b]	6.48 [a]	6.32 [a]	0.12	0.001
Total phenolic content	51.33 [b]	60.34 [a]	60.29 [a]	2.11	0.008

C: Control; RRF: rosemary residues + faba bean; RRS: rosemary residues + soybean; SEM: standard error mean a, b: values within a row different superscript differ significantly at $p < 0.05$.

4. Discussion

4.1. Feed Intake and Growth Performance

The richness of both concentrates based on RR in total phenols and tocopherol fractions (α-Tocopherol and γ-Tocopherol) was previously shown when RR were used as basal diet in lambs feeding [4]. The higher amount of these nutrients leads to higher intake of total phenolic compounds and tocopherols by lambs. In the current study, all concentrates were iso-energetic and iso-nitrogenous which explains the similarity in DM and CP intakes among groups. In contrast, when used as roughage to totally substitute oat hay at higher levels of RR incorporation (60 and 87%), the DM and CP intakes were significantly higher for groups that receive forage based on RR because they were richer on CP than oat hay [4]. The similarity in growth performances may result from the same DM intake for all treatments. Furthermore, and irrespective of concentrate type, there were no significant differences in BW gain associated with nitrogen source (soybean and faba bean), which is in line with previous reported results [20,21]. The total daily feed cost was lower for RRF and RRS than that of the control diet; hereafter, the main target of using aromatic plant by-products, the reduction of feed cost, was reached. The inclusion of RR in concentrate decreased the cost of concentrate by 36.5 and 33.8% for RRF and RRS, respectively, compared to control. Consequently, the cost/kg of gain was reduced by 40 and 22% for RRF and RRS, respectively which was 6.47, 3.88 and 5.07 Tunisian Dinar for C, RRF and RRS, respectively.

4.2. Carcass Weights, Dressing Percentage and Non-Carcass Components

The absence of variation for EBW, HCW, CCW as well as for commercial and real DP was generated by the similarity of SBW among all groups. This strong correlation between these parameters and SBW was previously documented [22–25]. Similar results were recorded on animal yield and carcass weights when RR were used at a low (10 to 20%) or high (60–80%) rate [4] or when ewes received myrtle by-products [26]. On the other hand, the similar concentrate proportion for all groups could be at the origin of this similarity, given that increasing dietary energy concentration or concentrate level affected significantly these parameters [27]. In addition, Asadollahi et al. [14] showed that supplementation of sugar beet pulp and roasted canola seed in a concentrate diet altered carcass traits of fattening lambs. The recorded values of CDP (44%) are lower than previous reported results for more heavy lambs of the same breed [5,26]. The dietary supplementation with aromatic plants extracts, additives or by-products had no effects on dressing percentages for sheep [4,26,28]. Regarding the nitrogen source, the substitution of soybean meal by faba bean leads to similar weights of carcasses and DP. These results are consistent with those reported in previous works where lambs received diets containing faba bean and soybean meal [21].

All lambs have similar age, sex, SBW and belonging to the same breed, which accounts for the absence of difference between dietary treatments for the non-carcass components' weights. These facts are the main factors that influence the non-carcass components rather than intake level or diet composition [23,29,30]. The weight of offal components high in bone content (head) and/or with a low metabolic activity was similar for all lambs given these components are early maturing parts [4,23,30,31] and are less affected by dietary treatments [29]. The skin, characterized by a high metabolic activity, is related to the EBW, then, the similarity in EBW leads to similar proportions of the skin [26,32]. The similarity of the gut weight and proportion in the EBW is originated by the similarity of intake for all groups given the digestive tract weight and activity increases with feed intake [30,33]. It is well established that, in young animals, some parts of the alimentary tract and particularly the rumen continue to develop as the animals become older and heavier [21,32]. Similarly, the weights and proportions in EBW of red organs and of liver were not affected by the dietary treatments given that the liver weight did not vary under the same DM intake [30].

4.3. Carcass Sectional and Tissular Composition

The result on constancy of joints' weight and proportions in the carcass confirmed the theory of anatomic harmony firstly established by [34,35] and then confirmed by other authors for fat- and thin-tailed breeds [3,4,22,24]. The dietary treatment did not affect the carcass joint's weight and proportions and the average percentage of leg and shoulder in the tailed and untailed carcass are close to those previously reported for the same breed [4,25,26] and for other sheep from thin-tail breeds [36]. The dietary supplementation with aromatic plants by-products did not affect the carcass sectional composition as previously reported [26,32]. The carcass tissue composition depends on breed, sex, age [25] and growth rate [23], which were similar for all lambs in the current study. Moreover, the diets are iso-nitrogenous which explain the production of the same amount of muscle given a higher protein level leads to higher muscle amount [2]. In addition, the same amount of fat can be explained by the fact that lambs had the same age and the same weight. In this context, it was shown that fat depot depends on SBW, nutritional level and nutrient utilization [30,37]. The constancy of bone tissue for all groups is explained by the precocity of this tissue, which had an early development regardless of the nutrition and which depends mostly on breed and age [25,28]. The similarity of subcutaneous, inter-muscular and fat tail among groups is originated by the same SBW [32]. The subcutaneous fat deposition depends more on carcass weight than on growth level or breed [23] and occurs late, hence its proportion increased when total body fat increased while for inter-muscular fat, an early maturating depot, the inverse occurred. The similarity of tissular carcass composition irrespective of nitrogen source should encourage the use of faba bean as substitute to soybean in concentrate for growing lamb.

4.4. Meat Quality

The initial pH value was significantly higher for control group than experimental ones (6.33 vs. 6.18 and 5.96 for C, RRF and RRS, respectively). This trend was maintained even after 24 h with 5.86, 5.59 and 5.51 for C, RRF and RRS, respectively. The groups fed RR presented ultimate pH lower than that of control group and which are close to pH values previously reported for sheep [4,32]. However, the pH value presented by control was considered as slightly high, which could be the result of an altered utilization of dietary energy or a different reaction to the stress of slaughter [38] that leads to low muscle glycogen reserve and then a higher pH value. Furthermore, the higher phenolic compounds intake for experimental groups could probably be at the origin of this difference in pH. However, although the slight difference in ultimate pH among groups, WCL and meat color parameters were not affected and were similar among groups. In previous studies, it was shown that the intake of myrtle or rosemary essential oils as additive, or rosemary residues did affect neither the pH nor the cooking loss for lambs, ewes and goats [4,32,39]. The meat lightness (L*) values presented by all groups averaged 43.3 indicating a light-colored meat, being in the range of average acceptability of meat given a meat with lightness equal or above 34 is acceptable and close to 44 which is considered the value of acceptability by 95% of consumers [40]. Similar results were reported after a dietary intake of rosemary extracts, rosemary and myrtle by-products [4,26]. The meat redness averaged 17, which is close to the result reported in previous works showing that RR intake did not affect redness [4]. Some reports have demonstrated that natural antioxidants can retard meat color loss by extending the red color, such as oregano essential oil supplementation which increased redness (a*) and yellowness (b*) of meat [41]. In addition, similar pH and the same slaughter age are the major factors determining meat color compared to diet effect [21]. The nitrogen source did not influence the meat physical properties as previously reported [21] where, meat color, pH, WCL were similar. Meat α-Tocopherol content was significantly doubled in both groups receiving RR in concentrate compared to control. The α Tocopherol represent the principal component of vitamin E although the presence of other tocopherols in its activity [42–44]. Several authors have reported the great action of diet on muscle's tocopherol content given α-tocopherols are not degraded in the rumen but are deposited in muscle cell membranes where their antioxidant action is more effective [45,46]. Therefore, the mechanism of absorption of vitamin E is the key to maximize its benefits in meat quality [47]. Anyway, it was observed in several studies assessing the effect of ruminal microbiota and fermentation in vitamin E absorption that vitamin E was not degraded during in vitro ruminal fermentation [48]. This result is in agreement of that noted by [49], who reported that neither in vitro no in vivo hydrolysis of α-Tocopherol in the rumen, given α-Tocopherol esters need to be hydrolyzed before their absorption, little if any absorption of vitamin E in the rumen should be expected. In contrast, tocopherol esters are largely hydrolyzed in the intestinal lumen where they are then absorbed in combination with lipid micelles. Then, once in the enterocytes, vitamin E is packed into chylomicrons and delivered to the liver in the form of chylomicron remnants [50]. In the liver, the hepatic α-tocopherol transfer protein, α-TTP, binds to the vitamin E to facilitate its incorporation into nascent VLDL and its secretion from hepatocytes. This lypoprotein has a central role in vitamin E metabolism as it regulates the body-wide levels of α-tocopherol [51]. The current result is in agreement with earlier results reporting the increase of muscle's α-tocopherol content when lambs received rosemary by-products as basal diet [5] or when sheep and goats received distillated myrtle leaves or myrtle essential oil [6,39]. The values found when animals received distillated rosemary or myrtle leaves exceed even those reported under grazing conditions although the richness of green herbs on vitamin and phenols [52,53]. The richness of meat on TPC could be explained by the richness of the diet on this component as previously shown [5,54]. In the same context, it was shown a positive transfer of phenolic compounds to lamb meat from pregnant ewes with the inclusion of rosemary by products in animal diet [54].

5. Conclusions

The results provide evidence that the use of RR as cereal substitute up to 30% in concentrate for sheep feeding did not alter animal performances. This smart strategy of using aromatic plant by-products could be effective especially in the Mediterranean region, where this by-product is available in an important amount and is free. The cost per kilogram of meat produced by Barbarine lambs was reduced until 40%. In addition, the carcass quality was not altered and the meat quality was enhanced seen the use of RR rich in tocopherols and phenolic contents. On the other hand, faba bean (*Vicia Faba*) could be used as a substitute to soybean without affecting carcass nor meat quality; this nitrogen source could potentially be produced, given its production is relatively cheap compared to the nutritional value, to reduce the import of soybean meal, which is still expensive. However, future studies with greater size should verify these results.

Author Contributions: Y.Y.: analyzed the data and wrote the first draft manuscript. H.S. and Y.Y.: Carried out animal feeding and weighting, sampling, etc.). I.M.: Participated to sampling and laboratory analysis. S.S.: Participated to design conception and data analysis. S.B.S.: Participated to sampling and laboratory analysis. N.A. and M.M.: Conceived and designed the experiments, Coordinated the research project. Y.Y., N.A. and M.M.: Wrote the paper. All authors have read and agreed to the published version of the manuscript.

Funding: This research received no external funding.

Institutional Review Board Statement: All procedures employed in this study (transport and slaughtering) meet ethical guidelines and adhere to Tunisian legal requirements (The Livestock Law No. 2005-95 of 18 October 2005, Chapter II; Sections 1 and 2 relative to the slaughter of animals).

Informed Consent Statement: Not applicable.

Data Availability Statement: The data presented in this study are available on request from the corresponding author.

Acknowledgments: The authors are grateful to the staff of ESA Kef for their collaboration in this study and Zina Taghouti, technician in Animal Production Laboratory in INRAT for her help in meat analyses. We wish also to acknowledge the doctoral school of the National Institute of Agronomy of Tunisia for the financial support.

Conflicts of Interest: The authors declare no conflict of interest.

References

1. Wreford, A.; Topp, C.F.E. Impacts of climate change on livestock and possible adaptations: A case study of the United Kingdom. *Agric. Syst.* **2020**, *178*, 102737.
2. Mahouachi, M.; Atti, N.; Hajji, H. Use of spineless cactus (*Opuntia ficus indica f. inermis*) for dairy goats and growing kids: Impacts on milk production, kid's growth and meat quality. *Sci. World J.* **2012**, *4*, 321567. [CrossRef]
3. Obeidat, B.S.; Mahmoud, K.Z.; Maswadeh, J.A.; Bsoul, E.Y. Effects of feeding *Atriplexhalimus* L. on growth performance and carcass characteristics of fattening Awassi lambs. *Small Rum. Res.* **2016**, *137*, 65–70.
4. Yagoubi, Y.; Hajji, H.; Smeti, S.; Mahouachi, M.; Kamoun, M.; Atti, N. Growth performance, carcass and non-carcass traits and meat quality of Barbarine lambs fed rosemary distillation residues. *Animal* **2018**, *12*, 2407–2414.
5. Yagoubi, Y.; Joy, M.; Ripoll, G.; Mahouachi, M.; Bertolin, J.R.; Atti, N. Rosemary distillation residues reduce lipid oxidation, increase alpha-tocopherol content and improve fatty acid profile of lamb meat. *Meat Sci.* **2018**, *136*, 23–29.
6. Tibaoui, S.; Smeti, S.; Essid, I.; Bertolin, J.R.; Joy, M.; Atti, N. Physicochemical characteristics, fatty acid profile, Alpha-Tocopherol content and lipid oxidation of meat from ewes fed different levels of distilled myrtle residues. *Molecules* **2020**, *25*, 4975. [CrossRef]
7. Haak, L.; Raes, K.; Van Dyck, S.; De Smet, S. Effect of dietary rosemary and alpha-tocopherol acetate on the oxidative stability of raw and cooked pork following oxidized linseed oil administration. *Meat Sci.* **2008**, *78*, 239–247.
8. Smeti, S.; Hajji, H.; Mekki, I.; Mahouachi, M.; Atti, N. Effects of dose and administration form of rosemary essential oils on meat quality and fatty acid profile of lamb. *Small Rum. Res.* **2018**, *158*, 62–68.
9. Ito, N.; Fukushinma, S.; Hasegawa, A.; Shibata, M.; Ogiso, T. Carcinogenicity of butylated hydroxy anisole in F344 rats. *J. Natl. Cancer Inst.* **1983**, *70*, 343–347.
10. Petrovska, B.B. Historical review of medicinal plants' usage. *Pharmacogn. Rev.* **2012**, *6*, 1–5.
11. APIA. *Aperçu Sur le Secteur Des Plantes Aromatiques et Médicinales "P.A.M"*; Agro-Services: Tunis, Tunisia, 2003; 122p.
12. Smeti, S.; Hajji, H.; Bouzid, K.; Abdelmoula, J.; Munoz, F.; Mahouachi, M.; Atti, N. Effects of Rosmarinus officinalis L. as essential oils or in form of leaves supplementation on goat's production and metabolic statute. *Trop. Anim. Health Prod.* **2015**, *47*, 451–457.
13. Ben Abdelmalek, Y.; Essid, I.; Smeti, S.; Atti, N. The anti-oxidant and antimicrobial effect of Rosmarinus officinalis L. distillation residues' intake on cooked sausages from ewes fed linseed. *Small Rumin. Res.* **2018**, *168*, 87–93.
14. Asadollahi, S.; Sari, M.; Erafanimajd, N.; Kiani, A. Supplementation of sugar beet pulp and roasted canola seed in a concentrate diet altered carcass traits, muscle (Longissimus Dorsi) composition and meat sensory properties of Arabian fattening lambs. *Small Rumin. Res.* **2017**, *153*, 95–102.
15. Bonanno, A.; Tornambè, G.; Di Grigoli, A.; Genna, V.; Bellina, V.; Di Miceli, G.; Giambalvo, D. Effect of legume grains as a source of dietary protein on the quality of organic lamb meat. *J. Sci. Food Agric.* **2012**, *92*, 2870–2875.

16. Atti, N.; Khaldi, G. Effects of slaughter weight of Barbary and noir of Thibar lambs on their carcass composition and meat qualities. *Ann. INRAT* **1988**, *61*, 24.
17. Chauveau-Duriot, B.; Doreau, M.; Nozière, P.; Graulet, B. Simultaneous quantification of carotenoids, retinol and tocopherols in forages, bovine plasma and milk: Validation of a novel UPLC method. *Anal. Bioanal. Chem.* **2010**, *397*, 777–790. [PubMed]
18. Vázquez, C.V.; Rojas, M.V.; Ramírez, C.A.; Chávez-Servín, J.L.; García-Gasca, T.; Martínez, R.A.F.; Montemayor, H.M.A. Total phenolic compounds in milk from different species. Design of an extraction technique for quantification using the Folin–Ciocalteu method. *Food Chem.* **2015**, *176*, 480–486.
19. SAS. *Statistical Analysis System, User's Guide. Statistical*, Version 7th ed.; SAS. Inst. Inc.: Cary, NC, USA, 2004.
20. Atti, N.; Mahouachi, M. Effects of rearing system and nitrogen source on lamb growth, meat characteristics and fatty acid composition. *Meat Sci.* **2009**, *81*, 344–348.
21. Polidori, P.; Quagliarini, C.; Vincenzetti, S. Use of faba bean as a replacer of soybean meal in diet of Fabrianese lambs. *J. Food Sci. Technol.* **2018**, *3*, 350–360.
22. Sents, A.E.; Walters, L.E.; Whiterman, J.V. Performance and carcass characteristics of ram lambs slaughtered at different weights. *J. Anim. Sci.* **1982**, *55*, 1360–1369.
23. Atti, N.; Ben Salem, H.; Priolo, A. Effects of polyethylene glycol in concentrate or feed blocks on carcass composition and offal weight of Barbarine lambs fed *Acacia cyanophylla* Lindl Foliage. *Anim. Res.* **2003**, *52*, 363–375. [CrossRef]
24. Atti, N.; Mahouachi, M. The effects of diet, slaughter weight, and docking on growth, carcass composition and meat quality of fat-tailed Barbarine lambs: A review. *Trop. Anim. Health Prod.* **2011**, *43*, 1371–1378. [CrossRef] [PubMed]
25. Hajji, H.; Smeti, S.; Ben Hammouda, M.; Atti, N. Effect of protein level on growth performance, non-carcass components and carcass characteristics of young sheep from three breeds. *Anim. Prod. Sci.* **2016**, *6*, 2115–2121. [CrossRef]
26. Tibaoui, S.; Hajji, H.; Smeti, S.; Mekki, I.; Essid, I.; Atti, N. Effects of distillated myrtle (*Myrtus communis* L.) leaves' intake on cull ewes' body weight gain, carcass composition and meat quality. *Span. J. Agric. Res.* **2020**, *18*. [CrossRef]
27. Papi, N.; Mostafa-Tehrani, A.; Amanlou, H.; Memarian, M. Effects of dietary forage-to-concentrate ratios on performance and carcass characteristics of growing fat-tailed lambs. *Anim. Feed. Sci. Technol.* **2011**, *163*, 93–98. [CrossRef]
28. Smeti, S.; Atti, N.; Mahouachi, M. Effects of finishing lambs in rich aromatic plant pasture or in feedlot on lamb growth and meat quality. *J. Appl. Anim. Res.* **2014**, *42*, 297–303. [CrossRef]
29. Kamalzadeh, A.; Koops, W.J.; van Bruchem Bangma, G.A.; Tamminga, S.; Zwart, D. Feed quality restriction and compensatory growth in growing sheep: Development of body organs. *Small Rum. Res.* **1998**, *29*, 71–82. [CrossRef]
30. Mahouachi, M.; Atti, N. Effects of restricted feeding and re-feeding of Barbarine lambs: Intake, growth and non-carcass components. *Anim. Sci.* **2005**, *81*, 305–312. [CrossRef]
31. Prud'hon, M. La croissance globale de l'agneau: Ses caractéristiques et ses lois. In *Deuxième Journées de la Recherché Ovine et Caprine: Croissance, Engraissement et Qualité des Carcasses D'agneaux et de Chevreaux, INRA-ITOVIC*; INRA: Paris, France, 1976; pp. 6–20.
32. Ben Abdelmalek, Y.; Smeti, S.; Mekki, I.; Hajji, H.; Essid, I.; Atti, N. Rehabilitation of Barbarine cull ewes using rosemary residues and linseed: Effect on weight gain, carcass and non-carcass characteristics and meat quality. *Animal Int. J. Anim. Biosci.* **2019**, *13*, 879–887. [CrossRef]
33. Reynolds, C.K. Quantitative aspects of liver metabolism in ruminants. In *Ruminant Physiology: Digestion, Metabolism, Growth and Reproduction*; Engelhardt, W.V., Leonhard-mark, S., Breves, G., Giesecke, D., Eds.; Ferdinand Enke Verlag: Stuttgart, Germany, 1995; pp. 351–371.
34. Kabbali, A.; Johnson, D.W.; Goodrich, R.D.; Allen, C.E. Effects of undernutrition and refeeding on weights of body parts and chemical components of growing Moroccan lambs. *J. Anim. Sci.* **1992**, *70*, 2859–2865. [CrossRef]
35. Boccard, R.; Dumont, B.L. Etude de la production de la viande chez les ovins. Variation de l'importance relative de différentes régions corporelles de l'agneau de boucherie. *Ann. Zootech.* **1960**, *9*, 385–398. [CrossRef]
36. Joy, M.; Ripoll, G.; Delfa, R. Effects of feeding system on carcass and non-carcass composition of Churra Tensina light lambs. *Small Rum. Res.* **2008**, *78*, 123–133. [CrossRef]
37. Murphy, T.A.; Loerch, S.C.; McClure, K.E.; Solomon, M.B. Effects of restricted feeding on growth performance and carcass composition of lambs. *J. Anim. Sci.* **1994**, *72*, 3131–3137. [CrossRef] [PubMed]
38. Simitzis, P.E.; Deligeorgis, S.G.; Bizelis, J.A.; Dardamani, A.; Theodosiou, I.; Fegeros, K. Effect of dietary oregano oil supplementation on lamb meat characteristics. *Meat Sci.* **2008**, *79*, 217–223. [CrossRef] [PubMed]
39. Smeti, S.; Tibaoui, S.; Bertolin, J.R.; Yagoubi, Y.; Mekki, I.; Joy, M.; Atti, N. Effects of myrtle (*Myrtus communis* L.) essential oils as dietary antioxidant supplementation on carcass and meat quality of goat meat. *J. Anim. Physiol. Anim. Nutr.* **2020**, in press. [CrossRef]
40. Khliji, S.; Van de Ven, R.; Lamb, T.A.; Lanza, M.; Hopkins, D.L. Relationship between consumer ranking of lamb colour and objective measures of colour. *Meat Sci.* **2010**, *85*, 224–229. [CrossRef] [PubMed]
41. Simitzis, P.E.; Symeon, G.K.; Charismiadou, M.A.; Bizelis, J.A.; Deligeorgis, S.G. The effects of dietary oregano oil supplementation on pigment characteristics. *Meat Sci.* **2010**, *84*, 670–676. [CrossRef]
42. Stinnett, J.D. *Nutrition and the Immune Response*; CRC Press: Boca Raton, FL, USA, 1983.
43. Cook-Mills, J.M. Vitamin E Isoform-Specific Functions in Allergic Inflammation and Asthma. In *Nutrition and Functional Foods for Healthy Aging*; Elsevier Inc: Cambridge, MA, USA, 2017; Chapter 17; pp. 167–188.

44. Traber, M.G. Vitamin E: Metabolism and Requirements. In *Encyclopedia of Human Nutrition*; Elsevier: Cambridge, MA, USA, 2013; pp. 383–389.
45. Gomez-Fernandez, J.C.; Villalain, J.; Aranda, F.J.; Ortiz, A.; Micol, V.; Coutinho, A.; Berberan-Santos, M.N.; Prieto, M.J. Localization of alpha-tocopherol in membranes. *Ann. N. Y. Acad. Sci.* **1989**, *570*, 109–120. [CrossRef]
46. Echegaray, N.; Gómez, B.; Barba, F.J.; Franco, D.; Estévez, M.; Carballo, J.; Marszałek, K.; Lorenzo, J.M. Chestnuts and by-products as source of natural antioxidants in meat and meat products: A review. *Trends Food Sci. Technol.* **2018**, *82*, 110–121. [CrossRef]
47. Bellés, M.; Del Mar Campo, M.; Roncalés, P.; Beltran, J.A. Supranutritional doses of vitamin E to improve lamb meat quality. *Meat Sci.* **2019**, *149*, 14–23. [CrossRef]
48. Weiss, W.P.; Smith, K.L.; Hogan, J.S.; Steiner, T.E. Effect of forage to concentrate ratio on disappearance of vitamins A and E during in vitro ruminal fermentation. *J. Dairy Sci.* **1995**, *78*, 1837–1842. [CrossRef]
49. Hymøller, L.; Jensen, S.K. Stability in the rumen and effect on plasma status of single oral doses of vitamin D and vitamin E in high-yielding dairy cows. *J. Dairy Sci.* **2010**, *93*, 5748–5757. [CrossRef] [PubMed]
50. Lauridsen, C.; Jensen, S.K. α-tocopherol incorporation in mitochondria and microsomes upon supranutritional vitamin E supplementation. *Genes Nutr.* **2012**, *7*, 475–482. [CrossRef] [PubMed]
51. Rigotti, A. Absortion, transport and tissue delivery of vitamin E. *Mol. Asp. Med.* **2007**, *28*, 423–436. [CrossRef] [PubMed]
52. Ponnampalam, E.N.; Burnett, V.F.; Norng, S.; Warmer, R.D.; Jacobs, J.L. Vitamin E and fatty acid content of lamb meat from perennial or annual pasture systems with supplements. *Anim. Prod. Sci.* **2012**, *52*, 255–262. [CrossRef]
53. Hopkins, D.L.; Lamb, T.A.; Kerr, M.J.; van de Ven, R.J.; Ponnampalam, E.N. Examination of the effect of ageing and temperature at rigor on colour stability of lamb meat. *Meat Sci.* **2013**, *95*, 311–316. [CrossRef]
54. Monino, M.I.; Martínez, C.; Sotomayor, J.A.; Lafuente, A.; Jordán, M.J. Polyphenolic transmission to segureño lamb meat from ewes dietary supplemented with the distillate from rosemary (*Rosmarinus officinalis*) leaves. *J. Agric. Food Chem.* **2008**, *56*, 3363–3367. [CrossRef]

Article

Effects of Dietary Protein Concentration on Lipid Metabolism Gene Expression and Fatty Acid Composition in 18–23-Month-Old Hanwoo Steers

Rajaraman Bharanidharan [1,†], Krishnaraj Thirugnanasambantham [2,3,†], Ridha Ibidhi [2], Geumhwi Bang [4], Sun Sik Jang [5], Youl Chang Baek [6], Kyoung Hoon Kim [2,7] and Yea Hwang Moon [8,*]

[1] Department of Agricultural Biotechnology, College of Agriculture and Life Sciences, Seoul National University, Seoul 08826, Korea; bharanidharan7@snu.ac.kr

[2] Department of Eco-Friendly Livestock Science, Institute of Green Bio Science and Technology, Seoul National University, Pyeongchang 25354, Korea; thiru_dna@yahoo.co.in (K.T.); ridha@snu.ac.kr (R.I.); khhkim@snu.ac.kr (K.H.K.)

[3] Pondicherry Centre for Biological Science and Educational Trust, Kottakuppam 605104, Tamil Nadu, India

[4] Department of Animal Science and Technology, Konkuk University, Seoul 05029, Korea; rharhadl1995@snu.ac.kr

[5] Hanwoo Research Institute, National Institute of Animal Science, RDA, Pyeongchang 25342, Korea; jangsc@korea.kr

[6] Division of Animal Nutritional and Physiology, National Institute of Animal Sciences, Wanju 55365, Korea; chang4747@korea.kr

[7] Department of International Agricultural Technology, Graduate School of International Agricultural Technology, Seoul National University, Pyeongchang 25354, Korea

[8] Division of Animal Bioscience and Integrated Biotechnology, Gyeongsang National University, Jinju 52828, Korea

* Correspondence: yhmoon@gntech.ac.kr; Tel.: +82-55-772-3265

† These authors contributed equally to this work.

Citation: Bharanidharan, R.; Thirugnanasambantham, K.; Ibidhi, R.; Bang, G.; Jang, S.S.; Baek, Y.C.; Kim, K.H.; Moon, Y.H. Effects of Dietary Protein Concentration on Lipid Metabolism Gene Expression and Fatty Acid Composition in 18–23-Month-Old Hanwoo Steers. *Animals* **2021**, *11*, 3378. https://doi.org/10.3390/ani11123378

Academic Editors: Guillermo Ripoll and Begoña Panea

Received: 27 October 2021
Accepted: 23 November 2021
Published: 25 November 2021

Publisher's Note: MDPI stays neutral with regard to jurisdictional claims in published maps and institutional affiliations.

Simple Summary: Intramuscular fat or fatty acids content was regarded as a quality index of meat as it increases meat tenderness and flavor. In Korea, commercial Hanwoo farms have been supplying concentrate feeds that are 2–3% higher in crude protein (CP) content rather than conventional feeds throughout the entire program and have tried to shorten the feeding period to less than 28 months of age. This has led to a probable stability in lean-to-fat ratio in Hanwoo steers from 19 to 21 months of age. This stability could be linked to the regulation of the gene expression mediating lipogenesis during this phase. However, there is a lack of data regarding the effects of the protein level in isoenergetic concentrate diets on transcriptional activity of such genes at this phase. The conventional feeding program and the current one with relatively higher CP content were compared in two groups of twenty Hanwoo steers. Results showed that higher CP endorsement during the growing phase increased the expression of intramuscular *PPARα* ($p < 0.1$) and *LPL* ($p < 0.05$) and decreased the genes, such as *VLCAD* ($p < 0.01$), *GPAT1* ($p = 0.001$), and *DGAT2* ($p = 0.016$), which are involved in lipogenesis and fatty acid esterification. This may result in a relatively lower lipid turnover which could be responsible for shortening the feeding period.

Abstract: The present study evaluated the influence of dietary protein level on growth performance, fatty acid composition, and the expression of lipid metabolic genes in intramuscular adipose tissues from 18- to 23-month-old Hanwoo steers, representing the switching point of the lean-to-fat ratio. Forty steers with an initial live weight of 486 ± 37 kg were assigned to one of two treatment groups fed either a concentrate diet with 14.5% CP and or with 17% CP for 6 months. Biopsy samples of intramuscular tissue were collected to analyze the fatty acid composition and gene expression at 23 months of age. Throughout the entire experimental period, all steers were restrained twice daily to allow individual feeding. Growth performance, blood metabolites, and carcass traits, according to ultrasonic measurements, were not affected by the experimental diets. The high-protein diet significantly increased the expression of intramuscular *PPARα* ($p < 0.1$) and *LPL* ($p < 0.05$) but did not affect genes involved in fatty acid uptake (*CD36* and *FABP4*) nor lipogenesis (*ACACA*, *FASN*, and *SCD*). In addition, it downregulated intramuscular *VLCAD* ($p < 0.01$) related to lipogenesis but also *GPAT1* ($p = 0.001$), *DGAT2* ($p = 0.016$), and *SNAP23* ($p = 0.057$), which are involved in fatty acid esterification and adipocyte size. Hanwoo steers fed a high-protein diet at 18–23 months of age resulted in a relatively lower lipid turnover rate than steers fed a low-protein diet, which could be responsible for shortening the feeding period.

Keywords: lipogenesis; *GPAT1*; *SNAP23*; fatty acid composition; Hanwoo steer

1. Introduction

In the Korean beef cattle (Hanwoo) grading system, intramuscular fat (IMF) content of the *Longissimus dorsi* is a key quality factor, and the current grading system (1++, 1+, 1, 2, 3) has been implemented since 2004 [1]. Specific feeding strategies after castration at about 7 months of age have been developed for Hanwoo steers until 30 months of age [2,3]. The conventional feeding program in Korea is based on restricted feeding of concentrates but a gradual increase in the dietary concentrate to forage ratio from 50% to 90%, which supplies sufficient energy for stimulating fat deposition during the 17-month fattening phase. Forage is usually provided *ad libitum* during the growth phase (7–13 months) and gradually decreases from 30% to 10% of total dry matter (DM) intake from the early (14–21 months) to late fattening (22–30 months) phases, respectively. The total digestible nutrient (TDN) content of concentrate feeds has gradually been increased from 70% to 74% DM, and crude protein (CP) content has gradually been decreased from 16% to 12% DM [3,4]. However, commercial farms have been supplying concentrate feeds that are 2–3% higher in CP content rather than conventional feeds throughout the entire program and have tried to shorten the feeding period to less than 28 months of age. Positive results on growth performance and carcass characteristics have been obtained for Hanwoo steers by shortening the feeding period, attributed not only to a 2.5% increase in CP content but also to an increase in TDN content compared to conventional diets [5].

In a comparative slaughter study, Hanwoo steers recorded their highest daily gain and the switching point of the ratio of daily retained energy to daily lean body weight gain at 19–21 months of age; thereafter, the retained energy ratio dramatically increased until 30 months of age [6]. This indicates that the lean-to-fat ratio is relatively stable due to a preferential increase in protein from 19 to 21 months of age. It can be assumed that regulation of the gene expression mediating lipogenesis during this phase would lead to relatively stable changes. Gene expression detected by quantitative real-time polymerase chain reaction using five consecutive biopsy samples at 2, 7, 12, 20, and 25 months of age revealed that the inflection point and peak expression of many genes associated with adipogenesis and lipogenesis were observed at 20–25 months of age [7]. Furthermore, studies have also demonstrated the roles of different genes in intramuscular fatty acid synthesis and growth performance in different cattle breeds [8,9].

However, there is a lack of data regarding the effects of the protein level in isoenergetic concentrate diets on transcriptional activity of genes related to fatty acid biosynthesis and fatty acid composition during the growth stage, which represents the switching point of the lean-to-fat ratio in Hanwoo steers. Therefore, the present study evaluated growth performance, blood metabolites, fatty acid composition, and the expression of 14 genes involved in lipid metabolism in intramuscular adipose tissues of Hanwoo steers fed isoenergetic concentrate diets with different protein levels for 6 months from 18 to 23 months of age.

2. Materials and Methods

2.1. Experimental Design, Animals, and Diet

All experiments were carried out at the animal farm of Seoul National University (Pyeongchang, Korea). Experiments were carried out according to the Guidelines for the

Care and Use of Experimental Animals of Seoul National University (SNU-171211–1). Forty Hanwoo steers were selected from steers that had been reared at a farm with the same concentrate and hay according to a conventional feeding program. After a 1-month adaptation to the new pens, steers with an initial live weight of 486 ± 37 kg (average age 18 months) were blocked by body weight (BW) and randomly allocated to two experimental groups (five steers/pen and four pens/treatment) fed either of the concentrate diets with 14.5% CP (LCP) or 17% CP (HCP) for 6 months. The concentrates were formulated to provide 75% TDN at equal metabolizable energy on a DM basis (Table 1). Steers in a pen were restrained in self-locking stanchions twice daily (0800 and 1800 h) for about 1 h to allow them to consume the concentrate and ryegrass hay individually. They were given full access to water and a mineral block *ad libitum* in the pen, and the monthly live BW of the steers was measured from the beginning of the experiment and used to calculate the average daily gain (ADG) during the experimental period. Daily feed refusal was noted to estimate the dry matter intake (DMI) and the feed conversion ratio (DMI/ADG).

Table 1. Ingredients and nutrient composition of the diets fed to steers during the experimental period.

Items	Concentrates		Ryegrass Hay
	LCP	HCP	
Ingredient (% DM)			
Broken corn	1.36	1.02	
Wheat bran	4.44	5.56	
Soy bean hull	7.08	2.12	
Urea	0.45	0.58	
NaCl	0.20	0.20	
Molasses	3.00	3.00	
Baking soda	1.07	0.94	
Steam flaked corn	22.00	22.00	
Ammonium chloride	0.15	0.15	
Corn flour	1.72	0.05	
Extruded palm seed	4.24	7.52	
CMS	1.50	1.50	
Wheat flour	20.02	20.00	
DDGS	10.00	13.00	
Corn gluten feed	20.00	19.92	
Limestone	2.34	2.00	
Amaferm [1]	0.01	0.01	
Palm oil	0.20	0.20	
Mineral/Vitamin premix [2]	0.23	0.23	
Nutrient composition (% DM)			
DM (%)	89.0	89.0	90.0
Ash	6.6	6.4	4.4
Crude Protein	14.5	17.0	5.7
Ether extract	3.4	3.7	1.2
NDF	22.0	22.0	67.1
ADF	8.4	8.1	40.8
TDN	75.1	75.2	-

LCP, low crude protein; HCP, high crude protein; CMS, condensed molasses solubles; DDGS, distiller's dried grain with solubles; NDF, neutral detergent fiber; ADF, acid detergent fiber; TDN, total digestible nutrients. [1] Fermentation extract of *Aspergillus oryzae* (Biozyme Enterprises Inc., St. Joseph, MO, USA). [2] Provided the following nutrients per kg additive (Grobic-DC, Bayer Health Care, Leverkusen, Germany): Vit. A, 2,650,000 IU; Vit. D, 3, 530,000 IU; Vit. E, 1050 IU; Niacin, 10,000 mg; Mn, 4400 mg; Zn, 4400 mg; Fe, 13,200 mg; Cu, 2200 mg; I, 440 mg; Co, 440 mg.

2.2. Carcass Evaluation, Blood Collection, and Analysis

After 180 day of feeding, the non-invasive and real-time assessment of the carcass in the live Hanwoo steers (23 months) were performed by ultrasonic scanning using Super-eye meat (FHK Co., Ltd., Tokyo, Japan) equipped with linear probe (2 MHz frequency:

27 × 147) between the 13th rib and lumbar vertebrae on the left side [10]. The estimates of back fat thickness, rib eye area, yield grade, marbling score, and quality grade were obtained using the ultrasound image.

One hour before the morning feeding on day 180 of the experiment, blood samples were collected from the jugular vein of two animals in each pen using a syringe (18G) and transferred to anticoagulant-free 6 mL yellow-capped BD Vacutainers® in an icebox. Serum was separated from the blood by centrifugation at 2500× g for 15 min at 4 °C (ScanSpeed 1580R, Labogene, Seoul, Korea) and transferred to 2 mL microtubes for storage at −80 °C until further analysis. Biochemical parameters, including total cholesterol, glucose, triglycerides, and non-esterified fatty acids, were analyzed using an automatic analyzer (BS-400, Mindray, Beijing, China).

2.3. Tissue Biopsies, RNA Extraction, and Real-Time Quantitative PCR

After 180 days of feeding, tissue samples (2 g/head) from eight animals in each group were collected by biopsy from the left-side rear of the third lumbar vertebra [11] using a spring-loaded biopsy instrument (Biotech, Karlova Ves, Slovakia), under intramuscular sedation (Xylazine 20 Inj., 20 mg/head, KEPRO B.V, Deventer, The Netherlands) and a line block (10 mL/head, 2% lidocaine injection, Cheil Pharma, Seoul, Korea) injection of local anesthesia. The harvested tissue samples were snap-frozen in liquid nitrogen and stored at −80 °C until further processing. The respective animals were intramuscularly injected with Procaine Penicillin G injection (4500 IU/kg, G.C. GPS Inj., Green Cross Veterinary Products, Seoul, Korea) immediately after collecting the intramuscular tissue. Then, for another 3 days, the animals were intramuscularly administrated 3 mg/kg Ketoprofen (New-Procop Inj., Shinil Biogen, Yesan-gun, Korea).

Total RNA was isolated from tissues using the RNeasy Lipid Tissue Mini Kit (Qiagen, Hilden, Germany) according to the manufacturer's instructions. The purity and concentration of the isolated RNA were determined using a NanoDrop™ 2000/2000c spectrophotometer (Thermo Fisher Scientific, Waltham, MA, USA). The integrity of isolated RNA was evaluated by visualizing the 28S and 18S bands using eco dye-stained (Biofact) agarose gel electrophoresis. The cDNA synthesis (1 µg) from the isolated RNA was performed using the QuantiTect Reverse Transcription Kit (Qiagen) according to the manufacturer's instructions. In the current study, 14 genes (Table 2) were investigated using real-time qPCR with SYBR Green real-time-PCR Master Mix (Bioneer, Seoul, Korea) and the CFX96 Touch™ Real-Time PCR Detection System (Bio-Rad Laboratories, Inc., Hercules, CA, USA). The 20-µL reaction mixture contained 30 ng cDNA, 10 µL SYBR Green Master Mix, and 1.0 µL each 10 µM primer. The PCR conditions were 95 °C for 15 min, followed by 40 cycles at 94 °C for 15 s, annealing temperature for 30 s, and 72 °C for 30 s. All primers were designed using the Primer-BLAST [12] tool based on the National Center for Biotechnology Information published sequences (www.ncbi.nlm.nih.gov, accessed on 5 January 2018). The $2^{-\Delta\Delta CT}$ method was used to determine relative-fold changes [13], and all data were normalized based on the β-actin housekeeping gene.

Table 2. Gene names, GenBank accession numbers, sequences, and amplicon sizes of the *Bos taurus* primers used in the real-time quantitative PCR analysis.

Gene	NCBI Acc. No.	Primer Name	Primer Sequence (5′–3′)	Product Size (bp)
Transcription factors:				
Peroxisome proliferator activated receptors	NM_001034036.1	PPAR FP PPAR RP	CAATGGAGATGGTGGACACA TTGTAGGAAGTCTGCCGAGAG	95
Sterol regulatory element-binding proteins	NM_001113302.1	SREBP FP SREBP RP	GAGCCACCCTTCAACGAA TGTCTTCTATGTCGGTCAGCA	88
Lipogenesis:				
Acetyl-CoA carboxylase	NM_174224.2	ACACA FP ACACA RP	CGCTCGGTGATTGAAGAGAA CGTCATGTGGACGATGGAAT	117

Table 2. *Cont.*

Gene	NCBI Acc. No.	Primer Name	Primer Sequence (5′–3′)	Product Size (bp)
Fatty acid synthase	NM_001012669.1	FASN FP FASN RP	ATCGAGTGCATCAGGCAAGT TGTGAGCACATCTCGAAAGCCA	92
Stearoyl-CoA desaturase	NM_173959.4	SCD FP SCD RP	TTATTCCGTTATGCCCTTGG TTGTCATAAGGGCGGTATCC	83
Fatty acid uptake:				
Lipoprotein lipase	NM_001075120.1	LPL FP LPL RP	CTCAGGACTCCCGAAGACAC GTTTTGCTGCTGTGGTTGAA	98
Fatty acid binding protein 4	NM_174314.2	FABP4 FP FABP4 RP	GGATGATAAGATGGTGCTGGA ATCCCTTGGCTTATGCTCTCT	80
Fatty acid translocase (CD36)	NM_174010.3	CD36 FP CD36 RP	GGTCCTTACACATACAGAGTTCG ATAGCGAGGGTTCAAAGATGG	115
Fatty acid esterification:				
Glycerol-3-phosphate acyltransferase-1	NM_001012282.1	GPAT1 FP GPAT1 RP	TGTGCTATCTGCTCTCCAATG CTCCGCCACTATAAGAATG	116
Diacylglycerol acyltransferase-2	NM_205793.2	DGAT2 FP DGAT2 RP	CATTGCCGTGCTCTACTTCA AGTTTCGGACCCACTGTGAC	86
Lipolysis:				
Adipose triglyceride lipase	NM_001046005.2	ATGL FP ATGL RP	TGACCACACTCTCCAACA AAGCGGATGGTGAAGGA	100
Very long chain acyl-CoA dehydrogenase	U30817.1	VLCAD FP VLCAD RP	TCTTCGAGGGGACAAATGAC AGCATTCCCAAAAGGGTTCT	116
Adipocyte size:				
Synaptosome-associated protein 23	BT030678.1	SNAP23 FP SNAP23 RP	GGAGGGGAGGCAAGAGATAA AAACCAAGCACTGGCCTAAA	148
Berardinelli-Seip congenital lipodystrophy2-seipin	BC105396.1	BSCL2 FP BSCL2 RP	CGAAAGGTCTCTGCCCATC GTTTTCTCCTCCTCGGACAG	140
Housekeeping:				
Beta-actin	BC142413.1	β-Actin FP β -Actin RP	GTCCACCTTCCAGCAGATGT CAGTCCGCCTAGAAGCATTT	90

PCR, polymerase chain reaction.

2.4. Chemical and Fatty Acid Analyses

The feed samples were dried in a forced-air oven at 65 °C for 72 h to estimate DM content and then ground to pass through a 1 mm screen (Model 4; Thomas Scientific, Swedesboro, NJ, USA). The feed samples were then analyzed for nutrient compositions, such as CP (Method 990.03 [14]), ash (Method 942.05 [14]), and ether extract (Method 960.39 [14]). Neutral detergent fiber content was assayed with heat-stable amylase (without sodium sulfite) and expressed exclusively as residual ash (aNDFom) using the method of Van Soest et al. [15]. The acid detergent fiber content, excluding residual ash (ADFom), was determined according to Van Soest [16].

The fatty acid composition of the biopsied intramuscular tissues was determined according to the method described by O'Fallon et al. [17]. A commercial 37-component fatty acid methyl ester (FAME) standard mixture (CRM47885) and the internal standard tridecanoic acid (C13:0) were obtained from Supelco Co. (Belafonte, PA, USA). The extracted fatty acids (1 μL) were injected with a split ratio of 30:1 into an Agilent 7890B GC system (Agilent Technologies, Santa Clara, CA, USA) equipped with a flame ionization detector (FID) and a 100 m × 0.25 mm × 0.20 μm SP-2560 biscyanopropyl polysiloxane capillary column (Cat. No: 24056, Supelco, Sigma-Aldrich, St. Louis, MO, USA). The inlet and detector temperatures were maintained at 250 °C and 260 °C, respectively. The helium carrier gas was set to a flow rate of 1.18 mL/min, and the initial oven temperature was set at 100 °C. The oven temperature was held constant at the initial temperature for 5 min and then increased by 4 °C/min to a final temperature of 240 °C, which was held for 14 min. The fatty acid contents were quantified after normalization with the internal standard using the fatty acid standards described by the AOCS Official Method Ce 1j-07 [18]. Theoretical FID correction factors reported in the AOCS Official Method Ce 1 h-05 [19] were applied to individual FAMEs. The concentrations of individual fatty acids are expressed in terms of total FAMEs instead of tissue content.

2.5. Statistical Analysis

Data were completely randomized, with steer as the experimental unit, using the MIXED procedure in SAS software, version 9.4 (SAS Institute, Cary, NC, USA). The fixed effects in the model included the dietary treatments. Animals within a treatment group were considered a random effect. Appropriate covariance structures were chosen based on Akaike's information criterion. Means were calculated using the LSMEANS statement, and treatment differences were considered significant at $p < 0.05$. Near significant trends were considered present at $0.05 < p < 0.10$ [20]. Pearson's correlation was used to test the correlations between fatty acid composition and genes, and among genes using the corr.test function in the 'psych' package of R-software, version 4.0.3 (The R Foundation for Statistical Computing, Vienna, Austria).

3. Results

3.1. Feed Intake, Daily Gain, Blood Metabolites, and Fatty Acid Composition of Biopsy Tissues

Feeding a high-protein concentrate during 18- to 23-month age period did not affect ($p > 0.05$) feed intake, ADG, blood metabolites, nor the ultrasonic evaluation of carcass traits in Hanwoo steers (Tables 3 and 4). However, a trend toward increased total fatty acids ($p = 0.07$) and oleic acid ($p = 0.09$) proportion and a significant increase in eicosenoic acid ($p < 0.01$) proportion in intramuscular tissue were noted (Table 5).

Table 3. Effects of dietary protein level on feed intake, growth performance ($n = 20$), and serum metabolites ($n = 8$) in Hanwoo steers.

Item	LCP	HCP	SEM	*p*-Value
DMI (kg/d)	9.5	9.5	0.02	0.993
Concentrate (% DMI)	81.6	82.0	0.17	0.111
Rye grass hay (% DMI)	17.5	17.1	0.31	0.379
Body weight (kg)				
Initial	484.6	487.5	9.04	0.822
Final	621.2	626.7	12.44	0.756
Average daily gain (kg/d)	0.76	0.77	0.04	0.815
Feed conversion ratio [1]	13.3	12.9	0.70	0.646
Serum metabolites				
Cholesterol, mg/dL	134.9	155.4	15.90	0.377
Triglycerides, mg/dL	17.3	20.5	1.47	0.141
Glucose, mg/dL	40.5	38.9	2.67	0.674
NEFA, mmol/L	0.1	0.1	0.01	0.226

LCP, low crude protein; HCP, high crude protein; NEFA, non-esterified fatty acids; SEM, standard error of the mean. [1] Feed conversion ratio = average daily intake/average daily gain.

Table 4. Effect of dietary protein level during the growing phase on carcass characteristics of Hanwoo steers evaluated using ultrasonic scanning at 23 months of age ($n = 20$).

Item	LCP	HCP	SEM	*p*-Value
Back fat thickness (mm)	7.08	6.90	0.48	0.799
Rib eye area (cm^2)	82.13	81.83	1.20	0.860
Yield grade (A:B:C, head) [1]	17:3:0	18:2:0		
Yield grade score [2]	2.85	2.90	0.08	0.643
Marbling score [3]	3.00	2.95	0.24	0.883
Quality grade (1$^+$:1:2:3, head) [4]	0:7:12:1	1:3:16:0		
Quality grade score [5]	2.35	2.25	0.13	0.582

LCP, low crude protein; HCP, high crude protein; SEM, standard error of the mean. [1] Carcass yield grades range from C (low yield) to A (high yield). [2] Yield grade score: A = 3, B = 2, and C = 1. [3] Marbling score ranges from 1 to 9, with higher numbers indicating better quality (1 = devoid, 9 = abundant). [4] Quality grades range from 3 (low quality) to 1$^+$ (high quality). [5] Quality grade score: 1$^+$ = 4, 1 = 3, 2 = 2, and 3 = 1.

Table 5. Effect of dietary protein level at growing stage on fatty acid composition in the intramuscular tissue of Hanwoo steers sampled by biopsy at 23 months of age (n = 8).

Fatty Acid (mg/100 g FAME)	LCP	HCP	SEM	p-Value
Myristic acid (C14:0)	3571	3476	283.7	0.815
Palmitic acid (C16:0)	26339	26921	1213.7	0.639
Palmitoleic acid (C16:1n7)	4062	4382	495.8	0.656
Stearic acid (C18:0)	10123	10606	572.7	0.560
Oleic acid (cis 9 C18:1)	37192	39747	990.7	0.090
Linoleic acid (C18:2n6c)	2283	2861	393.3	0.316
Gamma-linolenic acid (C18:3n6)	34	34	2.5	0.816
Alpha linolenic acid (C18:3n3)	114	129	7.9	0.218
Eicosenoic acid (C20:1n9)	147	217	19.0	0.008
Arachidonic acid (C20:4n6)	424	672	225.3	0.449
Others [1]	5843	6169	693.2	0.621
SFA [2]	41758	42705	1482.2	0.659
MUFA [3]	45201	48442	1261.0	0.091
PUFA [4]	3173	4067	666.3	0.359
Total fatty acids (mg/100 g FAME)	90132	95214	1842.4	0.071

SFA, saturated fatty acids; MUFA, monounsaturated fatty acids; PUFA, polyunsaturated fatty acids; FAME, fatty acid methyl ester. [1] Others = C4:0 + C6:0 + C8:0 + C10:0 + C11:0 + C12:0 + C14:1 + C15:0 + C15:1 + C17:0 + C17:1 + C18:1n9t + C18:2n6t + C20:0 + C20:2n6 + C20:3n6 + C20:3n3 + C20:5n3 + C21:0 + C22:0 + C22:1n9 + C22:2n6 + C23:0+ C24:0+C24:1n9 + C22:6n3. [2] SFA = C10:0 + C11:0 + C12:0 + C14:0 + C15:0 + C16:0 + C17:0 + C18:0 + C20:0 + C21:0 +C22:0 + C24:0. [3] MUFA = C14:1n5 + C16:1n7 + C17:1n7 + C18:1n7 + C18:1n9 + C20:1n9 + C22:1n9 + C24:1n9. [4] PUFA = C18:2n6 + C18:2c9,t11 + C18:3n3 + C18:3n6 + C20:2n6 + C20:3n3 + C20:3n6 + C20:4n6 + C20:5n3 + C22:2n6 + C22:4n6 + C22:5n3 + C22:6n3.

3.2. Intramuscular Lipid Metabolic Genes Expression and Their Associations

Figure 1 illustrates the effects of fattening diets with varying protein concentrations at 18–23 months of age on the expression patterns of lipid metabolic genes in intramuscular tissues. The dietary treatments had no effects (p > 0.05) on expression of genes, such as sterol regulatory element-binding protein (*SREBP*), fatty acid synthase (*FASN*), acetyl-CoA carboxylase (*ACACA*), or stearoyl-CoA desaturase (*SCD*), Berardinelli-Seip congenital lipodystrophy2-seipin (*BSCL*), and adipose triglyceride lipase (*ATGL*). The high-protein diet upregulated only intramuscular expression of peroxisome proliferator-activated receptor (*PPARα*) by 1.5-fold (p = 0.09) and lipoprotein lipase (*LPL*) by 3.5-fold (p < 0.05). However, a downregulation in glycerol-3-phosphate acyltransferase (*GPAT1*) (p < 0.005), diacylglycerol acyltransferase-2 (*DGAT2*) (p < 0.05), very long-chain acyl-CoA dehydrogenase (*VLCAD*) (p < 0.005), and synaptosome-associated protein 23 (*SNAP23*) (p = 0.057) genes were noted.

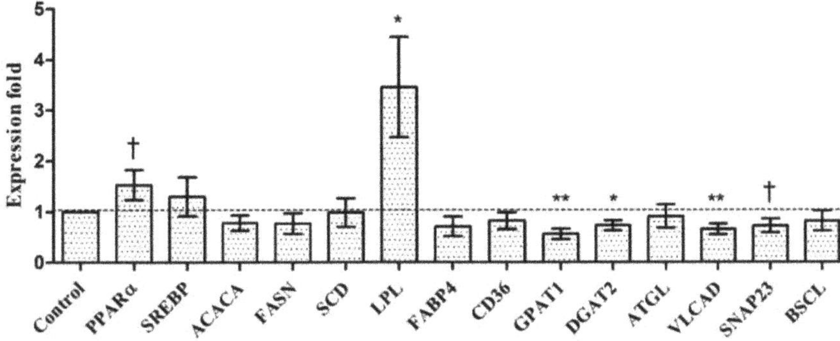

Figure 1. Effect of 6 months of an altered dietary protein level on the relative expression levels of genes in intramuscular tissue of Hanwoo steers sampled by biopsy at 23 months of age. [†] p < 0.1; [*]p < 0.05; [**] < 0.01. Values are the least-squares means with the standard error (n = 8).

Pearson's correlation analysis revealed a wide range of associations among different genes expressed (Supplementary Table S1). Most notably, *PPARα* expression was positively correlated with *LPL* ($p < 0.05$) and negatively with *VLCAD* ($p < 0.05$). Likewise, *SCD* expression was positively correlated with *DGAT2* ($p < 0.05$), *FASN* ($p < 0.001$), and *ACACA* expression ($p < 0.005$). Associations between genes and fatty acid composition are showed in Supplementary Table S2. Oleic acid content was positively and negatively associated with the expression level of *LPL* ($p = 0.060$) and *FABP4* ($p = 0.060$), respectively. A strong ($p < 0.05$) negative association of *FASN* and *ACACA* with alpha linoleic acid content was also noted.

4. Discussion

The observed effect of dietary protein concentration on growth performance of 18–23 months old Hanwoo steers are consistent with a previous study [21] that assessed the effects of different levels of dietary protein in 13- to 18-month-old finishing steers. Kamiya et al. [22] also did not find an effect of high CP level on the daily gain of Holstein steers. In addition, several studies have reported that diets containing more or less protein than the recommended amount do not lead to significant differences in marbling or IMF content [23,24]. Lee et al. [25] showed that diets with higher CP and non-degraded protein intake levels do not affect growth performance but tend to improve the carcass quality of Hanwoo steers. On the other hand, high total fatty acids and oleic acid contents are positively correlated with flavor [26], customer preference over imported beef [27], and favorable health characteristics for consumers [28,29]. Rats that eat Hanwoo beef fat show high feed intake but reduced lipogenic enzyme activities and increased high-density lipoprotein cholesterol, indicating a lower risk for cardiovascular disease compared to consuming Angus beef fat [30].

Peroxisome proliferator-activated receptors (*PPARs*) are fatty acids regulated transcription factors that control lipid metabolism [31]. Among the *PPAR* isoforms, *PPARα* binds to the *PPAR* response element in the *LPL* gene promoter and enhances *LPL* gene expression in adipose tissue [32] and human skeletal muscle [33]. The increased expression of *LPL* gene and high content of long-chain mono- and poly-unsaturated fatty acids in the HCP group may be supported by the functional role of *PPARα* binding to fatty acids with a general preference for long-chain unsaturated fatty acids [34]. This was further supported by the strong positive association between *LPL* and *PPARα* expression in the current experiment. In addition, adipose tissue *LPL* of young rats [35] and yaks [36] is downregulated by a low-protein diet and stimulated by a high-protein diet. *LPL* is the rate-limiting enzyme that hydrolyzes plasma triglycerides and plays key roles in lipoprotein metabolism through the efficient transfer of energy in the form of lipid from sites of synthesis to sites of storage or utilization [37]. In this sense, upregulation of *LPL* expression caused by the increase in dietary protein concentration at 18–23 months of age may facilitate uptake of fatty acids because Hanwoo steers typically start to retain energy in the form of fat into vascular, subcutaneous, and intramuscular tissues between 19 and 21 months of age [6]. Okumura et al. [38] observed that an additional fattening period of 6 months in 24- to 30-month-old Japanese black steers increases the rate of intramuscular fat deposition. Jeong et al. [9] reported a strong correlation between *LPL* mRNA abundance and IMF content in Hanwoo steers and suggested that *LPL* is a genetic marker for IMF deposition. In addition, the observed positive association between *LPL* and oleic acid ($p < 0.005$) in this experiment suggested that the expression of LPL might be responsible for increase in oleic acid content in intramuscular tissue.

Berardinelli-Seip congenital lipodystrophy2-seipin (*BSCL)* is an endoplasmic reticulum (ER) membrane protein involved in regulation of lipid droplet biogenesis. Payne et al. [39] strongly suggested that downregulation of *BSCL* causes a defect in lipid droplet morphology formation, indicating that it is crucial for normal adipogenesis of adipocytes. That study also showed that cells lacking *BSCL* fail to induce the expression of *SREBP* and the lipogenic enzymes *GPAT1* and *DGAT2*; our results are consistent with those findings. *GPAT1* is involved in the first step of triglyceride synthesis via acylation of glycerol 3-phosphate and direct incorporation of exogenous fatty acids into triglycerides rather than phospholipids [40]. Multiple regression analyses between IMF contents and the abundance of genes responsible for fat deposition and fat removal have revealed that *GPAT1* is a candidate gene for increasing IMF deposition in Hanwoo [9]. The *DGAT2* gene is an important contributor to triglyceride synthesis and storage and increases the total proportion of polyunsaturated fatty acids to saturated fatty acids within the adipocyte. Our results are similar to a previous study that reported downregulation of *DGAT2* expression in cattle fed a high-protein diet with fat [41]. Despite the observed downregulation of *GPAT1* and *DGAT2* genes involved in fatty acid esterification, the intramuscular fatty acid and oleic acid contents increased in response to the high-protein diet in this experiment. *DGAT* activity is not the only mechanism for triglyceride synthesis, as adequate triglyceride biosynthesis can be sustained by other enzymes via multiple mechanisms even under downregulation of *DGAT2* [42]. Yu et al. [43] demonstrated that *DGAT2* plays an important role in energy homeostasis through rigorous post-transcriptional gene expression in adipocytes. It is unclear if this type of triglyceride biosynthesis plays a significant role in mammalian cells, as another major triglyceride bypassing *DGAT* has been identified in yeast [44]. The interaction between fatty acid translocase (CD36) and long-chain fatty acids is important for the absorption and storage of dietary lipids [9]. Adipocyte fatty acid-binding protein 4 (*FABP4*) directly interacts with hormone-sensitive lipase, which is the first step in an organized lipid transfer process that leads to an increase in fatty acid hydrolysis. However, in the absence of the interaction, fatty acids are not efficiently released from the adipocyte and accumulate intracellularly [45,46]. These genes associated with fatty acid uptake are not affected by increasing the dietary protein concentration in this experiment. Stearoyl-CoA desaturase (SCD) is a membrane-bound enzyme that synthesize monounsaturated fatty acids from saturated fatty acids [47]. Although a strong positive association was noted between the expression of *DGAT2* and *SCD* in this experiment, they did not account for the physical interaction between both genes during fatty acid monounsaturation [48] and the incorporation of endogenously synthesized monounsaturated fatty acids into triglycerides [49]. Archibeque et al. [50] reported that total unsaturated fatty acid and oleic acid contents do not result from greater *SCD* gene expression in the adipose tissues of beef steer. In the present experiment, unchanged or suppressed genes related to fatty acid uptake, lipogenesis, and fatty acid esterification appeared to originate from the effects of dietary protein level and not from age-associated changes in lipogenesis of 18- to 23-month-old Hanwoo steers. A peak in expression of many genes associated with adipogenesis and lipogenesis is detected at 25 months of age; thereafter, cattle are beginning to deposit a substantial amount of IMF [7]. Additional increases in the expression of those genes are observed during the finishing phase [51,52].

Very long-chain acyl-CoA dehydrogenase (*VLCAD*) catalyzes the initial step in the mitochondrial β-oxidation of long-chain fatty acids. Adiposity, or the amount of triglyceride stored in adipocytes, is fundamentally a net result of lipogenesis and lipolysis. The observed downregulation of *VLCAD* by high protein diet in the current study indicated that the lipid turnover rate, i.e., the balance between the synthesis and degradation of triglycerides, was relatively lower in steers fed the high-protein diet than in those fed the low-protein diet, which could be responsible for the carryover effect on high marbling in later life. Jeong et al. [9] reported that the expression of the fat removal gene *VLCAD* is negatively correlated with IMF content. The negative correlation between *PPARα* and *VLCAD* can be explained by muscle-specific upregulation of *PPARα* in *VLCAD*-deficient mice [53].

The abundance of *SNAP23* transcripts is positively correlated with adipocyte size [54]. Therefore, downregulation of intramuscular *SNAP23* expression, mediated by the high-protein diet, may have contributed to reduced adipocyte size and increased fineness of the carcass marbling texture. Vierck et al. [55] reported that coarsely marbled steaks contain larger adipocytes than those of finely marbled steaks. This may serve as a pointer for future investigations to elucidate the mechanism of synthesis of fine marbling because Korean consumers prefer fine marbling and the quality grading system changes based on fineness. Although growth performance and final carcass traits after 23 months of age could not be included in this study, indirect identification of the functional links between gene regulation and adipose tissue occurred at a particular developmental stage for 6 months, which was made possible by the developmental processes related to the data [3] on the change in the lean-to-fat ratio or net energy content during daily gain of Hanwoo steers.

5. Conclusions

A high-protein diet fed to 18- to 23-month-old steers altered the switching point of the lean-to-fat ratio and could be responsible for less lipogenesis (*GPAT1* and *DGAT2*) and lipolysis (*VLCAD*) compared to a low-protein diet. These data reveal a relatively low lipid turnover rate, which could be responsible for shortening the feeding period. Furthermore, Hanwoo steers fed a high-protein diet during this period showed increased intramuscular fatty acid content, oleic acid, and fineness in the marbling texture during later life by downregulating *SNAP23*.

Supplementary Materials: The following are available online at https://www.mdpi.com/article/10.3390/ani11123378/s1, Table S1: Correlation coefficients for the expression of genes in intramuscular tissues; Table S2: Correlation coefficients for fatty acid composition and gene expression in intramuscular tissues.

Author Contributions: Conceptualization, K.H.K., Y.H.M.; methodology, K.H.K., R.B.; project administration, K.H.K.; formal analysis, R.B., K.T., R.I., G.B., S.S.J., Y.C.B.; data curation, R.B., K.T.; investigation, R.B., K.T.; validation, R.B., K.T.; writing—original draft preparation, R.B., K.T.; writing—review and editing, K.H.K.; software, R.B.; visualization, R.B.; funding acquisition, K.H.K.; supervision, Y.H.M. All authors have read and agreed to the published version of the manuscript.

Funding: This study was supported by the National Institute of Animal Science, Ministry of Rural Development and Administration, Korea (research project PJ014234022021).

Institutional Review Board Statement: This study was conducted in accordance with the guidelines of the Animal Care and Use Committee of Seoul National University, Korea (approval number SNU-171211-1).

Informed Consent Statement: Not applicable.

Data Availability Statement: All data generated in this study are included in the manuscript and Supplementary Files.

Conflicts of Interest: The authors declare no conflict of interest.

References

1. Chung, K.Y.; Lee, S.H.; Cho, S.H.; Kwon, E.G.; Lee, J.H. Current situation and future prospects for beef production in South Korea—A review. *Asian-Australas. J. Anim. Sci.* **2018**, *31*, 951–960. [CrossRef] [PubMed]
2. Park, S.J.; Beak, S.-H.; Jung, D.J.S.; Kim, S.Y.; Jeong, I.H.; Piao, M.Y.; Kang, H.J.; Fassah, D.M.; Na, S.W.; Yoo, S.P.; et al. Genetic, management, and nutritional factors affecting intramuscular fat deposition in beef cattle—A review. *Asian-Australas. J. Anim. Sci.* **2018**, *31*, 1043–1061. [CrossRef] [PubMed]
3. Kim, K.H.; Lee, J.H.; Oh, Y.G.; Kang, S.W.; Lee, S.C.; Park, W.Y.; Ko, Y.D. The Optimal TDN Levels of Concentrates and Slaughter Age in Hanwoo Steers. *J. Anim. Sci. Technol.* **2005**, *47*, 731–744. [CrossRef]
4. NIAS. National Institute of Animal Science. Rural Development Administration. Korean Feeding Standard for Hanwoo. 2017. Available online: https://www.nias.go.kr/english/indexList.do (accessed on 10 July 2017).
5. Reddy, K.E.; Jeong, J.Y.; Ji, S.Y.; Baek, Y.-C.; Lee, S.; Kim, M.; Oh, Y.K.; Lee, H.-J. Effects of High Levels of Nutrients on Growth Performance and Carcass Characteristics of Hanwoo Cattle. *J. Korean Soc. Grassl. Forage Sci.* **2018**, *38*, 180–189. [CrossRef]
6. Kim, K.H.; Oh, Y.G.; Lee, S.C.; Shin, K.J.; Chung, W.T.; Kang, S.W.; Hong, S.K.; Ju, J.C.; Baek, B.H. Determination of Net Energy and Protein Requirements for Growth in Hanwoo Steers by Comparative Slaughter Experiment. *J. Anim. Sci. Technol.* **2007**, *49*, 41–50. [CrossRef]

7. Wang, Y.H.; Bower, N.I.; Reverter, A.; Tan, S.H.; De Jager, N.; Wang, R.; McWilliam, S.M.; Cafe, L.M.; Greenwood, P.L.; Lehnert, S.A. Gene expression patterns during intramuscular fat development in cattle. *J. Anim. Sci.* **2009**, *87*, 119–130. [CrossRef] [PubMed]

8. Liu, L.; Cao, P.; Zhang, L.; Qi, M.; Wang, L.; Li, Z.; Shao, G.; Ding, L.; Zhao, X.; Zhao, X.; et al. Comparisons of adipogenesis- and lipid metabolism-related gene expression levels in muscle, adipose tissue and liver from Wagyu-cross and Holstein steers. *PLoS ONE* **2021**, *16*, e0247559. [CrossRef] [PubMed]

9. Jeong, J.; Kwon, E.G.; Im, S.K.; Seo, K.S.; Baik, M. Expression of fat deposition and fat removal genes is associated with intramuscular fat content in longissimus dorsi muscle of Korean cattle steers1. *J. Anim. Sci.* **2012**, *90*, 2044–2053. [CrossRef]

10. Song, Y.H.; Kim, S.J.; Lee, S.K. Evaluation of Ultrasound for Prediction of Carcass Meat Yield and Meat Quality in Korean Native Cattle (Hanwoo). *Asian-Australas. J. Anim. Sci.* **2002**, *15*, 591–595. [CrossRef]

11. Cheah, K.S.; Cheah, A.M.; Just, A. Simple and rapid biopsy technique for obtaining fat and muscle samples from live animals for predicting meat quality. *Dan. Vet.* **1997**, *80*, 775–777.

12. Ye, J.; Coulouris, G.; Zaretskaya, I.; Cutcutache, I.; Rozen, S.; Madden, T.L. Primer-BLAST: A tool to design target-specific primers for polymerase chain reaction. *BMC Bioinform.* **2012**, *13*, 134. [CrossRef]

13. Livak, K.J.; Schmittgen, T.D. Analysis of relative gene expression data using real-time quantitative PCR and the $2^{-\Delta\Delta C_T}$ method. *Methods* **2001**, *25*, 402–408. [CrossRef]

14. AOAC. *AOAC Official Methods of Analysis*, 18th ed.; Revision 1; Association of Official Agricultural Chemists: Gaithersburg, MD, USA, 2006. Available online: http://www.eoma.aoac.org (accessed on 10 November 2017).

15. Van Soest, P.J.; Robertson, J.B.; Lewis, B.A. Methods for Dietary Fiber, Neutral Detergent Fiber, and Nonstarch Polysaccharides in Relation to Animal Nutrition. *J. Dairy Sci.* **1991**, *74*, 3583–3597. [CrossRef]

16. Van Soest, P.J. Collaborative Study of Acid-Detergent Fiber and Lignin. *J. Assoc. Off. Anal. Chem.* **1973**, *56*, 781–784. [CrossRef]

17. O'Fallon, J.V.; Busboom, J.R.; Nelson, M.L.; Gaskins, C.T. A direct method for fatty acid methyl ester synthesis: Application to wet meat tissues, oils, and feedstuffs. *J. Anim. Sci.* **2007**, *85*, 1511–1521. [CrossRef]

18. AOCS. *Official Methods and Recommended Practices of the AOCS. American Oil Chemists' Society, Champaign, IL, Method Ce 1j-07. Determination of Cis-, Trans-, Saturated, Monounsaturated and Polyunsaturated Fatty Acids in Dairy and Ruminant Fats by Capillary GLC*, 7th ed.; AOCS: Urbana, IL, USA, 2018.

19. AOCS. *AOCS Official Method Ce 1h-05: Determination of Cis-, Trans-, Saturated, Monounsaturated and Polyunsaturated Fatty Acids in Vegetable or Non-Ruminant Animal Oils and Fats by Capillary GLC. Official Methods and Recommended Practices of the AOCS*, 6th ed.; AOCS Press: Champaign, IL, USA, 2006.

20. Gelman, A. Commentary: P values and statistical practice. *Epidemiology* **2013**, *24*, 69–72. [CrossRef] [PubMed]

21. Li, L.; Zhu, Y.; Wang, X.; He, Y.; Cao, B. Effects of different dietary energy and protein levels and sex on growth performance, carcass characteristics and meat quality of F1 Angus × Chinese Xiangxi yellow cattle. *J. Anim. Sci. Biotechnol.* **2014**, *5*, 21. [CrossRef] [PubMed]

22. Kamiya, M.; Yamada, T.; Higuchi, M. Influence of dietary crude protein content on fattening performance and nitrogen excretion of Holstein steers. *Anim. Sci. J.* **2020**, *91*, e13438. [CrossRef] [PubMed]

23. Oddy, V.H.; Smith, C.; Dobos, R.; Harper, G.S.; Allingham, P. *Effect of Dietary Protein Content on Marbling and Performance of Beef Cattle*; Final Report of Project FLOT 210; Meat and Livestock Australia: North Sydney, Australia, 2000.

24. Pethick, D.W.; McIntyre, B.L.; Tudor, G.E. *The Role of Dietary Protein as a Regulator of the Expression of Marbling in Feedlot Cattle (WA)*; Meat & Livestock Australia: North Sydney, Australia, 2000.

25. Lee, Y.H.; Ahmadi, F.; Lee, M.; Oh, Y.-K.; Kwak, W.S. Effect of crude protein content and undegraded intake protein level on productivity, blood metabolites, carcass characteristics, and production economics of Hanwoo steers. *Asian-Australas. J. Anim. Sci.* **2020**, *33*, 1599–1609. [CrossRef] [PubMed]

26. Garmyn, A.J.; Hilton, G.G.; Mateescu, R.G.; Morgan, J.B.; Reecy, J.M.; Tait, R.G., Jr.; Beitz, D.C.; Duan, Q.; Schoonmaker, J.P.; Mayes, M.S.; et al. Estimation of relationships between mineral concentration and fatty acid composition of longissimus muscle and beef palatability traits. *J. Anim. Sci.* **2011**, *89*, 2849–2858. [CrossRef]

27. Moon, H.; Seok, J.H. Price relationship among domestic and imported beef products in South Korea. *Empir. Econ.* **2021**, *61*, 3541–3555. [CrossRef]

28. Troy, D.J.; Tiwari, B.K.; Joo, S.-T. Health Implications of Beef Intramuscular Fat Consumption. *Korean J. Food Sci. Anim. Resour.* **2016**, *36*, 577–582. [CrossRef]

29. Gilmore, L.A.; Walzem, R.L.; Crouse, S.F.; Smith, D.R.; Adams, T.H.; Vaidyanathan, V.; Cao, X.; Smith, S.B. Consumption of High-Oleic Acid Ground Beef Increases HDL-Cholesterol Concentration but Both High- and Low-Oleic Acid Ground Beef Decrease HDL Particle Diameter in Normocholesterolemic Men. *J. Nutr.* **2011**, *141*, 1188–1194. [CrossRef] [PubMed]

30. Nogoy, K.M.C.; Kim, H.J.; Lee, D.H.; Smith, S.B.; Seong, H.A.; Choi, S.H. Oleic acid in Angus and Hanwoo (Korean native cattle) fat reduced the fatty acid synthase activity in rat adipose tissues. *J. Anim. Sci. Technol.* **2021**, *63*, 380–393. [CrossRef] [PubMed]

31. Varga, T.; Czimmerer, Z.; Nagy, L. PPARs are a unique set of fatty acid regulated transcription factors controlling both lipid metabolism and inflammation. *Biochim. Biophys. Acta (BBA) Mol. Basis Dis.* **2011**, *1812*, 1007–1022. [CrossRef] [PubMed]

32. Schoonjans, K.; Peinado-Onsurbe, J.; Lefebvre, A.-M.; Heyman, R.A.; Briggs, M.; Deeb, S.; Staels, B.; Auwerx, J. PPARalpha and PPARgamma activators direct a distinct tissue-specific transcriptional response via a PPRE in the lipoprotein lipase gene. *EMBO J.* **1996**, *15*, 5336–5348. [CrossRef]

33. Zhang, J.; Phillips, D.I.W.; Wang, C.; Byrne, C.D. Human skeletal muscle PPARα expression correlates with fat metabolism gene expression but not BMI or insulin sensitivity. *Am. J. Physiol. Endocrinol. Metab.* **2004**, *286*, E168–E175. [CrossRef]
34. Murakamia, K.; Idea, T.; Suzukia, M.; Mochizuki, T.; Kadowaki, T. Evidence for Direct Binding of Fatty Acids and Eicosanoids to Human Peroxisome Proliferators-Activated Receptor α. *Biochem. Biophys. Res. Commun.* **1999**, *260*, 609–613. [CrossRef]
35. Boualga, A.; Bouchenak, M.; Belleville, J. Low-protein diet prevents tissue lipoprotein lipase activity increase in growing rats. *Br. J. Nutr.* **2000**, *84*, 663–671. [CrossRef] [PubMed]
36. Zhang, H.-B.; Wang, Z.S.; Peng, Q.-H.; Tan, C.; Zou, H.-W. Effects of different levels of protein supplementary diet on gene expressions related to intramuscular deposition in early-weaned yaks. *Anim. Sci. J.* **2014**, *85*, 411–419. [CrossRef]
37. Wang, H.; Eckel, R.H. Lipoprotein lipase: From gene to obesity. *Am. J. Physiol. Endocrinol. Metab.* **2009**, *297*, E271–E288. [CrossRef]
38. Okumura, T.; Saito, K.; Sakuma, H.; Nade, T.; Nakayama, S.; Fujita, K.; Kawamura, T. Intramuscular fat deposition in principal muscles from twenty-four to thirty months of age using identical twins of Japanese Black steers. *J. Anim. Sci.* **2007**, *85*, 1902–1907. [CrossRef]
39. Payne, V.A.; Grimsey, N.; Tuthill, A.; Virtue, S.; Gray, S.L.; Nora, E.D.; Semple, R.K.; O'Rahilly, S.; Rochford, J.J. The Human Lipodystrophy Gene BSCL2/Seipin May Be Essential for Normal Adipocyte Differentiation. *Diabetes* **2008**, *57*, 2055–2060. [CrossRef]
40. Igal, R.A.; Wang, S.; Gonzalez-Baró, M.; Coleman, R.A. Mitochondrial Glycerol Phosphate Acyltransferase Directs the Incorporation of Exogenous Fatty Acids into Triacylglycerol. *J. Biol. Chem.* **2001**, *276*, 42205–42212. [CrossRef]
41. Segers, J.R.; Loor, J.J.; Moisá, S.J.; Gonzalez, D.; Shike, D.W. Effects of protein and fat concentration in coproduct-based growing calf diets on adipogenic and lipogenic gene expression, blood metabolites, and carcass composition. *J. Anim. Sci.* **2017**, *95*, 2767–2781. [CrossRef] [PubMed]
42. Smith, S.J.; Cases, S.; Jensen, D.R.; Chen, H.C.; Sande, E.; Tow, B.; Sanan, D.A.; Raber, J.; Eckel, R.H.; Farese, R.V. Obesity resistance and multiple mechanisms of triglyceride synthesis in mice lacking Dgat. *Nat. Genet.* **2000**, *25*, 87–90. [CrossRef]
43. Yu, Y.-H.; Zhang, Y.; Oelkers, P.; Sturley, S.L.; Rader, D.J.; Ginsberg, H.N. Posttranscriptional Control of the Expression and Function of Diacylglycerol Acyltransferase-1 in Mouse Adipocytes. *J. Biol. Chem.* **2002**, *277*, 50876–50884. [CrossRef]
44. Oelkers, P.; Tinkelenberg, A.; Erdeniz, N.; Cromley, D.; Billheimer, J.T.; Sturley, S.L. A Lecithin Cholesterol Acyltransferase-like Gene Mediates Diacylglycerol Esterification in Yeast. *J. Biol. Chem.* **2000**, *275*, 15609–15612. [CrossRef] [PubMed]
45. Shen, W.-J.; Liang, Y.; Hong, R.; Patel, S.; Natu, V.; Sridhar, K.; Jenkins, A.; Bernlohr, D.A.; Kraemer, F. Characterization of the Functional Interaction of Adipocyte Lipid-binding Protein with Hormone-sensitive Lipase. *J. Biol. Chem.* **2001**, *276*, 49443–49448. [CrossRef] [PubMed]
46. Shen, W.-J.; Sridhar, K.; Bernlohr, D.A.; Kraemer, F. Interaction of rat hormone-sensitive lipase with adipocyte lipid-binding protein. *Proc. Natl. Acad. Sci. USA* **1999**, *96*, 5528–5532. [CrossRef]
47. Ntambi, J.M. Regulation of stearoyl-CoA desaturase by polyunsaturated fatty acids and cholesterol. *J. Lipid Res.* **1999**, *40*, 1549–1558. [CrossRef]
48. Enoch, H.G.; Catalá, A.; Strittmatter, P. Mechanism of rat liver microsomal stearyl-CoA desaturase. Studies of the substrate specificity, enzyme-substrate interactions, and the function of lipid. *J. Biol. Chem.* **1976**, *251*, 5095–5103. [CrossRef]
49. Man, W.C.; Miyazaki, M.; Chu, K.; Ntambi, J. Colocalization of SCD1 and DGAT2: Implying preference for endogenous monounsaturated fatty acids in triglyceride synthesis. *J. Lipid Res.* **2006**, *47*, 1928–1939. [CrossRef]
50. Archibeque, S.L.; Lunt, D.K.; Gilbert, C.D.; Tume, R.K.; Smith, S.B. Fatty acid indices of stearoyl-CoA desaturase do not reflect actual stearoyl-CoA desaturase enzyme activities in adipose tissues of beef steers finished with corn-, flaxseed-, or sorghum-based diets1. *J. Anim. Sci.* **2005**, *83*, 1153–1166. [CrossRef]
51. Graugnard, D.E.; Berger, L.L.; Faulkner, D.B.; Loor, J.J. High-starch diets induce precocious adipogenic gene network up-regulation in longissimus lumborum of early-weaned Angus cattle. *Br. J. Nutr.* **2010**, *103*, 953–963. [CrossRef] [PubMed]
52. Moisá, S.J.; Shike, D.W.; Faulkner, D.B.; Meteer, W.T.; Keisler, D.; Loor, J.J. Central Role of the PPARγ Gene Network in Coordinating Beef Cattle Intramuscular Adipogenesis in Response to Weaning Age and Nutrition. *Gene Regul. Syst. Biol.* **2014**, *8*, 17–32. [CrossRef]
53. Zhang, D.; Christianson, J.; Liu, Z.-X.; Tian, L.; Choi, C.S.; Neschen, S.; Dong, J.; Wood, P.A.; Shulman, G.I. Resistance to High-Fat Diet-Induced Obesity and Insulin Resistance in Mice with Very Long-Chain Acyl-CoA Dehydrogenase Deficiency. *Cell Metab.* **2010**, *11*, 402–411. [CrossRef]
54. Kociucka, B.; Jackowiak, H.; Kamyczek, M.; Szydlowski, M.; Szczerbal, I. The relationship between adipocyte size and the transcript levels of SNAP23, BSCL2 and COPA genes in pigs. *Meat Sci.* **2016**, *121*, 12–18. [CrossRef] [PubMed]
55. Vierck, K.R.; O'Quinn, T.G.; Noel, J.A.; Houser, T.A.; Boyle, E.A.E.; Gonzalez, J.M. Effects of Marbling Texture on Muscle Fiber and Collagen Characteristics. *Meat Muscle Biol.* **2018**, *2*, 75–82. [CrossRef]

Article

Effects of Bypass Fat on Buffalo Carcass Characteristics, Meat Nutrient Contents and Profitability

Amirul Faiz Mohd Azmi [1], Fhaisol Mat Amin [1], Hafandi Ahmad [1], Norhariani Mohd Nor [1], Goh Yong Meng [1], Mohd Zamri Saad [2], Md Zuki Abu Bakar [1], Punimin Abdullah [3], Agung Irawan [4,5,6], Anuraga Jayanegara [6] and Hasliza Abu Hassim [1,6,7,*]

[1] Department of Veterinary Preclinical Sciences, Faculty of Veterinary Medicine, Universiti Putra Malaysia (UPM), Serdang 43400, Malaysia; amirulfaizazmi@gmail.com (A.F.M.A.); fhaisol@gmail.com (F.M.A.); hafandi@upm.edu.my (H.A.); norhariani@upm.edu.my (N.M.N.); ymgoh@upm.edu.my (G.Y.M.); zuki@upm.edu.my (M.Z.A.B.)

[2] Department of Veterinary Laboratory Diagnosis, Faculty of Veterinary Medicine, Universiti Putra Malaysia (UPM), Serdang 43400, Malaysia; mzamri@upm.edu.my

[3] Faculty of Science and Natural Resources, Universiti Malaysia Sabah, Jalan UMS, Kota Kinabalu 88400, Malaysia; puniminabdullah@ums.edu.my

[4] Vocational School, Universitas Sebelas Maret, Surakarta 57126, Indonesia; a.irawan@staff.uns.ac.id

[5] Department of Animal and Rangeland Sciences, Oregon State University, Corvallis, OR 97333, USA

[6] Animal Feed and Nutrition Modelling (AFENUE) Research Group, Department of Nutrition and Feed Technology, Faculty of Animal Science, IPB University, Bogor 16680, Indonesia; anuraga.jayanegara@gmail.com

[7] Laboratory of Sustainable Animal Production and Biodiversity, Institute of Tropical Agriculture and Food Security, Universiti Putra Malaysia (UPM), Serdang 43400, Malaysia

* Correspondence: haslizaabu@upm.edu.my; Tel.: +603-9769-3417

Citation: Mohd Azmi, A.F.; Mat Amin, F.; Ahmad, H.; Mohd Nor, N.; Meng, G.Y.; Zamri Saad, M.; Abu Bakar, M.Z.; Abdullah, P.; Irawan, A.; Jayanegara, A.; et al. Effects of Bypass Fat on Buffalo Carcass Characteristics, Meat Nutrient Contents and Profitability. *Animals* **2021**, *11*, 3042. https://doi.org/10.3390/ani11113042

Academic Editors: Guillermo Ripoll and Begoña Panea

Received: 18 August 2021
Accepted: 13 October 2021
Published: 24 October 2021

Publisher's Note: MDPI stays neutral with regard to jurisdictional claims in published maps and institutional affiliations.

Simple Summary: Bypass fat supplementation has been shown to influence the carcass and meat qualities of large ruminants, especially cattle. However, limited information is available on the influence of bypass fat on carcass characteristics and the meat proximate and fatty acid compositions of buffaloes. The objective of this study was to evaluate both the effects of bypass fat on carcass traits and meat proximate and fatty acid compositions, and the profitability of Murrah cross and swamp buffaloes. Bypass fat supplementation improved the proximate and fatty acid compositions of buffalo meat without affecting the carcass characteristics. Although the mixture of the concentrate and bypass fat supplement (26:4) used in this study was found to increase the feed cost, the eventual overall returns resulted in a greater profit.

Abstract: The deposition and distribution of buffalo body fats play a vital role in the quality of the buffalo carcass and are of great commercial value, since the carcass quality influences the profitability and consumer acceptability of ruminant meat. The current study examined the effect a mixture of 4% bypass fat and 26% concentrate supplementations in buffalo basal diet had on both the carcass characteristics and the proximate and fatty acid composition in *longissimus thoracis et lumborum* (LTL), *supraspinatus* (SS) and *semitendinosus* (ST) muscles of Murrah cross and swamp buffaloes. In addition, profit and loss analyses were performed to determine the profitability. This study employed a completely randomized 2×2 factorial arrangement with two diets, two breeds and four replicates per treatment. A total of sixteen buffaloes (eight buffaloes per breed, bodyweight 98.64 ± 1.93 kg) were randomly assigned into two dietary groups. The first group was given Diet A, which consisted of 70% *Brachiaria decumbens* + 30% concentrate, whereas the second group was given Diet B, which consisted of 70% *Brachiaria decumbens* + 26% concentrate + 4% bypass fat. The buffaloes were fed for 730 days before slaughter. The results showed that supplemented bypass fat significantly ($p < 0.05$) increased the pre-slaughter weight, hot and cold carcass weights, meat:fat ratio, pH at 24 h, moisture and crude protein of LTL, ST and SS, the ether extract of LTL and ST and the meat fatty acid of C16:0, C16:1, C18:1, PUFA n-6/n-3 and total MUFA. The carcass yield and carcass fat percentages, the ash content in ST, the EE in the SS muscle and the meat fatty acid of C18:3, total PUFA n-3, UFA/SFA and PUFA/SFA were significantly ($p < 0.05$) decreased. Furthermore, Murrah cross showed a significantly

($p < 0.05$) higher pre-slaughter weight, hot and cold carcass weights, carcass bone percentage and total fatty acid, but a lower ($p < 0.05$) meat:bone ratio, ash of LTL and CP of LTL and ST when compared to swamp buffaloes. No significant changes were found in the proximate composition of different types of muscle, but the ST muscle revealed significantly high C14:0, C16:0 and C18:1, and the SS muscle had high C18:2 and total fatty acid ($p < 0.05$). Supplementing using bypass fat increased the cost of buffalo feeding but resulted in a higher revenue and net profit. In conclusion, the concentrate and bypass fat supplementations in the buffalo diet could alter the nutrient compositions of buffalo meat without a detrimental effect on carcass characteristics, leading to a higher profit.

Keywords: buffalo; carcass; costs; meat; supplementation

1. Introduction

Traditionally, buffalo meat was obtained from retired draft animals of more than 10 years of age [1]. Therefore, public perception remains that buffalo meats are tough and of low quality. However, when slaughtered at body weights equivalent to those of cattle, the carcass compositions and meat quality are comparable [2,3]. In fact, buffalo meat has a better tenderness compared to beef due to its high calpain activity in early post-mortem [4]. Therefore, buffalo meat has gained interest in recent decades due to its favored nutrient characteristics, such as a lower cholesterol and high protein contents [5].

Several factors contribute to the carcass quality of ruminants, including the feeding regimen, breed of animals and farm management [6]. Feed supplementation using commercialize concentrate and/or rumen bypass fat products from the local industry provide an inexpensive alternative to fulfil the nutritional demands of buffaloes [7–9]. Furthermore, enriching buffalo meat with good fatty acid through dietary supplementation allows buffalo meat to be more acceptable, particularly by health-conscious consumers. In fact, the fatty acid compositions of buffalo fat has been reported to affect the nutritional value and the various aspects of buffalo meat quality traits, including flavor and shelf-life [10], and the compositions of meat fatty acids [11]. An improvement in the ratio of polyunsaturated:saturated fatty acids (PUFA:SFA) in favor of the former is essential in maintaining the consumers' health [12]. Fat supplements, such as palm, coconut and corn oils, given together with pasture, was shown to increase the percentage of fat in the carcass, thickness of meat-covering fat and marbling of the meat [7].

Crossbreeding of pure Murrah and swamp breeds is a common practice developed by buffalo farmers in Malaysia and other Asian countries, as the crossbreeding buffaloes were reported to inherit superior traits (e.g., growth, meat and milk production) possessed by their parents [9,13,14]. Indeed, the study by Mohd Azmi et al. [13] showed that the crossbreeds had a greater growth performance with a significantly heavier body weight than swamp buffaloes from birth until 24 months old. Other than nutritional factors, heterosis might have an impact on the growth and performance of buffaloes, which reflects the total production and profit of a farm [13,15].

Despite these superior traits and the potential use of crossbreeding Murrah and swamp buffaloes for meat purpose, it has received little attention. Indeed, the effects of bypass fat supplementation on carcass and meat quality traits in Murrah cross and swamp buffaloes are also scanty. Therefore, given the increasing importance of buffalo farming in developing countries, this study aimed to evaluate the impact of the dietary bypass fat supplements on the carcass quality and meat nutrient composition of Murrah crossbred and swamp buffaloes, as well as the profit and loss analysis.

2. Materials and Methods

2.1. Statement of Animal Rights

This study was performed and managed according to the Animal Utilization Protocol (AUP), Institutional Animal Care and Use Committee (IACUC) and Universiti Putra

Malaysia (Approval No. UPM/IACUC/AUP-017/2018, on 8 January 2018). Samplings from the experimental animals were strictly conducted under veterinary supervision.

2.2. Study Area and Experimental Animals

This study was conducted at the Buffalo Breeding and Research Centre, Sabah, Malaysia (Coordinate 5°30′ N, 117°7′ E). A total of sixteen buffaloes, consisting of swamp ($n = 8$) and Murrah cross ($n = 8$) buffaloes of approximately 3 months old and with an average body weight of 98.64 ± 1.93 kg, were each randomly divided into two groups of 4 animals per group. The buffaloes were individually confined in collective stalls with an area of 30 m^2 per animal, separated according to the dietary treatments. The feedlot facility had a compacted dirt floor, whereas the area close to the feeder was covered with concrete. The feeders were vinyl type and were placed transversely on the upper part of the pens, whereas the drinkers were located at the divider between two pens. Before the onset of the experiment, a proper physical examination was conducted on each buffalo. Then, all buffaloes were weighed and treated against ecto and endoparasites.

2.3. Experimental Design

This study was a completely randomized block design according to 2 × 2 factorial arrangement with two diets, two breeds and four replicates per treatment. Daily feed supply was calculated at 3% body weight (based on dry matter (DM) of total mixed ration), given in two equal portions at 07:00 h and 17:00 h [16]. Diets were adjusted to ensure that refusals were around 5% of the total supplied. The buffaloes were allowed a 14-day adaptation period to the respective diet before the start of the experiment. Two total mixed rations (TMR) were prepared to contain three ingredients; namely, *Brachiaria decumbens* grass (G), commercial concentrate (C) (composition: corn grain (25.0%), palm kernel cake (32.0%), rice bran (18.0%), soya bean meal (19.7%), calcium carbonate (1.0%), molasses (2.8%), vitamin–mineral premix (0.3%), sodium chloride (0.6%), dicalcium phosphate (0.6%)) and bypass fat (B) (OPTI-FAT F8016RXP—rumen bypass supplement sources from calcium salt fractionated palm fat without trans-fat). All diets were formulated similar to the previous study by Mohd Azmi et al. [13] in order to meet the nutritional requirements of growing buffalo [17–19]. The fresh *Brachiaria decumbens* grass (basal diet) was collected from the pastureland area of the farm at Telupid Buffalo Breeding and Research Centre, Sabah, Malaysia. The concentrate and bypass fat supplements were obtained from the authorized supplier, Lipidchem Sdn Bhd, Masai, Johor, Malaysia. Buffaloes of Group A were fed with Diet A, which consisted of 70% *Brachiaria decumbens* grass and 30% concentrate (Control) (DM: 90.31%, ash: 5.69% DM, crude fiber (CF): 23.73% DM, ether extract (EE): 2.92% DM, crude protein (CP): 8.08% DM, neutral detergent fiber (NDF): 57.96% DM, acid detergent fiber (ADF): 28.70% DM, acid detergent lignin (ADL): 3.32% DM, non-fiber carbohydrate (NFC): 24.84% DM, gross energy (GE): 12.10 MJ/kg, hemicellulose: 29.25% DM and cellulose: 25.38% DM), which is the common practice of buffalo feeding by farmers [13]. Group B was fed with Diet B, which consisted of 70% *Brachiaria decumbens* grass supplemented with 26% concentrate and 4% bypass fat (Treatment) (DM: 91.60%, ash: 5.93% DM, CF: 21.65% DM, EE: 6.66% DM, CP: 6.56% DM, NDF: 49.63% DM, ADF: 26.65% DM, ADL: 2.96% DM, NFC: 20.81% DM, GE: 14.59 MJ/kg, hemicellulose: 22.98% DM and cellulose: 23.69% DM) [13]. The fatty acid (FA) compositions of the feedstuffs and diets are presented in Tables 1 and 2, respectively. The trial lasted 2 years before the buffaloes proceeded to slaughtering process.

2.4. Slaughtering Procedure and Sample Collection

Prior to slaughter, the buffaloes were placed at lairage for 12 to 16 h, but had access to drinking water. The buffaloes were weighed and slaughtered at a commercial government abattoir (Sabah Meat Technology Centre) according to the standard procedure (Muslim law; MS15000:2009) of the Department of Standard, Malaysia. After slaughtering, the weight of the warm carcass was immediately recorded before the head, fore and hind limbs were removed by cutting the atlantooccipital, carpal and tarsal joints, respectively [20]. Then, all carcasses were conditioned at 4 °C for 24 h before they were split longitudinally into two halves from the neck to pelvis along the vertebral column using a carcass splitting saw (Jarvis, CT, USA). Each primal cut was then physically dissected into lean, bone, fat and trimmable tissues, which consisted of major blood vessels, tendons, thick connective tissue sheets and glands [21].

Approximately 40 g of sample of *longissimus thoracis et lumborum* (LTL), *supraspinatus* (SS) and *semitendinosus* (ST) muscles were taken for the analyses. The samples were kept on ice during sampling, immediately vacuum-packed and stored at −20 °C until further analyses [22].

2.5. Carcass Traits

The parameters of carcass characteristics that were measured included the pH, the weight of the warm carcass, the weight of the cold carcass, the lean meat, the fat, the bone, the offal, the carcass yield and the shrinkage percentages. The pH of each carcass was measured using a portable pH meter (HANNA Hi8314 with an INGOLD type electrode Metrohm, Herisau, Switzerland) involved the left *longissimus* muscle, caudal to the 12th rib at 30 min, 24 h and 7 days post-mortem. Carcass yield was estimated using carcass weight over live weight and was expressed as a percentage [23]. The warm carcass weight was measured when dressed carcasses were quickly weighed within 1 h post-mortem, and the cold carcass weight was the weight of chilled carcass, which was kept at 4 °C for 24 h. Shrinkage percentage was calculated as the difference between warm and cold carcass weights and was expressed as a percentage [20]. The physical carcass composition was referred to as the ratio of major body tissues, muscle, fat and bone. The weights of these tissues were determined separately upon carcass dissection. The relationship between the weight of each carcass components and the carcass weight was computed and expressed as a percentage of the cold carcass weight.

Table 1. Fatty acid compositions of feedstuffs.

	Feedstuffs		
Fatty Acids Composition (% Total FA) [†]	**Grass**	**Concentrate**	**Bypass Fat**
C12:0 Lauric acid	0.75	6.70	0.51
C14:0 Myristic acid	0.3	1.96	0.16
C16:0 Palmitic acid	2.03	51.47	51.23
C18:0 Stearic acid	8.43	10.12	0.43
C18:1 n-9, Oleic acid	45.87	19.90	47.44
C18:2 n-6, Linoleic acid	40.44	8.10	0.10
C18:3 n-3, Alpha linoleic acid	1.93	1.37	0.10
C20:0 Eicosanoic acid	0.24	0.38	0.05
Total fatty acid (μg/mg)	192.35	556.40	2861.57
Σ SFA [1]	11.75	70.63	52.39
Σ UFA [2]	88.25	29.37	47.65
Σ PUFA n-3 [3]	1.93	1.37	0.10
Σ PUFA n-6 [4]	40.44	8.10	0.10
Σ PUFA [5]	42.38	9.47	0.20
PUFA n-6/n-3 ratio	20.91	5.90	1.05
Σ MUFA [6]	45.87	19.90	47.44

Note: [1] SFA (saturated fatty acids) = C12:0 + C14:0 + C16:0 + C18:0 + C + 20:0; [2] UFA (unsaturated fatty acids) = C18:1 + C18:2 + C18:3; [3] PUFA n-3 = C18:3 n-3; [4] PUFA n-6 = C18:2 n-6; [5] Total PUFA = PUFA n-3 + PUFA n-6; [6] MUFA (monounsaturated fatty acids) = C18:1. [†] The result of individual fatty acid was expressed as percentage from total fatty acid in each sample.

Table 2. Fatty acid compositions of total mixed ration.

	Diets	
Ingredient (%)	**Diet A**	**Diet B**
Brachiaria decumbens (G)	70	70
Concentrate (C)	30	26
Bypass fat (B)	-	4
Total	100	100
Fatty acids composition (% total FA) [†]		
C12:0 Lauric acid	2.60	2.13
C14:0 Myristic acid	1.01	0.76
C16:0 Palmitic acid	11.02	59.60
C18:0 Stearic caid	55.90	18.83
C18:1 n-9, Oleic acid	13.26	9.58
C18:2 n-6, Linoleic acid	13.67	8.65
C18:3 n-3, Alpha linoleic acid	1.04	0.46
C20:0 Eicosanoic acid	0.22	0.04
Total fatty acid (μg/mg)	543.10	674.09
Σ SFA [1]	70.75	81.35
Σ UFA [2]	29.25	18.39
Σ PUFA n-3 [3]	1.04	0.46
Σ PUFA n-6 [4]	13.67	8.65
Σ PUFA [5]	14.71	9.11
PUFA n-6/n-3 ratio	13.19	18.87
Σ MUFA [6]	13.26	9.58

Note: [1] SFA (saturated fatty acids) = C12:0 + C14:0 + C16:0 + C18:0 + C + 20:0; [2] UFA (unsaturated fatty acids) = C18:1 + C18:2 + C18:3; [3] PUFA n-3 = C18:3 n-3; [4] PUFA n-6 = C18:2 n-6; [5] Total PUFA = PUFA n-3 + PUFA n-6; [6] MUFA (monounsaturated fatty acids) = C18:1. [†] The result of individual fatty acid was expressed as percentage from total fatty acid in each sample.

2.6. Proximate Composition

The proximate compositions of feed samples and different type of muscles included the moisture, the ash, the ether extract, the crude protein and the carbohydrates. They were determined according to the method of the Association of Official Analytical Chemist [24], whereas the gross energy was determined using bomb calorimeter and expressed as MJ/kg sample.

2.7. Fatty Acid (FA) Analysis

The total fatty acids in the feed samples and the different type of muscles were extracted in chloroform:methanol (2:1, *v/v*) mixture according to the method of Folch et al. [25], as modified by Rajion et al. [26]. The fatty acids were transmethylated into their fatty acid methyl esters (FAME) using 0.66 N KOH in methanol and 14% methanolic boron trifluoride (BF3) following the method of AOAC [24]. The FAME was separated in a gas chromatograph (Agilent 7890A) equipped with a flame ionization detector (FID) and a split-less injector. The column used was fused silica capillary (Supelco SP-2560, 100 m, 0.25 mm ID, 0.20 mm film thickness). High purity helium was used as the carrier gas at 40 mL/min. Compressed air and high purity hydrogen were used for the FID in the chromatograph. To facilitate the optimal separation, the oven temperature was set at 100 °C for 2 min and warmed to 170 °C at 10 °C/min, held for 2 min, warmed to 230 °C at intervals of 5 °C/min and then held for 20 min. Identification of sample fatty acids was carried out by comparing the relative retention time of FAME peaks from samples with those of standards. Identification of fatty acids was carried out by comparing with the relative FAME peak retention times of fatty acid methyl standards using henecosanoic acid (Sigma, St. Louis, MO, USA) as the internal standard. The fatty acid was expressed as percentage from total fatty acid in each sample.

2.8. Profit and Loss Analysis of Buffalo Production

The economic aspect of buffalo meat production was calculated on the hot carcass weight at an exchange rate of USD 1.00 = MYR 4.07. To calculate the total operational cost, it was assumed that the cost of feeding represented 87% of the total cost of the activity [27], and the cost of feeding comprised the costs of basal diet and supplementations, which were concentrate and bypass fat [28]. At the time of the study, the values of the feed stuffs (MYR/kg) were MYR 0.23 (USD 0.06) for *Brachiaria decumbens*, MYR 1.11 (USD 0.27) for concentrate and MYR 3.82 (USD 0.94) for bypass fat. Previous study by Mohd Azmi et al. [13] reported that the average feed intake of Murrah cross buffaloes fed with Diet A was 6.57 kg/day and, for Diet B, it was 7.41 kg/day, whereas, for swamp buffaloes fed with Diet A, it was 6.15 kg/day and, for Diet B, it was 6.37 kg/day. The information of feed intake was used to calculate the total cost of average daily dry matter intake (MYR/day/animal). The 2-year management cost of MYR 158.50 (USD 38.94) per animal was added, which included the MYR 0.50 (USD 0.12) cost of deworming, MYR 2.00 (USD 0.49) for ID tag, MYR 156.00 (USD 38.33) for fertilizer, MYR 83.33 (USD 20.48) for transportation and MYR 152.00 (USD 37.35) for labor cost. Additional fixed cost imposed by the abattoir for slaughter services was priced at MYR 5.00/animal (USD 1.23/animal). The average price of fresh meat buffalo in Malaysia was MYR 31.03/kg (USD 7.62/kg), whereas the average price of bones was MYR 26.00/kg (USD 6.39/kg). The average price of head per animal was MYR 70.00 (USD 17.20/animal) and the average price of skin per animal was MYR 65.00 (USD 15.97/animal). The average price of tail per animal was MYR 10.00 (USD 2.46/animal) and butcher imposed overall offal (heart, liver, lung, limp, renal, intestine and rumen) price with value of MYR 35.00/kg (USD 8.60/kg). The gross profit of selling meat without the cost of aggregators (person who collects the animals from buffalo farm and transports them to livestock market for sale to sub traders) and traders or sub traders (registered person who supplies buffaloes to abattoirs, which incurs fees, such as transport cost, abattoir fees, miscellaneous expenditure and payments to sub-traders) was then calculated [29]. The income through meat price, the cost of feed per day for the 2-year rearing and the estimation of net income after selling fresh meat, bone and offal was calculated as below [28,30,31].

Meat sales (MYR/kg) = Meat weight (kg) × current price of fresh meat (MYR)
Bone sales (MYR/kg) = Bone weight (kg) × current price of bone (MYR)
Offal sales (MYR/kg) = Offal weight (kg) × current price of offal (MYR)
Total cost of average daily DMI (MYR/day/animal) = Current price of feed × dry matter intake (based on 3% body weight)
Total feed cost in 2 years (MYR/animal) = Total cost of average daily DMI × 720 days

2.9. Statistical Analysis

The data for the carcass characteristics and the profit and loss analyses were analyzed using analysis of variance (ANOVA) following a completely randomized block design in a 2 × 2 factorial arrangement (diet × breed) using MIXED command in SPSS. For these variables, dietary treatments and breeds were set as fixed effects and animal units within the breeds were declared as random effect. In addition, to test the differences in fatty acid profiles among types of muscles, one way ANOVA was performed. The data were presented as means and standard errors of the means. Tukey's test was employed to test the differences between the means of the treatments and the differences were declared significant at $p < 0.05$ [32]. In addition, data of proximate composition of buffaloes' muscle and fatty acid composition of the meat under different dietary treatments and breeds were compared using independent *t*-test.

3. Results

3.1. Carcass Traits

Table 3 shows the slaughter data of buffaloes that were fed with different diets. Significant ($p < 0.05$) interactions between the diet and breed were observed in the live weight and weights of the hot and cold carcass of the buffaloes. Buffaloes fed with Diet B were significantly higher in the live weight, the weights of the hot and cold carcass, the meat:fat ratio and the pH at 24 h, and were lower in the carcass fat percentage when compared to those fed with Diet A ($p < 0.05$). However, the carcass meat percentage was higher in the swamp buffalo fed with Diet B than that of Diet A ($p < 0.05$), whereas no such change was recorded in the Murrah cross. Furthermore, the breed significantly ($p < 0.05$) influenced the carcass quality, including the hot and cold carcass weights, the carcass bone percentage and the meat:bone ratio. The Murrah cross buffaloes were significantly ($p < 0.05$) higher in the live weight, the hot and cold carcass weights and the carcass bone percentage, but were lower in the meat:bone ratio than the swamp buffaloes. Moreover, there was a significant ($p < 0.05$) interaction between the diet and breed on the live weight (0.035), the hot carcass weight (0.047) and the cold carcass weight traits (0.041).

Table 3. Effect of supplementation and breed, and their impact on buffaloes' carcass traits.

Breed	Murrah Cross		Swamp			*p*-Value		Interaction
Diets	Diet A [1]	Diet B	Diet A	Diet B	SEM [2]	Diet	Breed	Diet × Breed
Pre-slaughter weight (kg)	330.75 [aY]	451.70 [bY]	281.75 [aZ]	353.78 [bZ]	16.15	*	*	0.035
Hot carcass (kg)	157.08 [aY]	204.10 [bY]	131.33 [aZ]	150.28 [bZ]	7.23	*	*	0.047
Cold carcass (kg)	144.53 [aY]	192.60 [bY]	120.28 [aZ]	139.40 [bZ]	7.19	*	*	0.041
Carcass yield (%)	47.50 [a]	45.17 [b]	46.66 [a]	42.41 [b]	0.70	*	ns	ns
Shrinkage (%)	7.99	5.62	8.40	7.30	0.46	ns	ns	ns
Fat (%)	12.35 [a]	6.49 [b]	8.23 [a]	5.54 [b]	0.78	*	ns	ns
Bone (%)	21.85 [Y]	23.85 [Y]	19.02 [Z]	18.12 [Z]	1.01	ns	*	ns
Meat (%)	59.01 [a]	59.20 [b]	56.91 [a]	60.75 [b]	0.81	*	ns	ns
Meat:bone ratio	2.74 [Y]	2.45 [Y]	3.22 [Z]	3.43 [Z]	0.18	ns	*	ns
Meat:fat ratio	5.05 [a]	9.43 [b]	6.98 [a]	11.33 [b]	0.75	*	ns	ns
Offal (% BW)	6.87	6.92	6.79	6.81	0.16	ns	ns	ns
pH (0 h)	6.77	6.69	6.67	6.81	0.06	ns	ns	ns
pH (24 h)	5.51 [a]	5.87 [b]	5.43 [a]	5.56 [b]	0.06	*	ns	ns
pH (48 h)	5.48	5.38	5.43	5.35	0.03	ns	ns	ns

Note: [1] Diet A = 70% *Brachiaria decumbence* + 30% concentrate, Diet B = 70% *Brachiaria decumbence* + 26% concentrate + 4% bypass fat; [2] SEM = standard error of mean; [a,b] values with different superscripts indicate significant difference between dietary treatments at $p < 0.05$; [Y,Z] values with different superscripts indicate significant difference between breeds at $p < 0.05$; % = percentage based on the carcass weight; % BW = percentage based on pre-slaughter body weight; * = $p < 0.05$, ns = not significant.

3.2. Proximate Composition of Different Types of Muscle

The effects of dietary treatments on the proximate composition of the buffalo LTL, ST and SS muscles are shown in Table 4. The inclusion of bypass fat in Diet B significantly ($p < 0.05$) increased the moisture, EE and CP contents of the LTL and SS, but lowered the ash content of ST. In fact, the increment of moisture content of ST was recorded only in swamp buffaloes fed with Diet B ($p < 0.05$). Furthermore, the breed significantly ($p < 0.05$) affected the ash and the CP of different muscles, with Murrah cross having a significantly ($p < 0.05$) lower ash content in the LTL and the CP of the LTL and the ST than the swamp buffaloes.

3.3. Fatty Acid Composition of Meat

The effects of different dietary treatments on the fatty acid composition of buffaloes' muscle are shown in Table 5. The diet was found to significantly ($p < 0.05$) influence the fatty acid composition of meat. This was because buffaloes fed with Diet B had a higher ($p < 0.05$) C16:0, C16:1, C18:1, n6/n3 ratio and MUFA composition in the meat. On the other hand, the buffaloes fed with Diet A showed an increase in the meat fatty acid of C18:3,

PUFA n-3, UFA/SFA ratio and PUFA/SFA ratio. This study also revealed that the buffaloes fed with Diet A had an increased ALA composition in the meat at 1.88-fold compared to the buffaloes fed with Diet B.

Table 4. Proximate composition of different type of muscles in Murrah cross and swamp buffaloes under different dietary treatments.

Nutrient Composition of Different Type of Muscles	Breeds				Dietary Treatments			
	Murrah Cross	Swamp	SEM	*p*-Value	Diet A [1]	Diet B	SEM [2]	*p*-Value
Longissimus thoracis et lumborum (LTL)								
Moisture (%)	73.75	73.61	0.05	ns	71.37 [b]	75.99 [a]	1.63	*
Ash (%)	1.39 [Z]	1.78 [Y]	0.14	*	1.64	1.53	0.04	ns
Ether extract (%)	2.82	2.90	0.03	ns	2.13 [b]	3.59 [a]	0.52	*
Crude protein (%)	24.31 [Z]	25.58 [Y]	0.45	*	24.39 [b]	25.49 [a]	0.39	*
Carbohydrates (%)	nd	nd	-	ns	nd[3]	Nd	-	ns
Gross energy (MJ/kg)	21.18	20.66	0.18	ns	20.35	21.48	0.40	ns
Semitendinosus (ST)								
Moisture (%)	70.27	73.08	1.00	ns	70.64 [b]	72.71 [a]	0.73	*
Ash (%)	1.29	1.28	0.01	ns	1.64 [a]	0.93 [b]	0.25	*
Ether extract (%)	3.04	2.00	0.37	ns	1.52 [b]	3.52 [a]	0.71	*
Crude protein (%)	22.62 [Z]	26.57 [Y]	1.40	*	23.42 [b]	25.77 [a]	0.83	*
Carbohydrates (%)	4.71	1.24	1.23	ns	2.98	nd [3]	-	ns
Gross energy (MJ/kg)	21.27	20.96	0.11	ns	20.80	21.44	0.23	ns
Supraspinatus (SS)								
Moisture (%)	73.96	74.71	0.27	ns	72.53 [b]	76.13 [a]	1.27	*
Ash (%)	1.81	1.60	0.08	ns	1.67	1.74	0.02	ns
Ether extract (%)	2.96	2.15	0.29	ns	2.38 [b]	2.73 [a]	0.12	*
Crude protein (%)	26.38	24.38	0.71	ns	23.76 [b]	27.00 [a]	1.14	*
Carbohydrates (%)	0.72	1.83	0.39	ns	1.28	nd [3]	-	ns
Gross energy (MJ/kg)	21.19	21.02	0.06	ns	21.24	20.97	0.10	ns

Note: [1] Diet A (control): 70% *Brachiaria decumbence* + 30% concentrate, Diet B: 70% *Brachiaria decumbence* + 26% concentrate + 4% bypass fat; [2] SEM = standard error of mean; [3] nd = not determined; * = $p < 0.05$, [a,b] values with different superscripts indicate significant difference between dietary treatments at $p < 0.05$; [Y,Z] values with different superscripts indicate significant difference between breeds at $p < 0.05$; n.s = not significant.

The fatty acid composition of the meat from different breeds of buffaloes are shown in Table 6. The Murrah cross buffaloes were significantly ($p < 0.05$) higher in linoleic acid (C18:2) and the total fatty acid composition than the swamp buffaloes (25.18% vs. 13.19% and 811.75 µg/mg vs. 610.48 µg/mg, respectively). The most abundant fatty acid in the buffaloes' meats was alpha linoleic acid (C18:3) (45.39%), followed by linoleic acid (C18:2) (16.00%), oleic acid (C18:1) (8.54%), palmitic acid (C16:0) (8.30%) and stearic acid (C18:0) (2.68%), which accounted for approximately 75% to 80% of the total identified fatty acids in both buffalo breeds.

The fatty acid compositions in different types of buffaloes' muscles (LTL, ST and SS) are shown in Table 7. Total fatty acid and C18:2 compositions were significantly ($p < 0.05$) higher in SS, followed by LTL and ST. Meanwhile, the compositions of C14:0 and C16:0 were higher ($p < 0.05$) in ST than SS and LTL. In addition, the composition of C18:1 was higher ($p < 0.05$) in ST and LTL, and lower in the SS muscle.

Table 5. Fatty acid composition of buffalo meat under different dietary treatments.

Fatty Acid Composition (% Total FA) [†]	Dietary Treatments		SEM [9]	*p*-Value
	Diet A [1]	Diet B		
C14:0 Myristic acid	0.56	2.33	0.63	ns
C15:0 Pentadecanoic acid	1.55	0.77	0.28	ns
C15:1 Pentadecanoic acid (cis-10)	1.29	2.36	0.38	ns
C16:0 Palmitic acid	3.36 [B]	15.04 [A]	4.13	*
C16:1 Palmitoleic acid	1.09 [B]	1.87 [A]	0.28	*
C17:0 Heptadecanoic acid	2.10	3.09	0.35	ns
C17:1 Heptadecenoic acid	2.47	2.07	0.14	ns
C18:0 Stearic acid	2.04	2.32	0.10	ns
C18:1 n9c, Oleic acid	4.01 [B]	13.18 [A]	3.24	*
C18:2 n6c, Linoleic acid	20.33	18.36	0.70	ns
C18:3 n3c, Alpha linoleic acid	53.85 [A]	28.63 [B]	8.92	*
C20:0 Eicosanoic acid	0.45	0.32	0.05	ns
C20:1 n9c, Eicosenoic acid	0.19	0.28	0.03	ns
C20:2 n6c, Eicosadienoic acid	1.20	0.62	0.21	ns
C20:3 n6c, Dihomo γ linoleic acid	2.55	4.22	0.59	ns
C20:4 n6c, Arachidonic acid	1.17	2.83	0.59	ns
C20:5 n3c, Eicosapentanoic acid	1.52	0.61	0.32	ns
C22:0 Docosanoic acid	0.11	0.17	0.02	ns
C22:1 n9c, Erucic acid	0.56	0.69	0.05	ns
C22:2 n6c, Docosadienoic acid	0.14	0.26	0.04	ns
C22:6 n3c, Docosahexanoic acid	0.38	0.25	0.05	ns
Total fatty acid (µg/mg)	766.75	727.90	13.74	ns
Total SFA [2]	9.65	22.96	4.71	ns
Total UFA [3]	81.10	72.48	3.05	ns
Total PUFA n-3 [4]	55.66 [A]	29.44 [B]	9.27	*
Total PUFA n-6 [5]	18.58	2 3.66	1.80	ns
Total PUFA n-9 [6]	3.41	13.43	3.54	ns
Total PUFA [7]	77.66	66.52	3.94	ns
PUFA n-6/n-3 ratio	0.35 [B]	0.93 [A]	0.21	*
UFA/SFA	10.19 [A]	4.17 [B]	2.13	*
PUFA/SFA	9.80 [A]	3.91 [B]	2.08	*
Total MUFA [8]	6.16 [B]	18.50 [A]	4.36	*

Note: [1] Diet A (control): 70% *Brachiaria decumbence* + 30% concentrate, Diet B: 70% *Brachiaria decumbence* + 26% concentrate + 4% bypass fat; [2] Total SFA (saturated fatty acids) = C14:0 + C15:0 + C16:0 + C17:0 + C18:0 + C + 20:0 + C22:0; [3] Total UFA (unsaturated fatty acids) = C15:1 + C16:1 + C17:1 + C18:1 + C18:2 + C18:3 + C20:4 + C20:5 + C22:6; [4] Total PUFA n-3 = C18:3 n-3 + C20:5 n-3 + C22:6 n-3; [5] Total PUFA n-6 = C18:2 n-6 + C20:2 n-6 + C20:3 n-6 + C20:4 n-6 + C22:2 n-6; [6] Total PUFA n-9 = C18:1 n9 + C20:1 n9 + C22:1 n-9; [7] Total PUFA = PUFA n-3 + PUFA n-6 + PUFA n-9; [8] Total MUFA (monounsaturated fatty acids) = C15:1 + C16:1 + C17:1 + C18:1; [9] SEM = standard error of mean. [†] The result of individual fatty acid was expressed as percentage from total fatty acid in each sample; [A,B] values with different superscripts within a row differ significantly at $p < 0.05$; * indicates that the value is significantly different at $p < 0.05$; ns indicates that the value is not significantly different at $p < 0.05$.

Table 6. Fatty acid composition of meat from different breed of buffaloes.

Fatty Acid Composition (% Total FA) [†]	Breed		SEM [9]	*p*-Value
	MC [1]	SW		
C14:0 Myristic acid	1.17	1.70	0.19	ns
C15:0 Pentadecanoic acid	0.74	1.36	0.22	ns
C15:1 Pentadecanoic acid (cis-10)	1.81	1.61	0.07	ns
C16:0 Palmitic acid	8.30	10.78	0.88	ns
C16:1 Palmitoleic acid	0.94	2.46	0.54	ns
C17:0 Heptadecanoic acid	1.85	3.54	0.60	ns
C17:1 Heptadecenoic acid	1.34	2.92	0.56	ns
C18:0 Stearic acid	3.04	2.63	0.14	ns
C18:1 n9c, Oleic acid	10.69	11.02	0.12	ns
C18:2 n6c, Linoleic acid	25.18 [y]	13.19 [z]	4.24	*
C18:3 n3c, Alpha linoleic acid	36.58	39.47	1.02	ns
C20:0 Eicosanoic caid	0.30	0.39	0.03	ns
C20:1 n9c, Eicosenoic acid	0.23	0.26	0.01	ns
C20:2 n6c, Eicosadienoic acid	0.72	1.24	0.18	ns
C20:3 n6c, Dihomo-γ-linoleic acid	2.95	3.78	0.29	ns
C20:4 n6c, Arachidonic acid	2.27	1.90	0.13	ns
C20:5 n3c, Eicosapentanoic acid	0.69	1.47	0.28	ns
C22:0 Docosanoic acid	0.15	0.16	0.00	ns
C22:1 n9c, Erucic acid	0.57	0.65	0.03	ns
C22:2 n6c, Docosadienoic acid	0.24	0.16	0.03	ns
C22:6 n3c, Docosahexanoic acid	0.42	0.22	0.07	ns
Total fatty acid (μg/mg)	811.75 [y]	610.48 [z]	71.16	*
Total SFA [2]	14.29	20.40	2.16	ns
Total UFA [3]	75.44	77.79	0.83	ns
Total PUFA n-3 [4]	37.55	41.14	1.27	ns
Total PUFA n-6 [5]	24.66	19.11	1.96	ns
Total PUFA n-9 [6]	9.45	11.91	0.87	ns
Total PUFA [7]	71.66	72.16	0.18	ns
PUFA n-6/n-3 ratio	0.81	0.67	0.05	ns
UFA/SFA	7.98	5.26	0.96	ns
PUFA/SFA	7.63	4.99	0.93	ns
Total MUFA [8]	12.56	16.65	1.45	ns

Note: [1] MC = Murrah cross buffalo, SW = swamp buffalo; [2] Total SFA (saturated fatty acids) = C14:0 + C15:0 + C16:0 + C17:0 + C18:0 + C + 20:0 + C22:0; [3] Total UFA (unsaturated fatty acids) = C15:1 + C16:1 + C17:1 + C18:1 + C18:2 + C18:3 + C20:4 + C20:5 + C22:6; [4] Total PUFA n-3 = C18:3 n-3 + C20:5 n-3 + C22:6 n-3; [5] Total PUFA n-6 = C18:2 n-6 + C20:2 n-6 + C20:3 n-6 + C20:4 n-6 + C22:2 n-6; [6] Total PUFA n-9 = C18:1 n9 + C20:1 n9 + C22:1 n-9; [7] Total PUFA = PUFA n-3 + PUFA n-6 + PUFA n-9; [8] Total MUFA (monounsaturated fatty acids) = C15:1 + C16:1 + C17:1 + C18:1; [9] SEM = standard error of mean; [†] The result of individual fatty acid was expressed as percentage from total fatty acid in each sample; [y,z] values with different superscripts within a row differ significantly at $p < 0.05$; * indicates that the value is significantly different at $p < 0.05$; ns indicates that the value is not significantly different at $p < 0.05$.

Table 7. Fatty acid composition of different types of buffalo muscle.

Fatty Acid Composition (% Total FA) [†]	Type of Muscle			SEM [9]	*p*-Value
	LTL [1]	ST	SS		
C14:0 Myristic acid	0.84 [b]	2.55 [a]	0.93 [b]	0.47	*
C15:0 Pentadecanoic acid	1.38	0.75	1.03	0.08	ns
C15:1 Pentadecanoic acid (cis-10)	1.85	0.91	2.36	0.42	ns
C16:0 Palmitic acid	6.76 [c]	12.80 [a]	9.06 [b]	1.08	*
C16:1 Palmitoleic acid	1.03	3.86	0.23	1.05	ns
C17:0 Heptadecanoic acid	2.69	2.93	2.47	0.13	ns
C17:1 Heptadecenoic acid	1.46	3.34	1.58	0.51	ns
C18:0 Stearic acid	4.91	2.14	1.45	0.20	ns
C18:1 n9c, Oleic acid	12.02 [b]	14.26 [a]	6.28 [c]	2.30	*
C18:2 n6c, Linoleic acid	19.14 [b]	10.54 [c]	27.87 [a]	5.00	*
C18:3 n3c, Alpha linoleic acid	40.17	36.08	37.82	0.50	ns
C20:0 Eicosanoic acid	0.43	0.34	0.28	0.02	ns
C20:1 n9c, Eicosenoic acid	0.22	0.28	0.23	0.01	ns
C20:2 n6c, Eicosadienoic acid	1.94	0.54	0.46	0.02	ns
C20:3 n6c, Dihomo γ linoleic acid	2.81	3.89	3.40	0.14	ns
C20:4 n6c, Arachidonic acid	1.66	1.86	2.74	0.25	ns
C20:5 n3c, Eicosapentanoic acid	1.17	1.32	0.75	0.16	ns
C22:0 Docosanoic acid	0.17	0.10	0.19	0.03	ns
C22:1 n9c, Erucic acid	0.44	1.06	0.32	0.21	ns
C22:2 n6c, Docosadienoic acid	0.19	0.16	0.25	0.03	ns
C22:6 n3c, Docosahexanoic acid	0.36	0.30	0.32	0.01	ns
Total fatty acid (μg/mg)	652.74 [b]	628.02 [b]	852.59 [a]	64.83	*
Total SFA [2]	17.00	20.92	14.11	1.97	ns
Total UFA [3]	83.00	71.37	75.49	1.19	ns
Total PUFA n-3 [4]	41.64	37.58	38.82	0.36	ns
Total PUFA n-6 [5]	24.86	14.65	26.14	3.32	ns
Total PUFA n-9 [6]	12.62	12.66	6.77	1.70	ns
Total PUFA [7]	79.11	64.89	71.73	1.97	ns
PUFA n-6/n-3 ratio	0.88	0.60	0.74	0.04	ns
UFA/SFA	6.29	5.91	7.65	0.50	ns
PUFA/SFA	6.01	5.54	7.37	0.53	ns
Total MUFA [8]	15.90	17.95	9.97	2.30	ns

Note: [1] LTL = *longissimus thoracis et lumborum*; ST = *semitendinosus*; SS = *supraspinatus*; [2] Total SFA (saturated fatty acids) = C14:0 + C15:0 + C16:0 + C17:0 + C18:0 + C + 20:0 + C22:0; [3] Total UFA (unsaturated fatty acids) = C15:1 + C16:1 + C17:1 + C18:1 + C18:2 + C18:3 + C20:4 + C20:5 + C22:6; [4] Total PUFA n-3 = C18:3 n-3 + C20:5 n-3 + C22:6 n-3; [5] Total PUFA n-6 = C18:2 n-6 + C20:2 n-6 + C20:3 n-6 + C20:4 n-6 + C22:2 n-6; [6] Total PUFA n-9 = C18:1 n9 + C20:1 n9 + C22:1 n-9; [7] Total PUFA = PUFA n-3 + PUFA n-6 + PUFA n-9; [8] Total MUFA (monounsaturated fatty acids) = C15:1 + C16:1 + C17:1 + C18:1; [9] SEM = standard error of mean; [†] The result of individual fatty acid was expressed as percentage from total fatty acid in each sample; [a,b,c] values with different superscripts within a row differ significantly at $p < 0.05$; * indicates that the value is significantly different at $p < 0.05$; ns indicates that the value is not significantly different at $p < 0.05$.

3.4. Profit and Loss Analysis of Buffalo Production

The effects of different dietary supplements on the cost of slaughtering buffaloes are shown in Table 8. The total cost, which is a summation of variable and fixed costs, was 20.28% higher ($p < 0.05$) for buffaloes fed with Diet B compared to Diet A. However, buffaloes fed with the diet supplementation had higher carcass, meat and bone weights, resulting in a higher total revenue, which ranged between MYR 4707.80 (USD 1156.71) and MYR 6260.13 (USD 1538.12) for the Murrah cross and between MYR 3782.76 (USD 929.43) and MYR 4521.50 (USD 1110.93) for the swamp buffaloes ($p < 0.05$). In fact, the concentrate and bypass fat supplements increased the total revenue for Murrah cross and swamp buffaloes by 24.80% and 16.34%, respectively, compared to the solely concentrate supplementation. Therefore, the total net profit was significantly ($p < 0.05$) higher in Diet B than Diet A by approximately 26.05% for Murrah cross buffaloes and 15.48% for swamp buffaloes. In addition, the Murrah cross buffaloes showed a significantly ($p < 0.05$) higher net profit than the swamp buffaloes.

Table 8. Profit and loss analysis of buffaloes after being fed with different dietary treatments.

Breed	Murrah Cross		Swamp		SEM	*p*-Value		
Diets	Diet A	Diet B	Diet A	Diet B		Diet	Breed	Interaction
Revenue								
A. Meat sales (MYR)	2874.50 [aY]	3754.31 [bY]	2321.33 [aZ]	2824.85 [bZ]	257.80	*	*	*
B. Bone sales (MYR)	893.10	1266.72	646.88	708.50	120.80	*	*	*
C. Offal sales (MYR)	795.2	1094.10	669.55	843.15	77.14	*	*	*
D. Head sales (MYR)	70.00	70.00	70.00	70.00	-	-	-	-
E. Skin sales (MYR)	65.00	65.00	65.00	65.00	-	-	-	-
F. Tail sales (MYR)	10.00	10.00	10.00	10.00	-	-	-	-
G. Total revenue (MYR) (A + B + C + D + E + F)	4707.80 [aY]	6260.13 [bY]	3782.76 [aZ]	4521.50 [bZ]	901.60	*	*	*
Operating Expenses (2 years)								
Variable cost								
Cost of feeding (MYR/day)								
Brachiaria grass	0.71 [a]	0.80 [b]	0.66 [a]	0.68 [b]	0.03	*	ns	ns
Concentrate	1.46	1.43	1.36	1.23	0.04	ns	ns	ns
Bypass fat	-	0.75	-	0.65	0.03	ns	ns	ns
H. Total cost of average daily DMI (MYR/day/animal)	2.17 [a]	2.98 [b]	2.02 [a]	2.56 [b]	0.19	*	ns	ns
I. Total feed cost in 2 years (MYR/animal)	1580.44 [a]	2171.62 [b]	1477.74 [a]	1867.33 [b]	135.10	*	ns	ns
Fixed cost								
Management cost								
Deworming	0.50	0.50	0.50	0.50	-	-	-	-
ID tag	2.00	2.00	2.00	2.00	-	-	-	-
Fertilizer	156.00	156.00	156.00	156.00	-	-	-	-
Transportation	83.33	83.33	83.33	83.33	-	-	-	-
Labor cost	152.00	152.00	152.00	152.00	-	-	-	-
J. Total management cost (MYR/2 year/animal)	393.83	393.83	393.83	393.83	-	-	-	-
K. Pre- and post-inspection of slaughter services/animal (MYR)	5.00	5.00	5.00	5.00	-	-	-	-
L. Total cost (I + J + K)	1979.27 [a]	2570.45 [b]	1876.57 [a]	2266.16 [b]	135.10	*	ns	ns
Total net profit (MYR) (G-L)	2728.53 [aY]	3689.68 [bY]	1906.19 [aZ]	2255.34 [bZ]	335.04	*	*	*

Note: USD 1.00 = MYR 4.07 currency conversion 5 March 2021, MYR = Malaysian Ringgit. Estimations: income from fresh meat, MYR 31.03/kg; income from bone sales, MYR 26.00/kg, income from offal sales, MYR 35.00/kg, *Brachiaria* grass, MYR 0.23/kg dry matter; Concentrate mixture, MYR 1.11/kg; Bypass fat, MYR 3.82/kg. Diet A (control): 70% *Brachiaria decumbens* + 30% concentrate; Diet B (treatment): 70% *Brachiaria decumbens* + 26% concentrate + 4% bypass fat; SEM: standard error of means; [a,b,Y,Z]: means with different superscript letters in the same column are significantly different at $p < 0.05$; * indicates that the value is significantly different at $p < 0.05$; ns indicates that the value is not significantly different at $p < 0.05$.

4. Discussion

4.1. Carcass Quality

In the current study, certain carcass quality characteristic of buffaloes had a significant interaction between the diet and breed of buffaloes. Supporting these results, previous studies also reported that the diet and breed interaction influenced the livestock performance, carcass and meat quality characteristics [10,33,34]. This study was comparable with our finding that showed that the diet and breed had an impact on the pre-slaughter weight, as well as on the hot and cold carcass weight. Thus, the breed and the type of diets demonstrated a crucial role in the improvement and the optimization of the productivity and carcass traits of these buffaloes.

This study revealed that all parameters for the buffalo carcass characteristics were improved following feed supplementations. In fact, there was a significant interaction between the carcass weights and the diet, as well as the breed of buffalo when a heavier pre-slaughter weight and hot and cold carcass weights were recorded among the Murrah cross buffaloes fed with Diet B. Similarly, Bakker et al. [35] showed that the hot and cold carcass yield of cattle were greatly influenced by the type of supplements and breeds, while Di Stasio and Brugiapaglia [36] reported the same finding in buffaloes. It seems that a high energy diet produces a better live weight at slaughter, carcass yield and weights of internal organs and body fat [36–38].

Mohd Azmi et al. [13] reported that a supplementation of bypass fat was able to improve the growth performance, such as the average feed intake, the body weight and the body condition score of Murrah cross and swamp buffaloes, which might further influence the carcass quality traits. Indeed, Jones [39] reported that the carcass yield improvement was mainly due to the gains in the weight of muscle and bone, and the decrease in fat deposition, thereby increasing the weight of the bone and muscle, and decreasing the fat weight. Similarly, calves fed on high concentrate tended to have more fat than grazing, indicating that the short period of concentrate feeding before slaughter increased the fat depots in finisher calves [40]. On the contrary, there was no difference in the carcass yield and the body chemical composition [41], whereas a higher carcass yield and fat percentage by adding fat to the diet of lambs was reported [42]. It was reported that a lower level of fat supplementation (<5%) did not affect the carcass yield and body composition [43], whereas the lipid constituted the largest component of the carcass gain when cull cows were re-alimented with a high energy density diet [44].

The carcass yield, meat percentage and carcass meat weight are indicators used to evaluate slaughter performance [45]. The carcass yield is closely related to nutrient levels, breeds, age and feeding management [45]. In this study, the carcass fat and the carcass yield of buffaloes were slightly higher when they were fed with Diet A, due to the high percentage of concentrate in this diet. Earlier, Lambertz et al. [10] reported a higher carcass fat percentage in buffaloes following supplementation with a 2.0% body weight of concentrate, while Pimpa et al. [37] failed to improve the warm and cold carcass yield of the cattle when fed with 5% fat. Nevertheless, the average carcass yield of the Murrah cross and the swamp buffaloes in this study was markedly lower than the pure-breed Murrah buffaloes, as reported by Biswas and Rajkumar [3] and Arshadullah et al. [46], likely due to breed differences and the higher proportion of non-edible parts [16,47]. In fact, the Murrah cross in this study recorded a high carcass bone percentage ($p < 0.05$). Similar studies reported that Mediterranean pure Murrah, swamp and Murrah cross produce a high bone percentage, with an average of 25.63%, 18.50% and 16.37%, respectively [7,10,45]. The percentage of buffalo carcass bone, however, is not influenced by the diet, according to Bakker et al. [35].

The meat percentage is one of the most significant yields in carcass composition. It is influenced by the high-energy dietary energy intake [36]. Indeed, this study showed a significant increase in the carcass meat percentage and meat:fat ratio with Diet B. This was due to the adequate and continuous supply of a high energy density from the bypass fat and optimum protein provided by Diet B for growth and tissue maintenance [13,48], whereas the high concentrate in Diet A usually produces more carcass fat [48].

The pH of the post-slaughter meat is one of the factors that determines the quality of the carcass and meat [10,49]. Following the slaughter, the pH of the meat starts to decrease. In this study, the meat pH of Murrah cross fed with Diet A was significantly lower at 24 h than Diet B, but the pH of the meat of swamp buffaloes remained similar. Nevertheless, the range of the meat pH of this study was similar to the findings of Calabrò et al. [50] and Gecgel et al. [51].

4.2. Proximate Composition

It is well known that the nutrient properties of meat influence the meat quality, particularly the crude fat and crude protein contents and the fatty acid composition [52]. In this study, meat nutrient contents were found to be generally influenced by the diet and breed. The buffaloes fed with Diet B tended to have meat moisture, protein and fat contents that were significantly higher than the buffaloes fed with Diet A, probably due to the fact that feeding diets with a readily fermentable carbohydrate (concentrate) with bypass fat increased propionate production through ruminal fermentation and increased the insulin concentrations, which increased the intramuscular fat and protein syntheses [53,54]. Furthermore, the average protein content in buffalo meat was recorded as between 17.33% and 23.30%, and was significantly affected by the diets, and, to some extent, by the breed [55]. Our study revealed that the swamp buffaloes had a higher CP value in LTL and ST than the Murrah cross. The higher CP and fat contents in LTL than those of other topside meats [10] were probably due to the high myosin content, thus making the LTL juicier than other topsides [56]. Furthermore, since LTL is a less active muscle, it contains less collagen; thus, it is much softer, with a higher fat content than other topsides [57,58]. This study revealed that the swamp buffaloes had juicer meat than the Murrah cross. Similarly, the meat of the swamp buffaloes recorded a higher ash content in LTL than the Murrah cross ($p < 0.05$). Considering the swamp buffaloes as a draft animal, they probably store more iron in order to ensure there is enough myoglobin for their heavy workloads [9,59].

4.3. Fatty Acid Composition

In general, fatty acid compositions vary in the meat of water and river buffaloes. The most common fatty acid found in this study was C18:3, followed by C18:2, C18:1, C16:0 and C18:0, which were comparable with the results of Rao and Kowale [60], who reported a high C18:0, C18:1 and C18:2 composition in buffalo meat. However, Gecgel et al. [51] concluded that water buffaloes have a high C18:1 fatty acid composition, followed by C18:0 and C16:0. Meanwhile, river buffaloes were shown to be high in the fatty acid composition of C18:1, C16:1, C18:3 n-3 and C20:4 n-6 [10]. In this study, the percentages of the fatty acid composition for both breeds were comparable, but the Murrah cross recorded a significantly higher C18:2 and total fatty acid than the swamp buffaloes. A variation in the fatty acid composition between breeds of lambs was also reported by Budimir et al. [61], but the Italian Merino and Soravissana lambs had a similar fatty acids composition [61].

The buffalo SS muscle had a significantly high total fatty acid, and the LTL and ST had a high C18:2 composition, while the ST had a significantly high C14:0, C16:0 and C18:1 composition, as reported by Wood et al. [62] and Tamburrano et al. [63]. Similarly, Calabro et al. [50] reported significant differences between three types of muscles: the *longisimus thoracis, semitendinosus and iliopsas plus psoas minor*. In this study, all muscles showed a significant difference in the fatty acids composition, regardless of the type of diet offered. The meat of buffaloes fed with Diet B had a higher percentage of C16:0 and C16:1 than that of Diet A. Similarly, a study showed that Murrah buffaloes fed with a palm-based supplement significantly increased palmitic acid (C16:0) in the LTL muscle [64]. Furthermore, C18:1 and MUFA compositions in buffalo muscle were three-fold higher when fed with Diet B, but there was no difference in the C18:1 and MUFA compositions of the LTL muscle of Droper sheep fed with bypass fat [65]. Another study also revealed that Murrah buffalo fed with 1% body weight of either a palm, coconut or corn-based oil supplement had high C18:1 and MUFA concentrations in the LTL muscle [64]. The results of our study were in agreement with da Silva Lima et al. [66] and Andrade et al. [67], who studied in cattle. They reported that cattle showed a higher proportion of UFA, namely C18:1, followed by SFA, such as C16:0. Therefore, the difference in the C18:1 composition in the muscle might be due to the different proportions of diet given to the animals. In addition, the biohydrogenation process within the rumen contributed a small proportion of fatty acid [68,69]. Therefore, buffaloes that were fed Diet A had a lower content of MUFA compared to those on Diet B.

The presence of n-3 fatty acids helps to determine the buffalo meat's nutrient quality [63]. The n-3 fatty acid, in particular, has the potential to reduce the serum concentration of triglycerides in humans [10]. In this study, n-3 fatty acid was present at a low level in the buffalo muscle fed with Diet B. The low level might be due to the low intensity of C18:3 n-3 ruminal biohydrogenation [70] in the rumen, which led to a low composition of C18:3 n-3 in the meat. Due to the low amount of C18:3 n-3, the subsequent level of total n-3 fatty acids was significantly low, leading to a significantly high n-6/n-3 ratio compared to Diet A ($p < 0.05$). Therefore, the ratio of n-6/n-3 fatty acid in the meat is influenced by the diet [51]. Even though there was a high n-6/n-3 ratio with Diet B, it is still considered to be a healthy meat [71], since it falls within the recommended ratio of the World Health [72].

Cifuni et al. [73] concluded that the fat mass percentage in the animal muscle could influence the fatty acid composition of both meat and PUFA/SFA and UFA/SFA fatty acid ratios. Buffaloes fed with Diet B in this study had a lower carcass fat percentage compared to Diet A ($p < 0.05$), and, subsequently, were low in UFA/SFA and PUFA/SFA ratios. Indeed, the PUFA/SFA ratio is recommended to be low, since it is considered to be beneficial for human health [63]. Nevertheless, the content of the fatty acid in this study was different from that reported by Gecgel et al. [51] and Lambertz et al. [10]. This might be due to the diet and breed [74], but Lima et al. [75] and Silva et al. [76] found little differences in the fatty acid composition of cattle meat that included protected fat in the diets. These differences were potentially due to differences in the animal study, type and ratio of feed offered and muscle functionality. Diets that contain high concentrations of fat were either partially or entirely protected from microbial action in the rumen, thus, causing changes in the fatty acid composition, which may lead to an increase in the intramuscular fat deposition [67,77].

4.4. Profit and Loss Analysis of Buffalo Production

In feedlot production, a farmer needs to sell the reared animal as soon as possible in order to ensure a consistent cash inflow. To reduce the costs of the animal meat production, farmers are increasingly seeking alternative feed additives, particularly cheaper feedstuffs with desired qualities that enable the animal to reach a market weight and slaughter weight within the given rearing period. An alternative that attracts a great deal of interest is the use of concentrate, either solely or in a mixture with bypass fat [10,66,78]. However, excessive amounts of dietary lipids might result in a negative effect on fiber digestion in the rumen and may influence the quality of the meat and the cost of animal feed [75], which makes the decision to change to the alternative difficult. With the right proportion of supplementations used in this study, the adverse effects on meat quality and costs could be controlled.

This study revealed that the 2-year operating expenses were significantly ($p < 0.05$) different between Murrah cross and swamp buffaloes. Diet B cost between MYR 2570.45 (USD 631.56) and MYR 2266.16 (USD 556.80), whereas Diet A cost between MYR 1979.27 (USD 486.31) and MYR 1876.57 (USD 461.07). In fact, additions of feed ingredients, such as concentrate or bypass fat, have been reported to increase the feed cost [79]. In this study, the net profit was significantly ($p < 0.05$) higher with Diet B than Diet A, at the rate of 26.05% for Murrah cross buffaloes and 15.48% for swamp buffaloes. However, a proper supplement ratio should be able to improve the growth performance and the carcass characteristics of the ruminant, thus becoming cost-effective with the potential of enhancing the profit [80–82]. Overall, diets with a supplement for the Murrah cross resulted in a higher net profit than that of swamp buffaloes. The result can improve the awareness on costs of production to farmers and policy makers and can enable them to make appropriate decisions when changing farm management. More research on the cost effectiveness of diets with a supplement on buffalo health should be performed in the future.

5. Conclusions

Supplementations of pasture with 26% concentrate and 4% bypass fat enhanced the carcass quality (i.e., hot and cold carcass weights and meat:fat ratio) and the proximate compositions of different buffaloes' muscles (i.e., moisture and crude protein of LTL, ST and SS and the ether extract of LTL and ST), as well as the meat fatty acid (i.e., C16:0, C16:1, C18:1, PUFA n-6/n-3 and total MUFA). The Murrah cross showed a significantly better carcass quality (hot and cold carcass weights and carcass bone yield), with a higher C18:2 and total fatty acid content, whereas the proximate composition of meat (i.e., ash of LTL and crude protein of LTL and ST) was better in swamp buffaloes. The supplementation of bypass fat in the buffalo diet significantly increased the cost of feeding, but eventually resulted in a significantly higher revenue and net profit. Further studies should address the meat quality and the nuclear magnetic resonance metabolic profiles of Murrah cross and swamp buffaloes supplemented with concentrate and bypass fat in order to explore the compounds that contribute to the physiochemical properties, sensory assessment and nutritional quality of food.

Author Contributions: Conceived and designed the experiment: H.A.H., M.Z.A.B., M.Z.S., G.Y.M. and H.A. Provided supervision for animal health assessment and feed formulation during experiment: P.A., F.M.A., H.A.H., M.Z.A.B. and M.Z.S. Conducted the experiment and analyzed data: A.F.M.A. Data interpretation and scientific discussion: A.F.M.A., F.M.A., N.M.N., A.I. and A.J. Contributed materials and reagents: H.A.H. and H.A. Wrote the manuscript: A.F.M.A., F.M.A. and H.A.H. All authors have read and agreed to the published version of the manuscript.

Funding: The whole research work and manuscript writing were fully funded by Ministry of Science, Technology, and Innovation (MOSTI), Malaysia under the FLAGSHIP Research Grant Scheme (Project Title: Enhancing the Productivity of Farm Buffalo (*Bubalus bubalis*) Through Modification of Rearing Practices; Reference number: FP05514B0020-2(DSTIN)/6300858). Amirul Faiz M. A. was a recipient of Graduate Research Fellowship from the Universiti Putra Malaysia.

Institutional Review Board Statement: The authors confirm that the ethical policies of the journal, as noted on the journal's author guidelines page, have been adhered to and the appropriate ethical review committee approval has been received. The animals were cared for in accordance with the animal ethics guidelines of the Animal Utilization Protocol approved by the Institution Animal Care and Use Committee (IACUC) (Approval No. UPM/IACUC/AUP-017/2018, on 8 January 2018), Universiti Putra Malaysia. The sampling from the experimented animals were strictly conducted under veterinary supervision.

Data Availability Statement: Availability of data and equipment used and analyzed during this study is available from the correspondence author on reasonable request.

Acknowledgments: The authors would like to thank the staff of the Department of Veterinary Service Sabah (JPHPT), the Buffalo Breeding and Research Centre, Telupid Sabah, the Sabah Meat Technology Centre (SMTC) and the postgraduates and staff from Nutrition Laboratory, Faculty of Veterinary Medicine, Universiti Putra Malaysia for their assistance in this study. This study was financially supported by the Flagship Research grant ABI/FS/2016/01 of the Ministry of Science, Technology and Innovation Malaysia, the Higher Institutions Centres of Excellence (HiCoE) grant of the Ministry of Higher Education Malaysia and Universiti Putra Malaysia.

Conflicts of Interest: The authors declare that they have no competing interests.

Abbreviations

FA	Fatty acid
SS	*Supraspinatus* muscle
ST	*Semitendinosus* muscle
LTL	*Longissimus thoracis et lumborum*
SFA	Saturated fatty acid
UFA	Unsaturated fatty acid
MUFA	Monounsaturated fatty acid
PUFA	Polyunsaturated fatty acid
kg	Kilogram
MJ	Milli joule
GE	Gross energy
µg	Microgram
mg	Milligram
USA	United State of America

References

1. Nanda, A.S.; Nakao, T. Role of buffalo in the socioeconomic development of rural Asia: Current status and future prospectus. *Anim. Sci. J.* **2003**, *74*, 113–155. [CrossRef]
2. Irurueta, M.; Cadoppi, A.; Langman, L.; Grigioni, G.; Carduza, F. Effect of aging on the characteristics of meat from water buffalo grown in the Delta del Paraná region of Argentina. *Meat Sci.* **2008**, *79*, 529–533. [CrossRef]
3. Biswas, S.; Rajkumar, R.S. Buffalo as a potential food animal. *Int. J. Livest. Prod.* **2009**, *1*, 1–5.
4. Neath, K.E.; Del Barrio, A.N.; Lapitan, R.M.; Herrera, J.R.V.; Cruz, L.C.; Fujihara, T.; Kanai, Y. Difference in tenderness and pH decline between water buffalo meat and beef during postmortem aging. *Meat Sci.* **2007**, *75*, 499–505. [CrossRef] [PubMed]
5. Murthy, T.R.; Devadason, I.P. Buffalo meat and meat products–An overview. In *Proceedings of the 4th Asian Buffalo Congress on Buffalo for Food, Security and Employment*; Asian Buffalo Association: New Delhi, India, 2003; pp. 193–199.
6. Guerrero, A.; Velandia Valero, M.; Campo, M.M.; Sañudo, C. Some factors that affect ruminant meat quality: From the farm to the fork. Review. *Acta Scien. Anim. Sci.* **2013**, *35*, 335–347.
7. Peixoto, M.R.; Lourenço Junior, J.B.; Faturi, C.; Garcia, A.R.; Nahúm, B.D.; Lourenço, L.F.; Meller, L.H.; Oliveira, K.C. Carcass quality of buffalo (Bubalus bubalis) finished in silvopastoral system in the Eastern Amazon, Brazil. *Arq. Bras. Med. Vet. Zootec.* **2012**, *64*, 1045–1052. [CrossRef]
8. Mohd Azmi, A.F.; Ahmad, H.; Mohd Nor, N.; Meng, G.Y.; Zamri-Saad, M.; Abu Bakar, M.Z.; Salleh, A.; Abdullah, P.; Jayanegara, A.; Abu Hassim, H. The Impact of Feed Supplementations on Asian Buffaloes: A Review. *Animals* **2021**, *11*, 2033. [CrossRef] [PubMed]
9. Mohd Azmi, A.F.; Abu Hassim, H.; Mohd Nor, N.; Ahmad, H.; Meng, G.Y.; Abdullah, P.; Zamri-Saad, M. Comparative Growth and Economic Performances between Indigenous Swamp and Murrah Crossbred Buffaloes in Malaysia. *Animals* **2021**, *11*, 957. [CrossRef]
10. Lambertz, C.; Panprasert, P.; Holtz, W.; Moors, E.; Jaturasitha, S.; Wicke, M.; Gauly, M. Carcass characteristics and meat quality of swamp buffaloes (Bubalus bubalis) fattened at different feeding intensities. *Asian-Australas. J. Anim. Sci.* **2014**, *27*, 551. [CrossRef]
11. Ekiz, B.; Yilmaz, A.; Yalcintan, H.; Yakan, A.; Yilmaz, I.; Soysal, I. Carcass and meat quality of male and female water buffaloes finished under an intensive production system. *Ann. Anim. Sci.* **2018**, *18*, 557–574. [CrossRef]
12. Demirel, G.; Ozpinar, H.; Nazli, B.; Keser, O. Fatty acids of lamb meat from two breeds fed different forage: Concentrate ratio. *Meat Sci.* **2006**, *72*, 229–235. [CrossRef] [PubMed]
13. Mohd Azmi, A.F.; Ahmad, H.; Mohd Nor, N.; Meng, G.Y.; Saad, M.Z.; Abu Bakar, M.Z.; Abu Hassim, H. Effects of concentrate and bypass fat supplementations on growth performance, blood profile, and rearing cost of feedlot buffaloes. *Animals* **2021**, *11*, 2105. [CrossRef] [PubMed]
14. Cruz, L.C. Recent developments in the buffalo industry of Asia. *Rev. Vet.* **2010**, *21*, 7–19.
15. Shahudin, M.S.; Ghani, A.A.A.; Zamri-Saad, M.; Zuki, A.B.; Abdullah, F.F.J.; Wahid, H.; Hassim, H.A. The Necessity of a Herd Health Management Programme for Dairy Goat Farms in Malaysia. *Pertanika J. Trop. Agric. Sci.* **2018**, *41*, 1–18.
16. Lapitan, R.M.; Del Barrio, A.N.; Katsube, O.; Ban-Tokuda, T.; Orden, E.A.; Robles, A.Y.; Fujihara, T.; Cruz, L.C.; Homma, H.; Kanai, Y. Comparison of carcass and meat characteristics of Brahman grade cattle (Bos indicus) and crossbred water buffalo (Bubalus bubalis). *Anim. Sci. J.* **2007**, *78*, 596–604. [CrossRef]
17. NRC National Research Council. *Nutrient Requirements of Dairy Cattle*; National Academy of Sciences: Washington, DC, USA, 2001.
18. Bulbul, T. Energy and nutrient requirements of buffaloes. *Kocatepe Vet. J.* **2010**, *3*, 55–64.
19. Basra, M.J.; Nisa, M.; Khan, M.A.; Riaz, M.; Tuqeer, N.A.; Saeed, M.N. Nili-ravi buffalo III. Energy and protein requirements of 12–15 months old calves. *Int. J. Agric. Biol.* **2003**, *5*, 382–383.

20. Adeyemi, K.D.; Ebrahimi, M.; Samsudin, A.A.; Sabow, A.B.; Sazili, A.Q. Carcass traits, meat yield and fatty acid composition of adipose tissues and Supraspinatus muscle in goats fed blend of canola oil and palm oil. *J. Anim. Sci. Technol.* **2015**, *157*, 42. [CrossRef]

21. Calheiros, F.; Neves, M.J.M. Rendimentos ponderais no borrego Merino Precoce. carcaça e 5° quarto. *Sep. Bol. Pecuário* **1968**, *37*, 117–126.

22. Candyrine, S.C.; Mahadzir, M.F.; Garba, S.; Jahromi, M.F.; Ebrahimi, M.; Goh, Y.M.; Samsudin, A.A.; Sazili, A.Q.; Chen, W.L.; Ganesh, S.; et al. Effects of naturally-produced lovastatin on feed digestibility, rumen fermentation, microbiota and methane emissions in goats over a 12-week treatment period. *PLoS ONE* **2018**, *13*, e0199840. [CrossRef]

23. Kadim, I.T.; Mahgoub, O.; Al-Ajmi, D.S.; Al-Maqbaly, R.S.; Al-Saqri, N.M.; Ritchie, A. An evaluation of the growth, carcass and meat quality characteristics of Omani goat breeds. *Meat Sci.* **2004**, *66*, 203–210. [CrossRef]

24. Association of Official Analytical Chemists. *Official Methods of Analysis of the Association of Official Analytical Chemists*, 18th ed.; Association of Official Analytical Chemists: Washington, DC, USA, 2007.

25. Folch, J.; Lees, M.; Stanley, G.S. A simple method for the isolation and purification of total lipides from animal tissues. *J. Biol. Chem.* **1957**, *226*, 497–509. [CrossRef]

26. Rajion, M.A.; McLean, J.G.; Cahill, R.N. Essential fatty acids in the fetal and newborn lamb. *Aust. J. Biol. Sci.* **1985**, *38*, 33–40. [CrossRef] [PubMed]

27. Lopes, L.S.; Ladeira, M.M.; Machado Neto, O.R.; da Silveira, A.R.M.C.; Reis, R.P.; Campos, F.R. Economical viability of finishing Nellore and Red Norte bulls in feedlot, in Lavras-MG region. *Ciênc. Agrotecnologia* **2011**, *35*, 774–780. [CrossRef]

28. Raval, A.J.; Sorathiya, L.; Kharadi, V.B.; Patel, M.D.; Tyagi, K.K.; Patel, V.R.; Choubey, M. Effect of calcium salt of palm fatty acid supplementation on production performance, nutrient utilization and blood metabolites in Surti buffaloes (Bubalus bubalis). *Indian J. Anim. Sci.* **2017**, *87*, 1124–1129.

29. Bardhan, D.; Kumar, S.; Kumar, S.; Kumar, N.; Singh, R.K.; Khan, R.; Mendiratta, S.K. Value chain analysis of buffalo meat (carabeef) in India. *Agric. Econ. Res. Rev.* **2019**, *32*, 149–163. [CrossRef]

30. Lapitan, R.M.; Del Barrio, A.N.; Katsube, O.; Tokuda, T.; Orden, E.A.; Robles, A.Y.; Kanai, Y. Comparison of feed intake, digestibility and fattening performance of Brahman grade cattle (Bos indicus) and crossbred water buffalo (Bubalus bubalis). *Anim. Sci. J.* **2004**, *75*, 549–555. [CrossRef]

31. Iqbal, Z.M.; Abdullah, M.; Javed, K.; Jabbar, M.A.; Ahmad, N.; Ditta, Y.A.; Mustafa, H.; Shahzad, F. Effect of varying levels of concentrate on growth performance and feed economics in Nili-Ravi buffalo heifer calves. *Turk. J. Vet. Anim. Sci.* **2017**, *41*, 775–780. [CrossRef]

32. Steel, R.G.R.; Torrie, J.H. *Principles and Procedures of Statistics*, 2nd ed.; McGraw-Hill Int. Book Company: New Delhi, India, 1980.

33. Cittadini, A.; Sarriés, M.V.; Domínguez, R.; Indurain, G.; Lorenzo, J.M. Effect of breed and finishing diet on growth parameters and carcass quality characteristics of navarre autochthonous foals. *Animals* **2021**, *11*, 488. [CrossRef]

34. Mwangi, F.W.; Charmley, E.; Gardiner, C.P.; Malau-Aduli, B.S.; Kinobe, R.T.; Malau-Aduli, A.E. Diet and genetics influence beef cattle performance and meat quality characteristics. *Foods* **2019**, *8*, 648. [CrossRef]

35. Bakker, C.E.; Blair, A.D.; Grubbs, J.K.; Taylor, A.R.; Brake, D.W.; Long, N.M.; Underwood, K.R. Effects of rumen-protected long-chain fatty acid supplementation during the finishing phase of beef steers on live performance, carcass characteristics, beef quality, and serum fatty acid profile. *Transl. Anim. Sci.* **2019**, *3*, 1585–1592. [CrossRef]

36. Di Stasio, L.; Brugiapaglia, A. Current knowledge on river buffalo meat: A critical analysis. *Animals* **2021**, *11*, 2111. [CrossRef]

37. Pimpa, O.; Binsulong, B.; Pastsart, U.; Pimpa, B.; Liang, J.B. Bypass fat enhances liveweight gain and meat quality but not profitability of smallholder cattle fattening systems based on oil palm frond. *Anim. Prod. Sci.* **2021**, *62*, 1–8.

38. Honig, A.C.; Inhuber, V.; Spiekers, H.; Windisch, W.; Götz, K.U.; Ettle, T. Influence of dietary energy concentration and body weight at slaughter on carcass tissue composition and beef cuts of modern type Fleckvieh (German Simmental) bulls. *Meat Sci.* **2020**, *169*, 108209. [CrossRef] [PubMed]

39. Jones, S.D.M. Tissue growth in young and mature cull Holsteincows fed a high energy diet. *J. Anim. Sci.* **1983**, *61*, 593–599.

40. Moreno, T.; Varela, A.; Portela, C.; Perez, N.; Carballo, J.A.; Montserrat, L. The effect of grazing on the fatty acid profile of longissimus thoracis muscle in Galician Blond calves. *Animal* **2007**, *1*, 1227–1235. [CrossRef]

41. Manso, T.; Castro, T.; Mantecon, A.R.; Jimeno, V. Effects of palm oiland calcium soaps of palm oil fatty acids in fattening diets ondigestibility, performance and chemical body composition of lambs. *Anim. Feed Sci. Technol.* **2006**, *127*, 175–186. [CrossRef]

42. Clinquart, A.; Micol, D.; Brundseaux, C.; Dufrasne, I.; Istasse, L. Utilisation des matierres grasses chez les bovines avec a l'engraisse-ment. *INRA Prod. Anim.* **1995**, *8*, 29–42. [CrossRef]

43. Salinas, J.; Ramirez, R.G.; Dominguez, M.M.; Reyes- Bernal, N.; Trinidad-Larraga, N.; Montano, M.F. Effect of calcium soaps of tallow ongrowth performance and carcass characteristics of Pelibuey lambs. *Small Rumin. Res.* **2006**, *66*, 135–139. [CrossRef]

44. Field, R.A.; Prasad, V.S.S.; Riley, M.L. Characteristics of lean fromculled breeding ewes. *J. Anim. Sci.* **1987**, *64*, 1648–1649. [CrossRef]

45. Li, Q.; Wang, Y.; Tan, L.; Leng, J.; Lu, Q.; Tian, S.; Shao, S.; Duan, C.; Li, W.; Mao, H. Effects of age on slaughter performance and meat quality of Binlangjang male buffalo. *Saudi J. Biol. Sci.* **2018**, *25*, 248–252. [CrossRef] [PubMed]

46. Arshadullah, M.; Suhaib, M.; Raheel Baber, M.U.; Badar-uz-Zaman, I.A.; Hyder, S.I. Growth of Chenopodium quiona wild under naturally salt affected soils. *Malays. J. Sustain. Agric.* **2017**, *1*, 1–3. [CrossRef]

47. Manafiazar, G.; Mohsenourazary, A.; Afsharihamidi, B.; Mahmoodi, B. Comparison carcass traits of Azeri buffalo, native and crossbred (native* Holstein) male calves in west Azerbaijan-Iran. *Ital. J. Anim. Sci.* **2007**, *6*, 1167–1170. [CrossRef]

48. Alemneh, T.; Getabalew, M. Factors influencing the growth and development of meat animals. *Int. J. Anim. Sci.* **2019**, *3*, 1048.

49. Ziauddin, K.S.; Mahendrakar, N.S.; Rao, D.N.; Ramesh, B.S.; Amla, B.L. Observations on some chemical and physical characteristics of buffalo meat. *Meat Sci.* **1994**, *37*, 103–113. [CrossRef]

50. Calabrò, S.; Cutrignelli, M.I.; Gonzalez, O.J.; Chiofalo, B.; Grossi, M.; Tudisco, R.; Infascelli, F. Meat quality of buffalo young bulls fed faba bean as protein source. *Meat Sci.* **2014**, *96*, 591–596. [CrossRef]

51. Gecgel, U.; Yilmaz, I.; Soysal, M.I.; Gurcan, E.K.; Kok, S. Investigating proximate composition and fatty acid profile of Longissimus dorsi from Anatolian Water Buffaloes (*Bubalus bubalis*) raised in similar conditions. *J. Food Sci. Technol.* **2019**, *39*, 830–836. [CrossRef]

52. Wood, J.D.; Richardson, R.I.; Nute, G.R.; Fisher, A.V.; Campo, M.M.; Kasapidou, E.; Sheard, P.R.; Enser, M. Effects of fatty acids on meat quality: A review. *Meat Sci.* **2004**, *66*, 21–32. [CrossRef]

53. Johnson, D.D.; Mitchell, G.E.J.; Tucker, R.E.; Hemken, R.W. Plasma glucose and insulin responses to propionate in preruminating calves. *J. Anim. Sci.* **1982**, *55*, 1224–1230. [CrossRef]

54. Behan, A.A.; Loh, T.C.; Fakurazi, S.; Kaka, U.; Kaka, A.; Samsudin, A.A. Effects of supplementation of rumen protected fats on rumen ecology and digestibility of nutrients in sheep. *Animals* **2019**, *9*, 400. [CrossRef] [PubMed]

55. Naveena, B.M.; Mendiratta, S.K.; Anjaneyulu, A.S. Tenderization of buffalo meat using plant proteases from Cucumis trigonus Roxb (Kachri) and Zingiber officinale roscoe (Ginger rhizome). *Meat Sci.* **2004**, *68*, 363–369. [CrossRef]

56. Lawrie, R.A.; Ledward, D. *Lawrie's Meat Science*; Woodhead Publishing: Sawston, UK, 2014.

57. Joo, S.T.; Kim, G.D.; Hwang, Y.H.; Ryu, Y.C. Control of fresh meat quality through manipulation of muscle fiber characteristics. *Meat Sci.* **2013**, *95*, 828–836. [CrossRef] [PubMed]

58. Oh, M.; Kim, E.K.; Jeon, B.T.; Tang, Y.; Kim, M.S.; Seong, H.J.; Moon, S.H. Chemical compositions, free amino acid contents and antioxidant activities of Hanwoo (Bos taurus coreanae) beef by cut. *Meat Sci.* **2016**, *119*, 16–21. [CrossRef] [PubMed]

59. Mancini, R.A.; Hunt, M.C. Current research in meat color. *Meat Sci.* **2005**, *71*, 100–121. [CrossRef]

60. Rao, V.K.; Kowale, B.N. Changes in phospholipids of buffalo meat during processing and storage. *Meat Sci.* **1991**, *30*, 115–129. [CrossRef]

61. Budimir, K.; Mozzon, M.; Toderi, M.; D'Ottavio, P.; Trombetta, M.F. Effect of breed on fatty acid composition of meat and subcutaneous adipose tissue of light lambs. *Animals* **2020**, *10*, 535. [CrossRef]

62. Wood, J.D.; Enser, M.; Fisher, A.V.; Nute, G.R.; Sheard, P.R.; Richardson, R.I.; Hughes, S.I.; Whittington, F.M. Fat deposition, fatty acid composition and meat quality: A review. *Meat Sci.* **2008**, *78*, 343–358. [CrossRef] [PubMed]

63. Tamburrano, A.; Tavazzi, B.; Callà, C.A.M.; Amorini, A.M.; Lazzarino, G.; Vincenti, S.; Laurenti, P. Biochemical and nutritional characteristics of buffalo meat and potential implications on human health for a personalized nutrition. *Ital. J. Food Saf.* **2019**, *8*, 174–179. [CrossRef]

64. Peixoto Joele, M.R.S.; Lourenço Júnior, J.B.; Lourenço, L.F.H.; Amaral Ribeiro, S.C.; Meller, L.H. Buffalo meat from animals fed with agro industrial in Eastern Amazon. *Arch. Zootec.* **2014**, *63*, 359–369. [CrossRef]

65. Behan, A.A.; Akhtar, M.T.; Loh, T.C.; Fakurazi, S.; Kaka, U.; Muhamad, A.; Samsudin, A.A. Meat quality, fatty acid content and nmr metabolic profile of dorper sheep supplemented with bypass fats. *Foods* **2021**, *10*, 1133. [CrossRef]

66. Da Silva Lima, E.; Valente, T.N.; de Oliveira Roça, R.; Cezário, A.S.; dos Santos, W.B.; Deminicis, B.B.; Ribeiro, J.C. Effect of whole cottonseed or protected fat dietary additives on carcass characteristics and meat quality of beef cattle: A review. *J. Agric. Sci.* **2017**, *9*, 175–189.

67. Andrade, E.N.; Neto, A.P.; Roça, R.D.; Faria, M.D.; Resende, F.D.; Siqueira, G.R.; Pinheiro, R.S. Beef quality of young Angus x Nellore cattle supplemented with rumen-protected lipids during rearing and fatting periods. *Meat Sci.* **2014**, *98*, 591–598. [CrossRef]

68. Lock, A.L.; Harvatine, K.J.; Drackley, J.K.; Bauman, D.E. Concepts in fat and fatty acid digestion in ruminants. In *Proc. Intermountain Nutr. Conf.*; Utah State Univ.: Logan, UT, USA, 2006; pp. 85–100.

69. Nieto, G.; Ros, G. Modification of fatty acid composition in meat through diet: Effect on lipid peroxidation and relationship to nutritional quality—A review. *Lipid Peroxid.* **2012**, *12*, 239–258.

70. Lanza, M.; Fabro, C.; Scerra, M.; Bella, M.; Pagano, R.; Brogna, D.M.R.; Pennisi, P. Lamb meat quality and intramuscular fatty acid composition as affected by concentrates including different legume seeds. *Ital. J. Anim. Sci.* **2011**, *10*, 18. [CrossRef]

71. Giuffrida-Mendoza, M.; Arenas de Moreno, L.; Huerta-Leidenz, N.; Uzcátegui-Bracho, S.; Valero-Leal, K.; Romero, S.; Rodas-González, A. Cholesterol and fatty acid composition of longissimus thoracis from water buffalo (*Bubalus bubalis*) and Brahman-influenced cattle raised under savannah conditions. *Meat Sci.* **2015**, *106*, 44–49. [CrossRef] [PubMed]

72. Department of Health. *Nutritional Aspects of Cardiovascular Disease*; Report on Health and Social Subjects, 46; HMSO: London, UK, 1994.

73. Cifuni, G.F.; Contò, M.; Amici, A.; Faill, S. Physical and nutritional properties of buffalo meat finished on hay or maize silage-based diets. *Anim Sci.* **2014**, *85*, 405–410. [CrossRef]

74. Woods, V.B.; Fearon, A.M. Dietary sources of unsaturated fatty acids for animals and their transfer into meat, milk and eggs: A review. *Livest. Sci.* **2009**, *126*, 1–20. [CrossRef]

75. Lima, E.D.; Morais, J.P.; Roça, R.D.; Costa, Q.P.; Andrade, E.N.; Vaz, V.P.; Valente, T.N.; Costa, D.P. Meat characteristics of Nellore cattle fed different levels of lipid-based diets. *J. Agric. Sci.* **2015**, *7*, 174–183. [CrossRef]

76. Silva, S.D.; Leme, P.R.; Putrino, S.M.; Pereira, A.S.; Valinote, A.C.; Nogueira Filho, J.C.; Lanna, D.P. Fatty acid composition of intramuscular fat from Nellore steers fed dry or high moisture corn and calcium salts of fatty acids. *Livest. Sci.* **2009**, *122*, 290–295. [CrossRef]
77. Souza, N.E.; Silva, R.R.; Prado, I.M.; Prado, J.M.; Wada, F.Y.; Prado, I.N. Grãos de linhaça e canola sobre a composição do músculo Longissimus de novilhas confinadas. *Arch. Zootec.* **2007**, *56*, 863–874.
78. Machado Neto, O.R.; Ladeira, M.M.; Chizzotti, M.L.; Jorge, A.M.; Oliveira, D.M.; Carvalho, J.R.; Ribeiro, J.D. Performance, carcass traits, meat quality and economic analysis of feedlot of young bulls fed oilseeds with and without supplementation of vitamin E. *Rev. Bras. Zootec.* **2012**, *41*, 1756–1763. [CrossRef]
79. Adegbola, A.A.; Okonkwo, A.C. Nutrient intake, digestibility and growth rate of rabbits fed varying levels of cassava leaf meal. *Niger. J. Anim. Prod.* **2000**, *29*, 21–26. [CrossRef]
80. Boughalmi, A.; Araba, A. Effect of feeding management from grass to concentrate feed on growth, carcass characteristics, meat quality and fatty acid profile of Timahdite lamb breed. *Small Rumin. Res.* **2016**, *144*, 158–163. [CrossRef]
81. Jenkins, G.P.; Miklyaev, M. Cost-Benefit analysis of small ruminants fattening with feed concentrates in the highlands of Ethiopia. *JDI Exec. Programs* **2014**, *1*, 2013-12.
82. Naik, P.K. Bypass fat in dairy ration-a review. *Anim. Nutr. Feed Technol.* **2013**, *13*, 147–163.

 *animals*

Article

Characterization of Three Different Mediterranean Beef Fattening Systems: Performance, Behavior, and Carcass and Meat Quality

Denise Sánchez [1], Sònia Marti [1,*], Marçal Verdú [2], Joel González [3], Maria Font-i-Furnols [3] and Maria Devant [1,*]

1   IRTA, Ruminant Production, 08140 Caldes de Montbui, Spain; denise.sanchez@irta.cat
2   Animal Nutrition and Feed Industry, bonÀrea Agrupa, 25210 Guissona, Spain; marsal.verdu@bonarea.com
3   IRTA, Food Quality and Technology Program, Finca Camps i Armet, 17121 Monells, Spain;
    joel.gonzalez@irta.cat (J.G.); maria.font@irta.cat (M.F.-i.-F.)
*   Correspondence: sonia.marti@irta.cat (S.M.); maria.devant@irta.cat (M.D.); Tel.: +34-93-467-40-40 (M.D.)

Citation: Sánchez, D.; Marti, S.;
Verdú, M.; González, J.;
Font-i-Furnols, M.; Devant, M.
Characterization of Three Different
Mediterranean Beef Fattening
Systems: Performance, Behavior, and
Carcass and Meat Quality. *Animals*
2022, *12*, 1960. https://doi.org/
10.3390/ani12151960

Academic Editors: Guillermo Ripoll
and Begoña Panea

Received: 24 June 2022
Accepted: 29 July 2022
Published: 2 August 2022

Publisher's Note: MDPI stays neutral
with regard to jurisdictional claims in
published maps and institutional affiliations.

Simple Summary: Beef fattening systems present a large diversity according to the effect of the type (genetics and gender) of animals fattened, the nutrition programs, the housing conditions, and days on feed, resulting in different carcass traits and meat qualities. New beef fattening systems are arising in Mediterranean countries raising crossbred Angus bulls seeking new marketing opportunities. One of the strengths of the present study is that all animals of the three compared different production systems, two conventional (crossbred heifers, Holstein bulls) and one innovative (crossbred Angus bulls), were raised following their own commercial program (days on feed, nutrition) under the same housing, care, and weather conditions. Furthermore, the carcass and meat quality parameters were analyzed by using a common methodology. With this experimental design, potential factors like the housing conditions or the methodology used to analyze carcass and meat quality (aging time, cooling temperatures, lab equipment) did not interfere in data interpretation. The results indicated that fattening crossbred Angus bulls is suitable in an intensive fattening program but technical data (performance, meat quality or consumer preferences) do not support it as a better alternative to the current Holstein bull production.

Abstract: The aim of this study was to characterize three different commercial dairy beef fattening systems for intensive Mediterranean fattening programs differing in gender, breed, nutrition, and days of feed in order to describe their performance, behavior, and carcass and meat quality when they were raised simultaneously under the same housing and care conditions. Treatments were three different production systems: (1) crossbred Holstein x beef breeds such as Charolais or Limousine heifers, slaughtered at 10 months of age (CBH10, $n = 41$); (2) Holstein bulls, slaughtered at 11 months of age (HB11, $n = 42$); and (3) crossbred Holstein x Angus bulls, slaughtered at 12 months of age (CAB12, $n = 37$). According to our results, moving from a production system based on Holstein bulls to the crossbred Angus production system has no technical support as no large production and meat quality improvements were observed, and only marketing strategies for meat differentiation and consumer trends could favor this decision.

Keywords: beef cattle; Angus bulls; growth rate; carcass; crossbred Holstein

1. Introduction

Beef fattening systems differ widely among European Union countries, with a large diversity in the type (genetics and gender) of animals fattened, in the nutrition programs, in the housing conditions, and in the days on feed, resulting in different carcass traits and meat qualities [1,2]. These beef fattening systems can range from extensive systems in Ireland to very intensive systems in Italy [3,4]. In Spain, the beef fattening system is traditionally an intensive system where calves are housed during the fattening period, commonly in partially open barns with straw bedding [5,6]. Briefly, 30% of the animals fattened are females (mainly crossbred) and from the remaining 70% of males, 30% are Holstein calves [7]. Dairy calves (dairy beef system) account for 40% of the fattened calves, and these calves are fed ad libitum concentrate from very young ages (before the weaning period) [7]. In this dairy beef fattening system, commercially, crossbred females are fattened for shorter periods of time (slaughtered around 10 months of age) to avoid the decrease in feed efficiency [8]. Commonly, these crossbred females are fed the same fattening concentrate with a moderate energy content during the growing and finishing periods [5]. Holstein bulls are usually slaughtered later, before 12 months of age, as the decline of the efficiency with age takes place later compared with the crossbred females. These males are traditionally fed a growing concentrate with moderate energy content until 9–10 months of age and a finishing concentrate with a greater energy content until slaughter [9]. The carcass and meat quality of females and males differ; bulls are sexually more active and more susceptible to suffering pre-slaughter stressors (handling, transportation, lairage) than heifers, and the meat of their carcasses is more prone to become DFD (dark, firm, and dry) [10] than females. In addition, heifer meat has greater intramuscular fat and is more tender than bull meat [11,12]. Recently, in Mediterranean countries, raising crossbred Angus bulls in the dairy beef fattening system has been suggested as a new marketing opportunity. The expectations (hypotheses) are that these crossbred Angus animals would perform close to the Holstein males in terms of growth and close to crossbred females in terms of meat quality [13]. However, there is a lack of data describing crossbred Angus raised in intensive production systems, where calves are fed high-concentrate diets from very young ages as is usual in Mediterranean production systems. Moreover, it is difficult to analyze and compare different production systems as they are the result of a combination of type of animal (breed, gender) and housing, management, nutrition programs (days on feed), and carcass and meat handling procedures [2]. Therefore, one of the strengths of the present study is the comparison of three different beef production systems following their own commercial fattening program (days on feed, nutrition) and raising animals under the same housing, care, and weather conditions; moreover, carcass and meat quality parameters were analyzed with the same methodology. The aim of this study was to characterize three different commercial beef fattening systems in intensive Mediterranean fattening programs differing by gender, breed, nutrition, and days on feed and to describe their performance, behavior, and carcass and meat quality when raised simultaneously under the same housing and care conditions. Data generated from this study are the first step for decision making and offer technical information to consider whether raising crossbred Angus bulls can be a good alternative to Holstein bulls in a Mediterranean dairy beef fattening system.

2. Materials and Methods

2.1. Experimental Design, Animals, Housing, and Diets

Animals were reared in a commercial farm of Agropecuària Montgai S.L. (Montgai, Spain) and were managed following the principles and guidelines from the Animal Care Committee of the Institute for Research and Technology in Agrifood (IRTA, Caldes de Montbui, Spain). This study was conducted in accordance with the Spanish guidelines for experimental animal protection (Royal Decree 53/2013 of 1 February on the protection of animals used for experimentation or other scientific purposes; Boletín Oficial del Estado, 2013).

The experiment was designed as a randomized, balanced design with covariance adjustment with 3 treatments. Pen was the experimental unit and animals were the sampling units. Treatments were the three different production systems: (1) crossbred Holstein x beef breeds such as Charolais or Limousine heifers, slaughtered at 10 months of age (CBH10); (2) Holstein bulls, slaughtered at 11 months of age (HB11); and (3) crossbred Holstein x Angus bulls, slaughtered at 12 months of age (CAB12). A total of 41 crossbred heifers (CBH10; 165 ± 24.8 kg BW and 141 ± 12.6 d age), 42 Holstein bulls (HB11; 176 ± 18.7 kg BW and 142 ± 4.2 d age) and

37 Angus crossbred bulls (CAB12; 192 ± 52.5 kg BW and 154 ± 34.0 d age) were used to evaluate 3 different production systems and their potential effects on performance, animal behavior, and carcass and meat quality. At arrival, animals belonging to each production system (treatment) were weighed and distributed in pens to equalize initial BW among pens, and then pens were randomly allocated to the 3 treatments (2 pens/treatment; 18–21 animals/pen). Thereafter, animals were weighed at the start of study (d 0) and every 14 d until the end. The CBH10 heifers were slaughtered on d 168 and d 174 of study (BW 425 ± 46.9 kg), HB11 bulls were slaughtered on d 209 and d 216 of study (BW 518 ± 45.7 kg), and CAB12 bulls were slaughtered on d 226 and d 230 of study (BW 554 ± 58.8 kg), simulating commercial criteria in intensive beef fattening in Mediterranean countries.

Animals were allocated in pens (12 m × 6 m) that were equipped with a single-space feeder (0.50 m long × 0.26 m wide × 0.15 m depth) with 10 kg of concentrate capacity, and it was protected by two lateral barriers (1.50 m length × 0.90 m height) forming a chute. Width of chute was regulated from 0.45 to 0.60 m throughout the study to accommodate the increase in animal size and provide sufficient space to allow only one animal to eat comfortably at a time [14], one water bowl (0.30 m length, 0.30 m width, 0.18 m depth) and straw was offered in a separated straw, five-space feeder (3.00 m length, 1.12 m wide, and 0.65 m depth). Animals were offered concentrate, straw, and water ad libitum. Each concentrate feeder was equipped with a scale that consisted of 4 load cells (Ultilcell, Barcelona, Spain), where the feeder was suspended and concentrate contained was continuously weighed, and its weight was displayed by digital screen reader. The scales were calibrated weekly. Every morning, concentrate refusals were recorded as final feed weight of day before; after that, all feeders were automatically refilled via refilling system, and concentrate offers were recorded as initial feed weight of the current day. All feeders were refilled daily by an auger conveying automated feeding distribution system and had a reservoir with storage capacity of 200 kg of concentrate to ensure continuous feed availability, which was dispensed slowly by gravity fall maintaining a continuous and low level of concentrate in the trough. Concentrate intake was considered as concentrate disappearance from feeder, which referred to both concentrate consumption and wastage without discriminating between them because feed spillage was not measured [14]. The amount of straw offered to each pen was recorded weekly to estimate the total amount of straw consumed; however, these data were only an approximation of straw intake because straw was also used for bedding. Before the beginning of the study, animals had a 1-month adaptation period by widening the chute to facilitate feeder access; from then, the width of the chute was adapted to the animal size to allow them to eat easily [15]. Pens were totally covered, i.e., a thick black curtain was installed from the roof to the floor of the barn to avoid eye contact and smelling between heifers and bulls, which could enhance sexual behavior of the bulls. Following commercial feeding practices, heifers (CBH10) were fed the growing concentrate (Table 1) throughout the study, whereas bulls (HB11 and CAB12) were fed the growing concentrate from d 1 to d 168 and the finishing concentrate (Table 1) from d 168 to d 230. Ingredient and nutrient composition from the growing and finishing concentrates (Table 1) were formulated according to FEDNA recommendations [16]. Main differences between the growing and finishing concentrates were metabolizable energy, CP, and fat content. Moreover, animals also had access to barley straw (35 g/kg CP, 16 g/kg EE, 796 g/kg NDF, and 61 g/kg ash; DM basis) and fresh water.

2.2. Feed Ingredient Analyses

Feed samples were collected every feed manufacturing for growing and finishing concentrate to analyze dry matter (DM) (method 925.04), ash (method 642.05), crude protein (CP) by the Kjeldahl method (method 988.05) [17], neutral detergent fiber (NDF) [18] using sodium sulfite and alpha-amylase, and ether extract EE by Soxhlet with a previous acid hydrolysis (method 920.39; [17]).

Table 1. Ingredient and nutrient composition of the growing and finishing concentrates.

Item		
Ingredient, g/kg	**Growing**	**Finishing**
Corn	421	398
Barley	107	149
Wheat middlings	103	67
Wheat	100	99
Corn DDG	120	99
Peas meal		59
Palm kernel	100	80
Palm oil	10	22
Calcium carbonate	18	14
Urea	8	3
Sodium bicarbonate	4	4
White salt	2	2
Vitamin–mineral premix [a,b]	3	2
Nutrient, per kg DM		
Metabolizable energy (ME), Mcal/kg	3.18	3.34
CP, g	158	144
Ether extract, g	55	65
Ash, g	54	47
NDF, g	220	198
NFC, g [c]	511	545

[a] Premix of the growing concentrate (Montgai, Spain). Vitamins and minerals contained per kg of DM: 3052 kIU of vitamin A, 610 kIU of vitamin D3, 10.2 g of vitamin E, 0.04 g of vitamin K, 10.2 g of vitamin B1, 0.34 g of vitamin B2, 0.04 g of vitamin B6, 0.007 g of vitamin B12, 1.7 g of vitamin B3, 0.2 g of Co, 1.7 g of Cu, 0.2 g of I, 15.3 g of Mn, 0.1 g of Se, 16.7 g of Zn, 200 g of sodium sulfate, 152 g of magnesium oxide, 42.4 g of etoxiquine, 1 kg of barley as excipient. [b] Premix of the finishing concentrate (CAG, Guissona, Spain). Vitamins and minerals contained per kg of DM: 3575 kIU of vitamin A, 858 kIU of vitamin D3, 101 g of vitamin E, 2.3 g of vitamin B1, 0.2 g of Co, 2.5 g of Cu, 0.3 g of I, 15.7 g of Mn, 0.2 g of Se, 20.6 g of Zn, 250 g of magnesium oxide, 75.7 g of etoxiquine, 1 kg of barley as excipient. [c] NFC = nonfiber carbohydrates calculated as 1000 − (CP + ash + NDF + ether extract).

2.3. Feed Ingredient Analyses

Feed samples were collected every feed manufacturing for growing and finishing concentrate to analyze dry matter (DM) (method 925.04), ash (method 642.05), crude protein (CP) by the Kjeldahl method (method 988.05) [17], neutral detergent fiber (NDF) [18] using sodium sulfite and alpha-amylase, and ether extract EE by Soxhlet with a previous acid hydrolysis (method 920.39; [17]).

2.4. Animal Behavior Evaluation

Animal behavior was recorded for general activities (standing, lying, eating, drinking, and ruminating) and social behavior (non-agonistic, agonistic, and sexual interactions) with a visual scan observation of 2 pens at the same time from 8:00 to 10:00 h [15] for each pen on d 13, 37, 49, 62, 76, 90, 104, 118, 132, 146, 160, 174, 188, 202, and 216, and the last sampling day was 160, 202, or 216 for CBH10, HB11, and CAB12, respectively. General activities were scored using 3 scan samplings of 10 s at 5 min intervals, and social behavior was scored during three continuous sampling periods of 5 min. This scanning procedure of 15 min was repeated twice consecutively in each pen, starting randomly in a different pen every scanning day [15].

2.5. Measurements and Sample Collection

Animals from CBH10, HB11, and CAB12 were transported to the slaughterhouse (La Closa, Guissona, Spain) by truck between on d 167 and 174, on d 209 to 216, and on d 226 and 230 of study, respectively, following the EU Regulation 1099/2009 using a captive-bolt pistol and dressed according to commercial practices. Animal transport was organized in six different loads without mixing animals of different treatments and pens. Transport distance was less than 35 km. The hot carcass weight (HCW) was recorded and degree of carcass fatness and conformation were graded according to the EU classification system into 1.2.3.4.5 (EU Regulation No. 1208/81) and into (S)EUROP categories (EU Regulation No. 1208/81, 1026/91), respectively. Dressing percentage was calculated dividing the HCW by the final BW. At 24 h post mortem of carcass chilling at 6.9 °C, pH was measured with a pH meter (pH 25 DL; Crison, Alella, Spain) by penetration of the probe equipped with a xerolyt electrode between the lumbar vertebrae L4 and L5 of the left side of the carcass.

Eighteen samples per treatment (total of 54 carcasses) were selected at random (avoiding the extreme HCW) for meat quality evaluation. At 48 h post mortem, samples from the central part of the *longissimus thoracis* (LT) from rib 12 to 13 were removed from each carcass and cut in 5 steaks of 2.5 cm. Thereafter, three steaks of 2.5 cm were individually packaged in modified atmosphere (MAP; 70% O_2:30% CO_2) with polypropylene trays (day 0 of MAP), and the remaining steaks were vacuum-packaged and frozen at −20 °C until determination of intramuscular fat and tenderness.

2.6. Meat Analyses

The MAP steaks were displayed for 9 d (11 d post mortem) in an illuminated cooling room (5 ± 0.5 °C) with a homogeneous fluorescent light (900 lx) activated for 12 h a day. From one steak, instrumental color was evaluated on d 2, 6, and 9 post packaging using a Minolta chromameter (CM600d, Minolta Inc., Osaka, Japan) in the CIE-Lab space (L*: lightness, a*: redness, and b*: yellowness; Commission Internationale de l'Éclairage, 1976) with illuminant D65 and 10° viewing angle. In addition, a group of 3 trained panelists evaluated daily a color perception and purchase decision until d 9 post packaging from Monday to Friday. The subjective perception of color and purchase decision were recorded using a 5-point scale (Color perception: (1), highly undesirable; (2), moderately undesirable; (3), slightly desirable; (4), moderately desirable; and (5), highly desirable. Purchase decision: (1), would not buy; (2), would probably not buy; (3), buy dubiously; (4), would probably buy; (5), would buy) [19]. Intramuscular fat was analyzed by using near infrared spectroscopy at wavelengths between 850 and 1048 nm (Foodscan; FOSS, Hillerød, Denmark), previously removing subcutaneous fat and connective tissue and homogenizing with a conventional meat grinder. Instrumental tenderness (Warner–Bratzler shear force (WBSF)) was measured using a texturometer (Stable Micro Systems, Godalming, United Kingdom); samples were thawed for 24 h at 2 °C and wrapped in aluminum foil and baked at 200 °C until the internal temperature reached 71 °C [19]. Cooked steaks were cut into six 1.25 cm diameter cores with a cork borer, parallel to the muscle fiber orientation. The Warner–Bratzer shear blade was perpendicularly oriented to the direction of the fibers [20].

2.7. Statistical Analysis

The pen was considered the experimental unit, as the pen was the unit on which all uncontrolled factors were occurring at random. In the case where data were registered individually, the animal was included in the analysis as a sampling unit. A power analysis was conducted to ensure that 2 replicates were appropriate for the statistical power [21].

Initial BW, initial age, final BW, days of study, HCW, and dressing percentage data were analyzed using a mixed-effects model (version 9.4, SAS Inst., Inc., Cary, NC, USA), including treatment as a main effect and pen as a random effect. Initial BW was used as a covariate.

Concentrate intake and average daily gain data were analyzed using a mixed-effect model with repeated measures. The model included initial BW as a covariate, treatment, period, and the interaction between the treatment and period as main effects and pen, animal within pen, and the interaction of treatment and pen as random effects. Period was considered a repeated factor, and the pen nested within treatment was subjected to 2 variance–covariance structures: compound symmetry and autoregressive order. The covariance structure that minimized Schwarz's Bayesian information criterion was considered the most desirable analysis.

In behavior data, due to the lack of normality of the data analyzed in a previous analysis, the non-parametric Kruskal–Wallis test was performed, and then the data were transformed. Percentage of general activities data were transformed into natural logarithm from behavioral performances, and data from social behavior was transformed into root-square to achieve a normal distribution. These data were analyzed with repeated measures, as described above.

Data from carcass conformation and fatness were analyzed with the FREQ procedure of SAS with the χ^2 distribution procedure (version 9.4, SAS Inst., Inc., Cary, NC, USA).

Meat quality data, such us pH ultimate, instrumental color and texture at 48 h, and intramuscular fat, were analyzed using a mixed-effects model (version 9.4, SAS Inst., Inc., Cary, NC, USA), including treatment as a main effect and the pen as a random effect. The variables of evolution of instrumental color, color preference, and purchase decision were analyzed using a mixed model with repeated measures as described above, where temperature of the refrigeration room was included as a covariate, treatment day and their interactions were included as fixed effects, and pen, animal within pen, and the interaction of treatment and pen were included as random effects. Least square mean values were compared with Tukey's HSD test.

Differences were declared significant at $p < 0.05$, and trends were discussed at $0.05 \leq p \leq 0.10$ for all models.

3. Results

3.1. Performance and Carcass Quality

Performance and carcass quality data are presented in Table 2. During the first 168 days, total concentrate consumption was greater ($p < 0.001$) only in the last period (Figure 1) for CAB12 and HB11 compared with CBH10 (interaction between production system and period, $p < 0.001$). However, during this 168-day period, ADG of CBH10 were less ($p < 0.05$) than HB11 and CAB12 in most of the periods studied (interaction production system and period was $p < 0.05$) as can be observed in Figure 2. Additionally, as a consequence, feed efficiency data were slightly lower in most of periods studied for CBH10 than for HB11 and CAB12 bulls (interaction production system and period were $p < 0.001$; data not shown). When global performance data were analyzed, as the days in the study differed, differences in total concentrate intake, ADG, and final BW and HCW were observed (Table 2). Total intake of CAB12, despite being slaughtered 16 days later, was close to HB11 (Table 2). Final global efficiency did not differ between both beef fattening systems (Table 2). Moreover, the global efficiency did not differ between raising crossbred heifers and slaughtering them at 10 months of age (CBH10) versus the tested bull production systems (HB11 and CAB12). The HCW of CAB12 slaughtered at 382 d of age was 6.4% greater ($p < 0.001$) than that for HB11 slaughtered at 354 d of age, and the HCW of the HB11 was 14.8% greater ($p < 0.001$) than that for CBH10 slaughtered at 311 d of age. When analyzing the global carcass efficiency (carcass yield expressed by total concentrate consumption) of the fattening bulls, both HB11 or CAB12 were 14% less efficient ($p < 0.001$) than the fattening crossbred heifers slaughtered at 10 months of age. Differences were observed in carcass conformation ($p < 0.001$), resulting CAB12 treatment with the best carcass conformation scoring rate (more "U" and "R" scoring percentages), followed by the carcasses of CBH10 heifers, while the carcasses of the HB11 bulls had the poorest conformation carcasses (greatest percentage of carcasses scored as "P"). However, no differences in the dressing percentage or carcass fatness among the production systems were observed.

3.2. Animal Behavior

In Table 3, animal behavior data are presented. A significant interaction ($p < 0.001$) between production system and periods was observed in general activities such as standing, lying, and rumination. Although standing and rumination differed among treatments, both behaviors did not follow a regular pattern, as can be observed in Figure 3a–c. Social behavior of CAB12 was less ($p < 0.05$) than CBH10 and HB11 in period 1, HB11 was less ($p < 0.05$) compared with CBH10 and CAB12 in period 4, and CBH10 was greater ($p < 0.05$) compared with HB11 and CAB12 in period 8 (data not represented in figures). Among the agonistic and sexual behaviors, the effect of the production system was more regular. Fighting was less ($p < 0.01$) for CBH10 compared with HB11 and CAB12. Displacement of HB11 was greater ($p < 0.05$) compared with CBH10 and CAB12 in period 8 (Figure 3d). Moreover, HB11 was greater ($p < 0.01$) in chasing compared with CBH10 and CAB12. Flehmen, attempt to mount and complete mount, was less ($p < 0.001$) for CBH10 compared with HB11 and CAB12. Finally, CBH10 performed more stereotyped behaviors ($p < 0.001$) compared with CAB12, and the latter performed more stereotyped behaviors than HB11 ($p < 0.001$).

Table 2. Performance and carcass characteristics for the first 168 days and whole study for different Mediterranean beef fattening systems.

Item	Production System [1]			SEM [2]	*p*-Value [3]		
	CBH10	HB11	CAB12		Production System	Time	Production System x Time
Numbers of animals	39	42	37	-	-	-	-
Initial age, days	140	141	154	17.2	0.83	-	-
Initial BW, kg	171	172	188	21.4	0.83	-	-
Performance from 0 to 168 d of study							
Concentrate intake, kg/d	6.24	6.49	6.38	0.138	0.45	<0.001	<0.001
ADG, kg/d	1.53	1.70	1.70	0.030	<0.04	<0.001	0.02
Efficiency, kg/kg	0.25	0.26	0.26	0.010	0.54	<0.001	<0.001
Global performance							
Days of study	171 [c]	212 [b]	228 [a]	0.3	<0.001	-	-
Final age, days	310 [b]	354 [ab]	382 [a]	16.3	<0.01	-	-
Final BW, kg	425 [c]	523 [b]	553 [a]	7.8	<0.001	-	-
ADG, kg/d	1.53 [b]	1.63 [a]	1.60 [ab]	0.027	<0.05		
Total concentrate consumption, kg	1064 [b]	1437 [a]	1490 [a]	22	<0.001		
Efficiency, kg/kg	0.24	0.24	0.24	0.003	0.45		
Carcass parameters							
HCW, kg	243 [c]	279 [b]	297 [a]	4.0	<0.001	-	-
Carcass efficiency, kg/kg	0.22 [a]	0.19 [b]	0.19 [b]	0.003	<0.001		
Dressing percentage, %	55.6	53.7	54.4	0.58	0.17	-	-
Conformation [4], %					<0.001		
E	2.6	-	-				
U	23.1	-	32.4				
R	33.3	-	67.6	-		-	-
O	35.9	54.8	-				
P	5.1	45.2	-				
Fatness [5], %					1.00		
1	-	-	-				
2	2.6	2.4	2.7	-		-	-
3	97.4	97.6	97.3				

[a,b,c] Rows with different superscripts differ ($p < 0.05$). [1] Treatments CBH10 = crossbred Holstein x beef breeds such as Charolais or Limousine heifers, slaughtered at 10 months of age; HB11= Holstein bulls, slaughtered at 11 months of age; CAB12 = crossbred Holstein with Angus bulls, slaughtered at 12 months of age. [2] SEM = standard error of the mean. [3] Production system effect; Time = time effect (period of 14 d); Production system x Time = production system by time interaction effect. [4] The conformation class designated by the letter "E" (excellent) describes carcasses with all profiles convex to super-convex with exceptional muscle development, and the conformation classified as "U" (very good) describes carcasses with profiles on the whole convex with very good muscle development. The carcasses classified as "R" (good) present profiles, overall, straight and with good muscle development. Carcasses classified as "O" (fair) present profiles straight to concave with average muscle development, and carcasses classified as "P" (poor) present all profiles concave to very concave with poor muscle development. [5] The carcass fat cover that is classified as 1 (low) describes none to low fat cover, the class of fat cover classified as 3 (very high) describes an entire carcass covered with fat and with heavy fat deposits in the thoracic cavity.

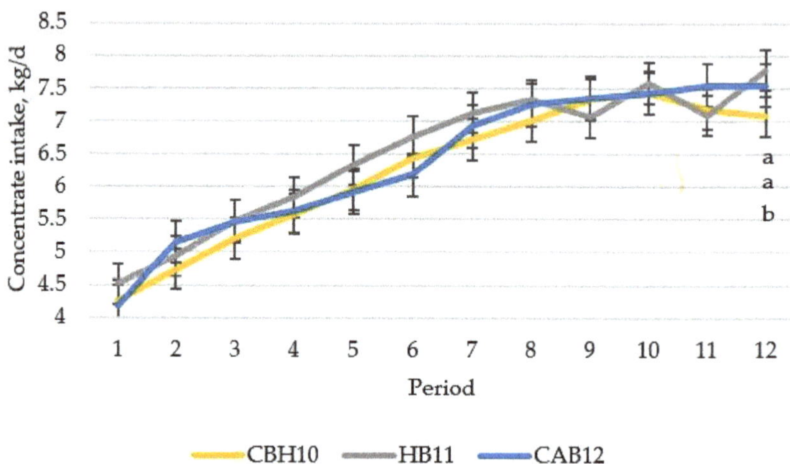

Figure 1. Daily concentrate intake over the time in crossbred Holstein x beef breeds such as Charolais or Limousine heifers, slaughtered at 10 months of age (CBH10); Holstein bulls, slaughtered at 11 months of age (HB11); and crossbred Holstein with Angus bulls, slaughtered at 12 months of age (CAB12) ([a,b] significant differences ($p < 0.05$) between production system within the same period).

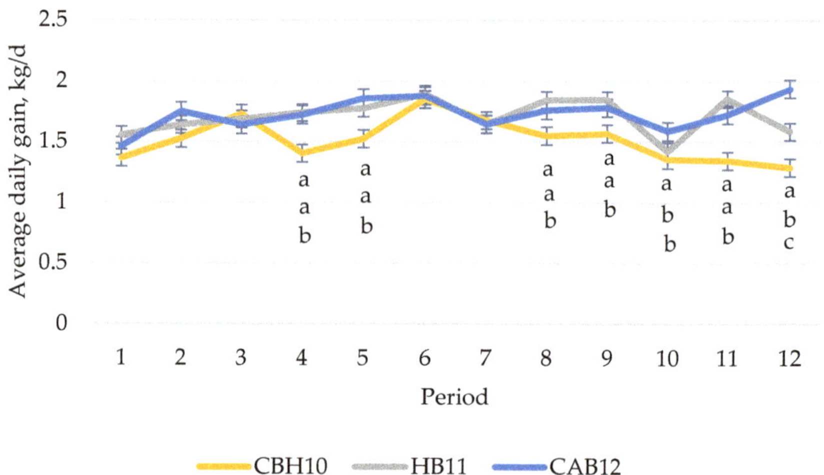

Figure 2. Average daily gain over the time in crossbred Holstein x beef breeds such as Charolais or Limousine heifers, slaughtered at 10 months of age (CBH10); Holstein bulls, slaughtered at 11 months of age (HB11); and crossbred Holstein with Angus bulls, slaughtered at 12 months of age (CAB12) ([a,b,c] significant differences ($p < 0.05$) between production system within the same period).

Table 3. Animal behavior for the first 168 days in different Mediterranean beef fattening systems.

Item	Production System [1]			SEM [2]	*p*-Value [3]		
	CBH10	HB11	CAB12		Production System	Period	Production System x Period
General activities, %							
Standing	67.7	76.9	77.9	0.85	<0.001	<0.001	<0.001
Lying	32.3	23.1	22.1	0.85	<0.001	<0.001	<0.001
Concentrate intake	5.0	4.7	5.3	0.05	<0.001	0.22	<0.01
Straw intake	9.7	9.5	12.2	0.89	0.06	<0.001	<0.01
Drinking water	1.1	1.7	1.5	0.28	0.32	0.93	0.48
Ruminating	13.8	10.1	10.5	0.31	<0.001	<0.01	<0.01
Behavior, each 15 min							
Self-grooming	16.6	9.9	10.5	0.20	<0.001	<0.001	<0.001
Social	3.8	2.9	2.7	0.13	0.10	<0.001	<0.001
Oral	5.1	6.4	4.8	0.13	0.22	<0.01	0.52
Fighting	0.7 [b]	2.9 [a]	2.1 [a]	0.21	<0.01	0.29	0.27
Butting	0.4 [b]	1.3 [a]	1.3 [a]	0.14	0.01	0.92	0.50
Displacement	0.2	0.2	0.2	0.05	0.37	0.03	0.03
Chasing	0.1 [b]	0.6 [a]	0.1 [b]	0.09	<0.01	<0.01	0.24
Chasing-up	0.0	0.1	0.0	0.01	0.26	<0.001	0.12
Flehmen	0.1 [b]	3.3 [a]	2.7 [a]	0.16	<0.001	<0.001	0.12
Attempt to mount	0.8 [b]	3.7 [a]	2.2 [a]	0.20	0.03	<0.001	0.15
Complete mount	0.4 [b]	2.8 [a]	2.3 [a]	0.24	<0.01	<0.001	0.98
Stereotype	1.4 [a]	0.1 [c]	0.4 [b]	0.06	<0.001	<0.001	0.05

[a,b,c] Rows with different superscripts differ ($p < 0.05$). [1] Treatments CBH10 = crossbred Holstein x beef breeds such as Charolais or Limousine heifers, slaughtered at 10 months of age; HB11 = Holstein bulls, slaughtered at 11 months of age; CAB12 = crossbred Holstein with Angus bulls, slaughtered at 12 months of age. [2] SEM = standard error of the mean. [3] Production system effect; Period = time effect (period of 14 d); Production system x Period = production system by period interaction effect.

3.3. Meat Quality

Meat pH was greater ($p < 0.001$) for CBH10 and CAB12 compared with HB11 (Table 4). The maximum force and the total area from WBSF were not affected by the production system. Only the slope was significant ($p = 0.03$), but the differences were not relevant. Intramuscular fat had no significant differences between production systems. Lightness (L*) of CBH10 were greater ($p < 0.001$) compared with HB11 and CAB12 during all the conservation time in MAP on days 2, 6, and 9 post-packaging (Figure 4a). Redness (a*) of CBH10 was less ($p < 0.05$) than HB11 and CAB12 at 2 and 9 d, and only less than HB11 at 6 d (Figure 4b). Yellowness (b*) of CBH10 was lower ($p < 0.05$) than HB11 and CAB12 on day 2, higher than CAB12 on day 6, and lower than HB11 at day 9 (Figure 4c). On days 6 and 9, color perception differed among treatments: CBH10 had a less preferred color ($p < 0.05$) than CAB12, which had a less preferred color than HB11. No differences in color perception by meat from different treatments were found at day 2 (Figure 4d). Similar results were obtained by purchase decision (Figure 4e).

Table 4. Meat pH, shear force, intramuscular fat and evolution of the shelf-life parameters (instrumental color, color perception, and purchase decision) over the time in MAP of meat of animals raised in different Mediterranean beef fattening systems.

Item	Production System [1]			SEM [2]	*p*-Value [3]		
	CBH10	HB11	CAB12		Production System	Days	Production System x Day
pH, 24 h	5.7 [a]	5.5 [b]	5.7 [a]	0.01	<0.001	-	-
WBSF							
Maximum force (kg)	6.6	7.3	6.6	0.38	0.27	-	-
Total area (kg.mm)	69.5	76.9	64.4	5.43	0.27	-	-

Table 4. *Cont.*

Item	Production System [1]			SEM [2]	p-Value [3]		
	CBH10	HB11	CAB12		Production System	Days	Production System x Day
Slope (kg.mm)	0.9 [a]	1.0 [b]	1.0 [b]	0.03	0.03	-	-
Intramuscular fat (%)	1.9	1.7	1.7	0.17	0.60	-	-
Instrumental color [4]							
L*	36.2	32.3	33.0	0.09	<0.001	<0.001	0.03
a*	14.9	16.7	16.9	0.53	<0.001	<0.001	<0.001
b*	15.6	16.2	16.3	0.36	0.03	<0.001	<0.001
Color perception [5]	2.9	4.2	3.5	0.10	<0.001	<0.001	<0.001
Purchase decision [6]	2.9	4.3	3.5	0.11	<0.001	<0.001	<0.001

[a,b] Rows with different superscripts differ ($p < 0.05$). [1] Treatments CBH10 = crossbred Holstein x beef breeds such as Charolais or Limousine heifers, slaughtered at 10 months of age; HB11 = Holstein bulls, slaughtered at 11 months of age; CAB12= crossbred Holstein with Angus bulls, slaughtered at 12 months of age. [2] SEM = standard error of the mean; [3] Production system effect; Day = time effect (2, 6, and 9 d after packaging in MAP); Production system x Day = production system by time interaction effect. [4] Instrumental color: L* = lightness, a* = redness, and b* = yellowness. [5] 5-point scale, color perception: 1 = highly undesirable, 2 = moderately undesirable, 3 = slightly desirable, 4 = moderately desirable, and 5 = highly desirable. [6] 5-point scale, purchase decision: (1) would not buy, (2) would probably not buy, (3) buy dubiously, (4) would probably buy, and (5) would buy.

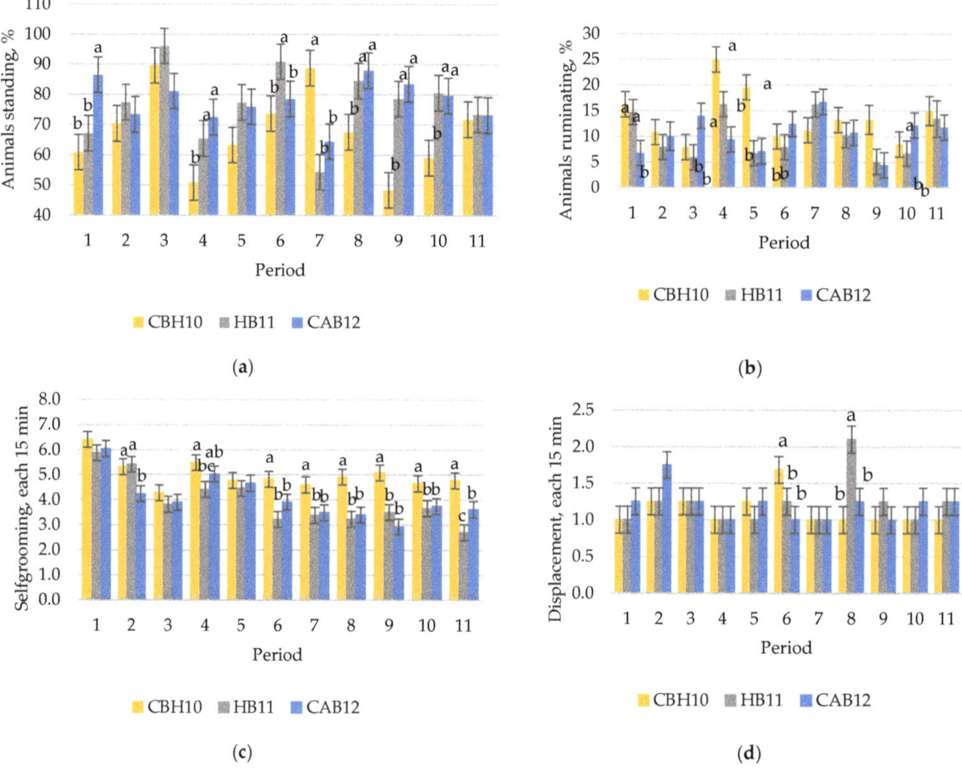

Figure 3. (**a**) Standing, (**b**) ruminating, (**c**) self-grooming, and (**d**) displacement behavior during the first 168 days of study in crossbred Holstein x beef breeds such as Charolais or Limousine heifers, slaughtered at 10 months of age (CBH10); Holstein bulls, slaughtered at 11 months of age; (HB11) and crossbred Holstein with Angus bulls, slaughtered at 12 months of age (CAB12) ([a,b,c] significant differences ($p < 0.05$) between production system within the same period).

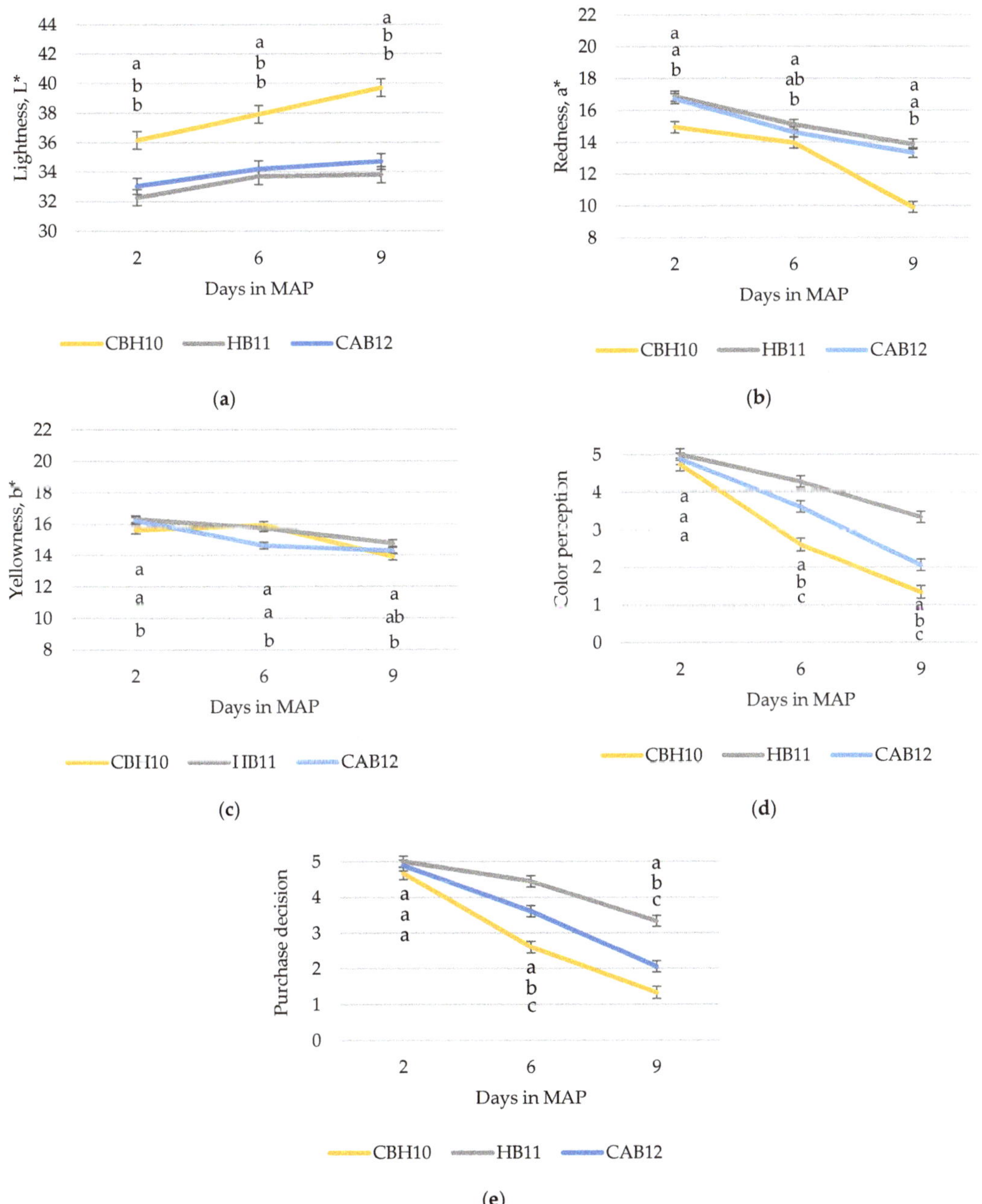

Figure 4. Evolution of the instrumental color ((**a**) lightness, (**b**) redness, (**c**) yellowness)), (**d**) color perception (5-point scale: 1 = highly undesirable, 2 = moderately undesirable, 3 = slightly desirable, 4 = moderately desirable, and 5 = highly desirable), and (**e**) purchase decision (5-point scale: (1) would not buy, (2) would probably not buy, (3) buy dubiously, (4) would probably buy, (5) would buy) of meat over the storage time in refrigeration and MAP conditions by production system (crossbred Holstein x beef breeds such as Charolais or Limousine heifers, slaughtered at 10 months of age (CBH10); Holstein bulls, slaughtered at 11 months of age (HB11); and crossbred Holstein with Angus bulls, slaughtered at 12 months of age (CAB12)) ([a,b,c] significant differences ($p < 0.05$) between production system within the same day).

4. Discussion

When crossbred heifers, Holstein bulls, and Angus–Holstein crossbred bulls were raised under the same housing and management conditions and fed similar diets (concentrate and straw) during the first 168 d of study, heifers, as expected, had a reduced growth and worse efficiency in several growing periods. Some studies have reported that bulls gain weight more rapidly and efficiently than heifers [8], which has been attributed to the anabolic properties of androgens, in particular testosterone [22].

Furthermore, meat from heifers has more intramuscular fat and is more tender than meat from bulls slaughtered at the same age [11,12]; however, in the present study, bulls were slaughtered 41 to 56 days later (Holstein or crossbred Angus, respectively) than heifers, and this could explain the lack of significant differences in intra-muscular fat content since it has been reported that in Holstein bulls intramuscular fat increases around 0.3% every 30 d [21]. It was expected that bulls would be more sexually active and more susceptible to stress than heifers, and they would be more susceptible to pre-slaughter stressors (handling, transportation, lairage), and the meat of their carcasses would be more prone to become DFD (dry, firm, and dark) [10]. However, in the present study, Holstein bulls had the lowest meat pH even if their sexual and agonistic behaviors were not significantly different than those of the crossbred Angus bulls. Some authors have reported differences in pH in bulls slaughtered at different ages and showed that pH was higher at 12 than at 10 and 14 months [21]. Nevertheless, comparisons between works are difficult since the production systems are, not the same, and there are many factors that could affect the pH. The meat pH plays a very important role in technological quality as it largely determines shelf-life and processability as well as water holding capacity and may affect sensory quality attributes, such as the visual perception of color or its tenderness. However, in the present study, differences observed among the three production systems were not considered as relevant since the mean pH was below 6.0 in all production systems. Maybe unrecorded factors such as temperature during transport, waiting time in lairage pens prior to slaughter, or other stressors before slaughtering [23], as well as cooling conditions, could have affected meat quality parameters, such as the pH. In addition, when analyzing the performance data of the global production system, no great differences in efficiency (expressed as growth divided by total concentrate intake) were observed; these results are probably related to the impairment of growth and efficiency with increasing slaughter age [21]. However, a greater impairment of global carcass efficiency (carcass yield expressed as percentage of total concentrate intake) was observed in Holstein or Angus crossbred bulls than in crossbred heifers. Meat quality data, such as intramuscular fat content and meat tenderness, were also very similar among production systems; these data are indicative that each production system had optimized the slaughter age and nutrition program and similar meat quality could be achieved. However, this affirmation is not fully supported by the meat color and purchase decision data. The evolution of the color of the heifer meat (lightness, redness), color perception, and purchase decision were indicative of an unexpected decrease in meat quality and duration of meat shelf-life. These data are not surprising as meat color is the first criterion for consumer appreciation of meat at the time of purchase [24]. Meat color can be affected by many factors (age, gender, type of muscle, intramuscular fat content), but in the present study, the intake of antioxidants via feed may be one of the potential reasons behind this impaired meat color in the heifer meat [25]. The vitamin E content of the growing concentrate was 30 UI per kg, less than the vitamin E content of the finishing concentrate, which was 202 UI per kg; in the present, study animals were fed the same concentrate from day 0 to 168 to be able to compare the different production systems, and the vitamin E concentration was not adapted for the females that were slaughtered at day 170 of the study and, consequently, were not fed finishing concentrate. The authors suggest that supplementing feedlot cattle, especially Holstein steers, with vitamin E extended the color display stability of fresh beef [26]. This was accomplished whether an additional 300 IU/d was supplemented for 9 months, 1140 IU/d for 67 d, or 1200 IU/d for 38 d. In the present study, all animals consumed less than 200 IU daily for 168 days; thereafter, heifers were slaughtered and bulls consumed the finishing concentrate, in which vitamin E was highly increased so that bulls consumed around 1400 IU daily before slaughter, either for 40 or 57 days before slaughter for Holstein bulls or crossbred Angus, respectively. To be able to confirm that vitamin E content was causing the lowest meat color perception scores and the lightness and redness meat, an additional study should be performed. Another unexpected result was that when comparing meat from Holstein bulls vs. crossbred Angus bulls after 6 and 9 days of display in MAP, Holstein bull meat had higher color preference and purchase decision scores than Angus bull meat, although the amount of Vitamin E consumed was higher in the Angus bulls. According to the results of the present study (performance and meat quality data, purchase decision), moving from a production system based on Holstein bulls slaughtered at 11 months to the crossbred Angus bulls slaughtered at 12 months has no technical (performance, animal behavior or meat quality) data support, and only marketing strategies for meat differentiation could support this decision. The decision of crossing dairy cows with Angus was based on the hypothesis that meat from Angus animals is perceived by the consumers as the meat with the best quality [27]; however, the Angus meat that consumers perceive as a good quality meat is usually coming from pure breeds, castrated, probably hormone-implanted animals with a different meat aging protocols, improving meat intramuscular fat, tenderness, and flavor [28]. At the point of purchase, aspects such as color, freshness, appearance, and fat quantity, as well as price and expiration date, are the most important factors for Spanish consumers [29], and, partly, they can be modified with the production system, including breed, age at slaughter, type of feeding, and transport conditions. Thus, any change in the production system should consider all these factors aiming to match consumer demands. As discussed previously, non-technical-based marketing strategies favoring specific claims, such as the production system type, could bias the consumer's choice, thus pushing the farmers to match market demands.

5. Conclusions

No great differences in efficiency (expressed as ADG divided by total concentrate intake), intramuscular fat, or meat tenderness among the three Mediterranean productions systems evaluated (crossbred Holstein x beef breeds such as Charolais or Limousine heifers slaughtered at 10 months of age, Holstein bulls slaughtered at 11 months of age and crossbred Holstein with Angus bulls slaughtered at 12 months of age) were observed. Surprisingly, Holstein bulls had the lowest meat pH even if their sexual and agnostic behavior was not significantly different from the crossbred Angus bulls. However, carcass conformation of crossbred Angus bulls was greater than in Holstein bulls. The evolution of color of heifer meat (lightness, redness), color perception, and purchase decision were indicative of an unexpected impairment in meat quality and meat shelf-life; maybe reduced antioxidant consumption could be one of the potential explanations. Additionally, meat shelf-life was less in crossbred Angus bulls than in Holstein bulls. In summary, according to the present study (performance and meat quality data), moving from a production system based on Holstein bulls to crossbred Angus has no technical support except if carcass conformation wants to be improved, and only marketing strategies for meat differentiation could support this decision.

Author Contributions: Conceptualization, M.D. and M.V.; methodology, M.D., M.F.-i.-F., J.G.; validation, S.M.; formal analysis, S.M., M.V., J.G.; investigation, D.S.; resources, M.D. and M.V.; data curation, D.S., S.M.; writing—D.S., M.D., S.M., M.F.-i.-F.; supervision, S.M.; project administration, M.D.; funding acquisition, M.D. and M.V. All authors have read and agreed to the published version of the manuscript.

Funding: This research was financially supported by bonÀrea Agrupa, 25210 Guissona, Lleida, Spain. IRTA also thanks the support of the Generalitat de Catalunya through the CERCA Program.

Institutional Review Board Statement: This study was conducted in accordance with the Spanish guidelines for experimental animal protection (Royal Decree 53/2013 of February 1st on the protection of animals used for experimentation or other scientific purposes; Boletín Oficial del Estado, 2013).

Informed Consent Statement: Not applicable.

Data Availability Statement: Not applicable.

Acknowledgments: Finally, the authors thank the collaboration of the personnel of Agropecuària Montgai S.L. (Lleida, Spain) for their assistance with animal care and the collaboration of their personnel during the development of this study.

Conflicts of Interest: The authors declare no conflict of interest.

References

1. Hocquette, J.F.; Ellies-Oury, M.P.; Lherm, M.; Pineau, C.; Deblitz, C.; Farmer, L. Current situation and future prospects for beef production in Europe—A review. *Asian-Australas. J. Anim. Sci.* **2018**, *31*, 1017–1035. [CrossRef] [PubMed]
2. Clinquart, A.; Oury, M.; Hocquette, J.-F.; Guillier, L.; Sante-Lhoutellier, V.; Prache, S. Review: On-farm and processing factors affecting bovine carcass and meat quality. *Animal* **2022**, *16*, 100426. [CrossRef]
3. Savoia, S.; Brugiapaglia, A.; Pauciullo, A.; Di Stasio, L.; Schiavon, S.; Bittante, G.; Albera, A. Characterisation of beef production systems and their effects on carcass and meat quality traits of Piemontese young bulls. *Meat Sci.* **2019**, *153*, 75–85. [CrossRef]
4. Serrapica, F.; Masucci, F.; De Rosa, G.; Calabrò, S.; Lambiase, C.; Di Francia, A. Chickpea Can Be a Valuable Local Produced Protein Feed for Organically Reared, Native Bulls. *Animal* **2021**, *11*, 2353. [CrossRef]
5. Albertí, P.; Panea, B.; Sañudo, C.; Olleta, J.L.; Ripoll, G.; Ertbjerg, P.; Christensen, M.; Gigli, S.; Failla, S.; Concetti, S.; et al. Live weight, body size and carcass characteristics of young bulls of fifteen European breeds. *Livest. Sci.* **2008**, *114*, 19–30. [CrossRef]
6. Devant, M.; Penner, G.B.; Marti, S.; Quintana, B.; Fábregas, F.; Bach, A.; Arís, A. Behavior and inflammation of the rumen and cecum in Holstein bulls fed high-concentrate diets with different concentrate presentation forms with or without straw supplementation. *J. Anim. Sci.* **2016**, *94*, 3902–3917. [CrossRef] [PubMed]
7. Ministerio de Agricultura, Pesca Y Alimentación. *Estudio del Sector Español de Cebo de Vacuno. Datos SITRAN*; Secretaría GeneralTécnica, Centro de Publicaciones: Madrid, Spain, 2019; pp. 1–40. Available online: http://publicacionesoficiales.boe.es/ (accessed on 1 August 2022).
8. Bureš, D.; Barton, L. Growth performance, carcass traits and meat quality of bulls and heifers slaughtered at different ages. *Czech J. Anim. Sci.* **2012**, *57*, 34–43. [CrossRef]
9. Mach, N.; Bach, A.; Realini, C.E.; Font i Furnols, M.; Velarde, A.; Devant, M. Burdizzo pre-pubertal castration effects on performance, behaviour, carcass characteristics, and meat quality of Holstein bulls fed high-concentrate diets. *Meat Sci.* **2009**, *81*, 329–334. [CrossRef]
10. Monin, G. Facteurs biologiques des qualités de la viande bovine. *INRA. Prod. Anim.* **1991**, *4*, 151–160. [CrossRef]
11. Modzelewska-Kapituła, M.; Nogalski, Z. Effect of gender on collagen profile and tenderness of infraspinatus and semimembranosus muscles of Polish Holstein-Friesian x limousine crossbred cattle. *Livest Sci.* **2014**, *167*, 417–424. [CrossRef]
12. Venkata Reddy, B.; Sivakumar, A.S.; Jeong, D.W.; Woo, Y.B.; Park, S.J.; Lee, S.Y.; Byun, J.Y.; Kim, C.H.; Cho, S.H.; Hwang, I. Beef quality traits of heifer in comparison with steer, bull and cow at various feeding environments. *J. Anim. Sci.* **2015**, *86*, 1–16. [CrossRef]
13. Sinclair, K.D.; Lobley, G.E.; Horgan, G.W.; Kyle, D.J.; Porter, A.D.; Matthews, K.R.; Commission, L.; Drive, S.; Mk, M.K. Factors influencing beef eating quality 1. Effects of nutritional regimen and genotype on organoleptic properties and instrumental texture. *J. Anim. Sci.* **2001**, *72*, 269–277. [CrossRef]
14. Verdú, M.; Bach, A.; Devant, M. Effect of feeder design and concentrate presentation form on performance, carcass characteristics, and behavior of fattening Holstein bulls fed high-concentrate diets. *Anim. Feed Sci. Technol.* **2017**, *232*, 148–159. [CrossRef]
15. Paniagua, M.; Crespo, J.; Arís, A.; Devant, M. *Citrus aurantium* flavonoid extract improves concentrate efficiency, animal behavior, and reduces rumen inflammation of Holstein bulls fed high-concentrate diets. *Anim. Feed Sci. Technol.* **2019**, *258*, 114304. [CrossRef]
16. Ferret, A.; Calsamiglia, S.; Bach, A.; Devant, M.; Fernández, C.; García-Rebollar, P.; Fundación española para el Desarrollo de la Nutrición Animal (FEDNA). Necesidades Nutricionales Para Ruminates de Cebo. 2008. Available online: http://www.fundacionfedna.org/sites/default/files/NORMAS_RUMIANTES_2008.pdf (accessed on 1 August 2022).
17. AOAC. *Official Methods of Analysis*, 17th ed.; Association of the Official Analytical Chemists: Arlington, VA, USA, 2005.
18. Van Soest, P.J.; Robertson, J.B.; Lewis, B.A. Methods for dietary fiber, neutral detergent fiber, and nonstarch polysaccharides in relation to animal nutrition. *J. Dairy Sci.* **1991**, *74*, 3583–3597. [CrossRef]
19. AMSA. *Research Guidelines for Cookery, Sensory Evaluation and Instrumental Measurements of Fresh Meat*; American Meat Science Association and National Livestock and Meat Board: Chicago, IL, USA, 1995.
20. Marti, S.; Realini, C.E.; Bach, A.; Pérez-Juan, M.; Devant, M. Effect of vitamin A restriction on performance and meat quality in finishing Holstein bulls and steers. *Meat Sci.* **2011**, *89*, 412–418. [CrossRef]

21. Marti, S.; Realini, C.E.; Bach, A.; Pérez-Juan, M.; Devant, M. Effect of castration and slaughter age on performance, carcass, and meat quality traits of Holstein calves fed a high-concentrate diet. *J. Anim. Sci.* **2013**, *91*, 1129–1140. [CrossRef]
22. Heitzman, R.J. The effectiveness of anabolic agents in increasing rate of growth in farm animals; report on experiments in cattle. *Environ. Qual. Saf. Suppl.* **1976**, *5*, 89–98.
23. Flores, A.; Linares, C.; Saavedra, F.; Serrano, A.B.; Lopez, E.S. Evaluation of changes in management practices on frequency of DFD meat in cattle. *J. Anim. Vet. Adv.* **2008**, *7*, 319–321.
24. Font-i-Furnols, M.; Guerrero, L. Consumer preference, behavior and perception about meat and meat products: An overview. *Meat Sci.* **2014**, *98*, 361–371. [CrossRef] [PubMed]
25. Castillo, C.; Pereira, V.; Abuelo, Á.; Hernández, J. Effect of supplementation with antioxidants on the quality of bovine milk and meat production. *Sci. World J.* **2013**, *2013*, 616098. [CrossRef]
26. Arnold, R.N.; Scheller, K.K.; Arp, S.C.; Williams, S.N.; Buege, D.R.; Schaefer, D.M. Effect of long- or short-term feeding of alpha-tocopheryl acetate to Holstein and crossbred beef steers on performance, carcass characteristics, and beef color stability. *J. Anim. Sci.* **1992**, *70*, 3055–3065. [CrossRef]
27. Bureš, D.; Bartoň, L. Performance, carcass traits and meat quality of Aberdeen Angus, Gascon, Holstein and Fleckvieh finishing bulls. *Livest. Sci.* **2018**, *214*, 231–237. [CrossRef]
28. Reichhardt, C.C.; Messersmith, E.M.; Brady, T.J.; Motsinger, L.A.; Briggs, R.K.; Bowman, B.R.; Hansen, S.L.; Thornton, K.J. Anabolic Implants Varying in Hormone Type and Concentration Influence Performance, Feeding Behavior, Carcass Characteristics, Plasma Trace Mineral Concentrations, and Liver Trace Mineral Concentrations of Angus Sired Steers. *Animal* **2021**, *11*, 1964. [CrossRef]
29. Magalhaes, D.R.; Maza, M.T.; Do Prado, I.N.; Fiorentini, G.; Kirinus, J.K.; del Mar Campo, M. An Exploratory Study of the Purchase and Consumption of Beef: Geographical and Cultural Differences between Spain and Brazil. *Foods* **2022**, *11*, 129. [CrossRef]

Article

Predicting Beef Carcass Fatness Using an Image Analysis System

José A. Mendizabal [1,*], Guillerno Ripoll [2], Olaia Urrutia [1], Kizkitza Insausti [1], Beatriz Soret [1] and Ana Arana [1]

[1] IS-FOOD Research Institute, Campus de Arrosadia, Universidad Pública de Navarra, 31006 Pamplona, Spain; olaia.urrutia@unavarra.es (O.U.); kizkitza.insausti@unavarra.es (K.I.); soret@unavarra.es (B.S.); aarana@unavarra.es (A.A.)

[2] Centro de Investigación y Tecnología Agroalimentaria de Aragón (CITA), Instituto Agroalimentario de Aragón–IA2 (CITA-Universidad de Zaragoza), Avda. Montañana 930, 50059 Zaragoza, Spain; gripoll@cita-aragon.es

* Correspondence: jamendi@unavarra.es

Simple Summary: The degree of conformation and the degree of fatness are the primary parameters taken by the European beef carcass classification system (the SEUROP system) for assessing carcass quality and pricing. Evaluations have conventionally been performed by graders suitably trained using photographic standards but in recent years new techniques have been developed to enhance grading accuracy and objectivity. This study reports a method that uses an image analysis to assess the degree of fatness of beef carcasses. The results obtained show that the accuracy significantly improves by using this image analysis method compared with the conventional method that assigns scores based on photographic standards. It would therefore be appropriate to implement this technique on slaughter lines to improve the beef carcass classification system.

Abstract: The amount and distribution of subcutaneous fat is an important factor affecting beef carcass quality. The degree of fatness is determined by visual assessments scored on a scale of five fatness levels (the SEUROP system). New technologies such as the image analysis method have been developed and applied in an effort to enhance the accuracy and objectivity of this classification system. In this study, 50 young bulls were slaughtered (570 ± 52.5 kg) and after slaughter the carcasses were weighed (360 ± 33.1 kg) and a SEUROP system fatness score assigned. A digital picture of the outer surface of the left side of the carcass was taken and the area of fat cover (fat area) was measured using an image analysis system. Commercial cutting of the carcasses was performed 24 h post-mortem. The fat trimmed away on cutting (cutting fat) was weighed. A regression analysis was carried out for the carcass cutting fat (y-axis) on the carcass fat area (x-axis) to establish the accuracy of the image analysis system. A greater accuracy was obtained by the image analysis ($R^2 = 0.72$; $p < 0.001$) than from the visual fatness scores ($R^2 = 0.66$; $p < 0.001$). These results show the image analysis to be more accurate than the visual assessment system for predicting beef carcass fatness.

Keywords: carcass fatness; image analysis; prediction; young bulls

Citation: Mendizabal, J.A.; Ripoll, G.; Urrutia, O.; Insausti, K.; Soret, B.; Arana, A. Predicting Beef Carcass Fatness Using an Image Analysis System. *Animals* **2021**, *11*, 2897. https://doi.org/10.3390/ani11102897

Academic Editor: John-Erik Haugen

Received: 14 September 2021
Accepted: 1 October 2021
Published: 5 October 2021

Publisher's Note: MDPI stays neutral with regard to jurisdictional claims in published maps and institutional affiliations.

1. Introduction

The EU's Beef Carcass Classification System (SEUROP) [1] is chiefly based on two parameters: conformation and the degree of carcass fat cover. These two parameters, plus weight, are used to determine the beef carcass quality and price.

Carcass conformation and fatness are evaluated using photographic standards depicting the different conformation (SEUROP scale) and fatness (scale from 1 to 5) grades. The grades are assigned by duly accredited slaughterhouse staff trained by official agencies and who have passed the corresponding grading test. However, this method has occasionally been faulted as being subject to a certain degree of subjectivity that could result in differences in grading by slaughterhouses and even within each individual slaughterhouse depending on the grader on duty [2].

For this reason, more objective tools for classifying beef carcasses have been developed in recent years [3–5]. These include instrumental methods such as ultrasound [6–9], computed tomography [10,11], dual energy X-ray absorptiometry (DEXA) [12,13], a bioelectrical impedance analysis (BIA) [14], near-infrared spectroscopy [15,16], microwave systems [17] and, above all, an image analysis [18–20]. Artificial intelligence-based techniques for modelling carcass classification parameters by official graders have even been developed [21].

These instrumental methods achieve highly accurate measurements for one of the two main classification measures used to grade carcass quality in the EU's classification system, namely, beef carcass conformation and have even shown to be capable of predicting commercial carcass meat yields [19–22]. However, the results for predicting carcass fatness levels do not reach the same satisfactory results because fat distribution is highly variable and not so easy to measure [19,20,23,24].

This article therefore presents an attempt to develop an image analysis application for obtaining accurate, objective measurements of the degree of carcass fatness capable of enhancing the official method based on the visual classification of beef carcasses currently in use.

2. Materials and Methods

2.1. Animals and Feeding

A total of 50 young bulls of the local Pirenaica (n = 25) and Asturiana (n = 25) breeds from northern Spain were weaned at a live weight of 271 ± 28.4 kg and 181 ± 30.2 days of age. During the feeding stage, at the CITA research center (Zaragoza, Spain), the animals were given ad libitum access to a commercial concentrate and barley straw. An initial concentrate (12.9 MJ metabolizable energy (ME) and 160 g crude protein (CP) per kg dry matter (DM)) was used up to a weight of 320 kg. Thereafter, a finishing concentrate (13.0 MJ ME and 150 g CP per kg DM) was used. The young bulls were slaughtered at 570 ± 52.5 kg live weight and 409.7 ± 27.04 days of age. Procedures involving animal care and use were conducted following the European guidelines [25].

2.2. Slaughter and Fatness Measurement

The slaughter of the animals was conducted at the General Refrigerated Slaughterhouse in Zaragoza according to EU regulations [26]. The hot carcass weight (HCW) was recorded after slaughter. Thereafter, the carcasses were chilled for 24 h at 4 °C and were classified using the SEUROP five-class fat cover notation system (from 15 for 5+, very high fat cover, to 1 for 1−, very low fat cover) [1]. The evaluations were carried out by two official graders.

2.3. Acquisition of VIA Images

A picture of each left half-carcass was taken (Figure 1) using a digital camera (Olympus E-300 SLR) with an 8 megapixel sensor (3264 × 2448) and an Olympus Zuiko 28 mm f/2.8 lens. The camera settings were: manual operation mode, aperture F/5.6, shutter 1/60, ISO 400, flash off. The camera was mounted perpendicularly to the carcass at a height of 1.75 m on a tripod placed 3 m from the carcass. The illumination intensity in the cold room where the photographs were taken was 500 lux.

2.4. Image Processing

The images acquired were processed using the ImageJ image analysis software (National Institutes of Health, USA). To that end, the images were binarized (8 bit format) and the contours of the area of the carcass were outlined. The carcass mean gray level was then calculated based on the gray scale, which assigns a value between 0 (pure black) and 255 (pure white) to each image pixel (gray level). The optimum gray level threshold value capable of accurately discriminating the whiter parts of the carcass (fat cover) from the darker parts (no fat cover) was then determined. The optimum threshold values for the individual beef carcasses ranged from 100 to 112 depending on the fat color and muscle color, which vary from one animal to another. It was decided to use a single threshold value for all the images in the interest of achieving a greater level of standardization for the method and, hence, the mean value of 106 was selected. Accordingly, the area obtained by grouping together all the pixels with a gray level value greater than 106 was taken as representing the carcass fat cover and given the parameter designation "carcass fat area". Figure 1 illustrates the different steps in the method.

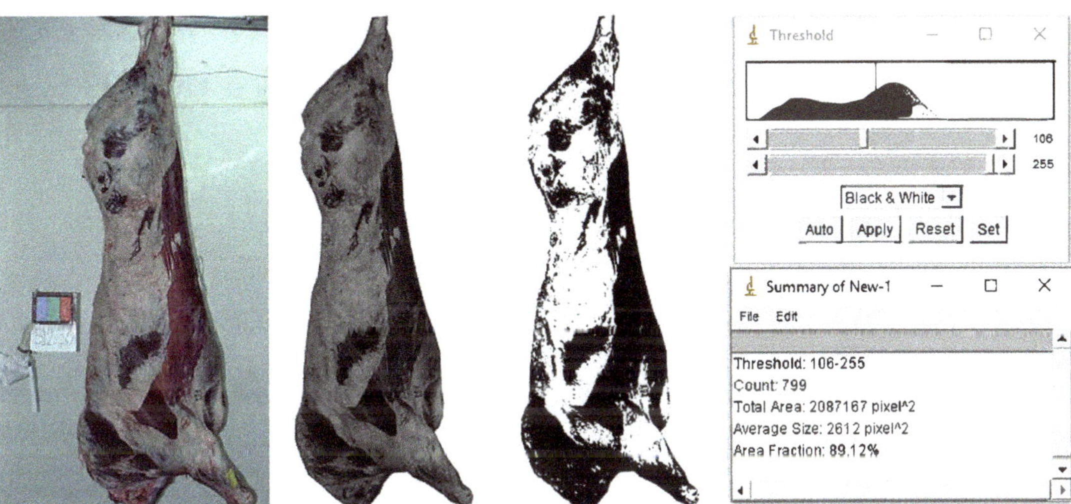

Figure 1. Image acquisition and processing for quantifying the carcass fat area.

2.5. Cutting Fat

The carcasses were dressed 24 h after slaughter using the method described by Panea et al. [27] to obtain the different commercial cuts from the carcasses. During the process, the weights of the different types of fat (basically kidney and pelvic fat, intermuscular fat and subcutaneous fat) not attached to the commercial cuts were recorded and the total was designated "cutting fat". This value was taken as an indicator of carcass fatness.

2.6. Statistical Analyses

An analysis of variance was used to examine the differences in the amount of cutting fat, the mean gray level and the carcass fat area values based on the fatness score assigned to the carcasses using the SEUROP system. Differences among the means were tested by Tukey's HSD test. Simple and multiple (stepwise) linear and non-linear regression analyses were used to study the relationships among the four variables considered. The multiple regressions also included the HCW. Available variables to the models were untransformed variables; the same variables squared and base 10 and e logarithms were also tested. The third level interactions between the variables were included as an option. The variables were retained in the models when $p < 0.05$. The coefficients of determination of each model obtained (R^2) and the standard error of the model were estimated and presented.

A statistical analysis of the data was performed using SPSS statistical software, version 27.0 [28].

3. Results

3.1. Carcass Fat Measurements

The young bull carcasses considered in this experiment were divided into four groups according to the fatness scores assigned using the SEUROP method (a scale of 1–5 with each score broken down into 3 sub-scores on a scale of 1 (low) to 15 (high)). The scores recorded were 3 (n = 9), 4 (n = 6), 5 (n = 28) and 6 (n = 7) (Table 1).

Table 1. Carcass fat area (mean ± se) based on the fatness scores (1–15) assigned to the carcasses.

Score (SEUROP: 1–15(1–5))	3 (1+) (n = 9)	4 (2−) (n = 6)	5 (2) (n = 28)	6 (2+) (n = 7)	*p*-Value
HCW (kg)	360 ± 12.2	358 ± 7.1	363 ± 6.8	347 ± 11.3	0.715
Cutting fat (%)	4.2 ± 0.3 [a]	6.4 ± 0.5 [b]	9.2 ± 0.4 [c]	12.8 ± 0.9 [d]	0.000
Mean gray level (0–255)	78 ± 1.6 [a]	84 ± 2.1 [a]	93 ± 2.1 [b]	99 ± 3.4 [b]	0.000
Carcass fat area (%)	25.0 ± 1.4 [a]	31.5 ± 2.4 [a]	41.7 ± 1.8 [b]	48.4 ± 2.8 [b]	0.000

Different letters (e.g., a, b, c) represent $p \leq 0.05$; the same or no letters represent $p > 0.05$.

The gray level values processed from the digitized carcass images yielded mean values between 78 and 99 (on a scale of 0–255) with values increasing in line with the fatness scores (Table 1).

Finally, an image analysis quantification of the degree of fat cover of the carcasses returned a mean value of 38.4 ± 1.6% (on a scale of 1–100) with values ranging from 19.2% to 62.3%. On that basis, a fatness score of 2+, for instance, represented carcasses with a fat cover over approximately half the surface area.

3.2. Predicting the Cutting Fat from the SEUROP Scores

Figure 2 plots the linear regression of the amount of carcass cutting fat (*y*-axis) on the fatness score assigned to the carcass using the SEUROP classification system (*x*-axis). The regression equation showed that the SEUROP fatness score explained 66.4% (R^2 = 0.66) of the variation recorded in the amount of carcass cutting fat (RMSE = 1.86%).

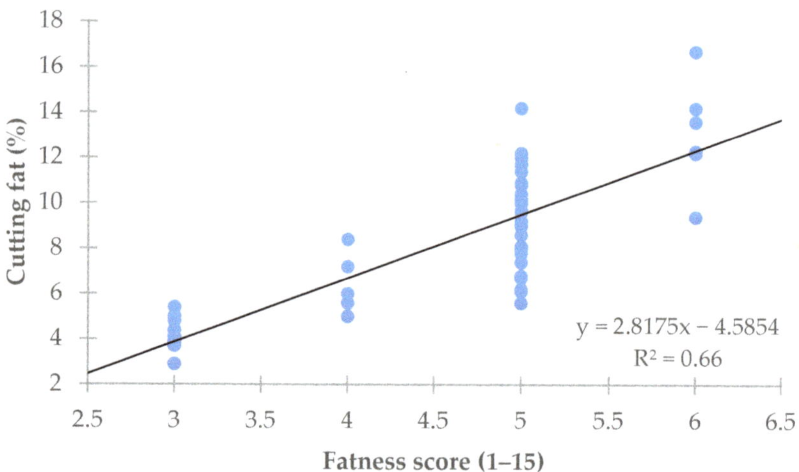

Figure 2. Linear regression of the cutting fat on the SEUROP fatness score.

3.3. Predicting the Cutting Fat from the Mean Gray Level Value

Figure 3 plots the regression of the amount of carcass cutting fat (*y*-axis) on the mean carcass gray level (*x*-axis). The regression equation showed that the mean gray level value explained 61.0% (R^2 = 0.61) of the variation recorded in the amount of carcass cutting fat (RMSE = 2.01%).

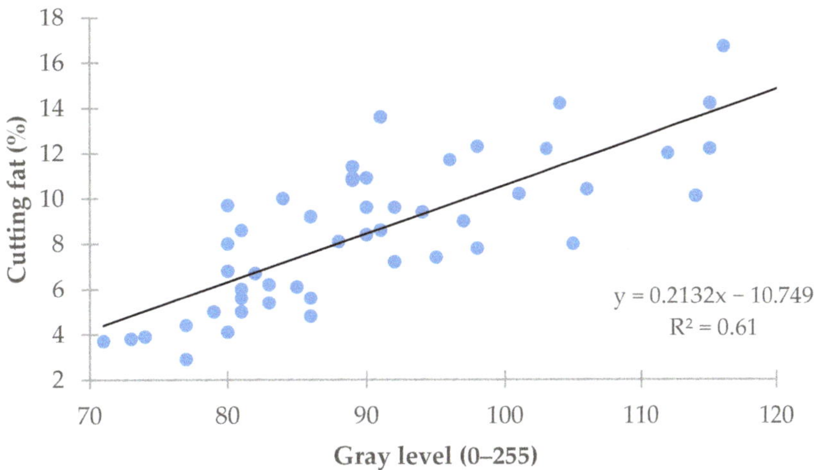

Figure 3. Regression of the cutting fat on the carcass gray level.

3.4. Predicting the Cutting Fat from the Carcass Fat Area

Figure 4 plots the regression of the amount of carcass cutting fat (*y*-axis) on the carcass fat area (*x*-axis) obtained using image analysis. Carcass fat area explained 72.0% ($R^2 = 0.72$) of the variation recorded in the amount of carcass cutting fat (RMSE = 1.70%).

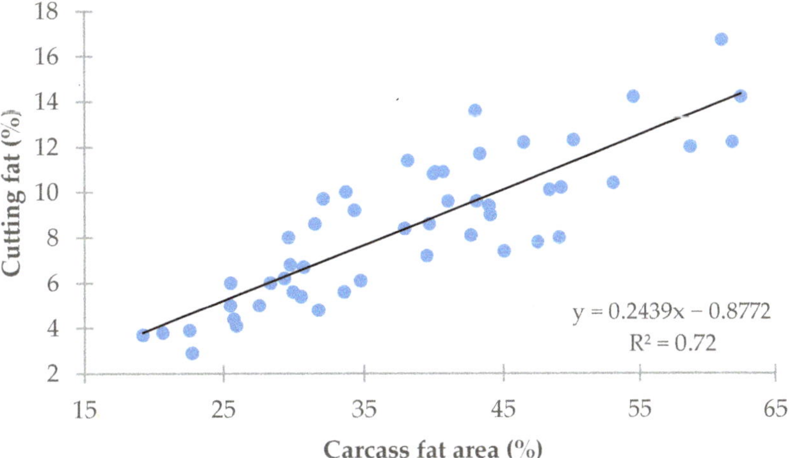

Figure 4. Regression of cutting fat on carcass fat area.

3.5. Predicting the Cutting Fat from the Set of Variables

Figure 5 depicts the most salient results found on running the various multiple regressions on the set of variables considered with the HCW (hot carcass weight) variable included in the models.

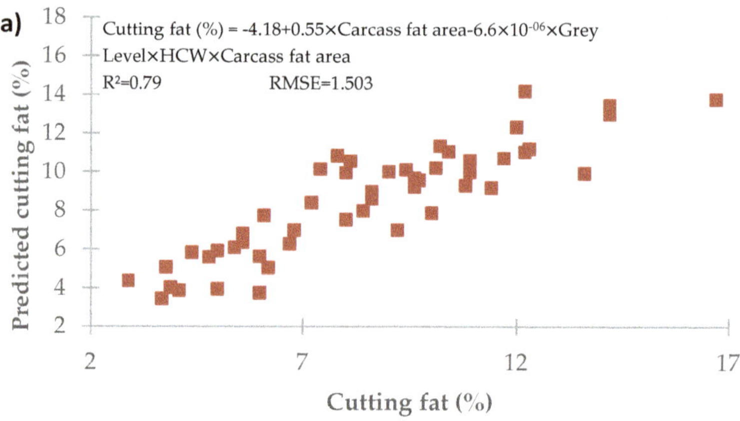

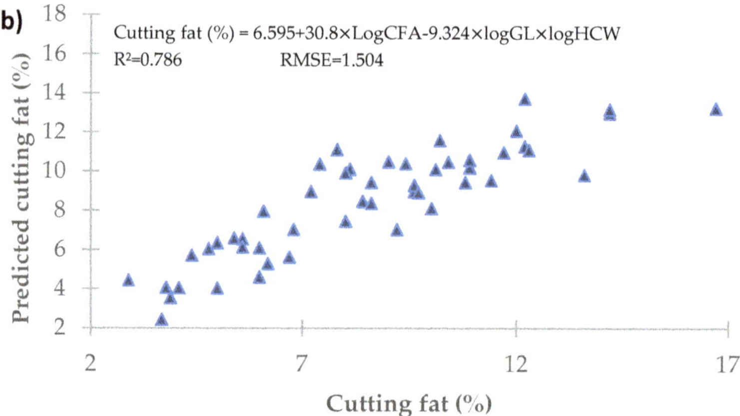

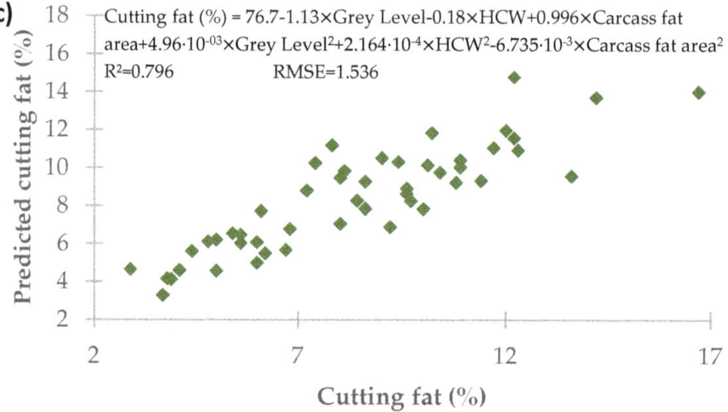

Figure 5. Multiple regressions. (**a**) Stepwise model based on the gray level, carcass weight and carcass fat area and their third-order interactions; (**b**) stepwise model based on the gray level, carcass weight and carcass fat area with their log transformations and fourth-order interactions; (**c**) stepwise model based on the gray level, carcass weight and carcass fat area with their log and square transformation and fourth-order interactions.

These models were significantly better at predicting the carcass cutting fat, attaining values of $R^2 > 0.79$ and RMSE < 1.54%.

4. Discussion

The EU's SEUROP beef carcass classification system is based on assigning visual scores for carcass conformation and carcass fatness. It is widely used at slaughterhouses across Europe; however, it is often faulted for a certain degree of subjectivity in scoring by official graders and scores can vary from one slaughterhouse to another [2]. This has substantial economic repercussions as grading scores have a major impact on beef carcass prices. Albertí et al. [29] estimated that a difference of one point in a carcass's conformation score may result in a 6 to 10% difference in the carcass's price per kg.

This being the case, a number of different techniques are in study in an effort to make carcass grading more objective. The greatest advances in this area have possibly been made for pork. Classification systems are officially approved by EU authorities and have been implemented following the adoption by the European Commission of regulations accepting and implementing new carcass classification technologies in 2017 [30]. Work in the field on beef carcasses has been under way for several years. A number of different methods are being tested and an image analysis is one of the more advanced technologies of this kind [5,19] (Craigie et al., 2012; Allen, 2021).

One of the two assessment parameters for carcass quality in the EU's SEUROP system, conformation assessment, has been comprehensively studied using an image analysis, obtaining very good results. The method has even been expanded to try to predict the commercial meat yields of beef carcasses, achieving very high coefficients of determination values greater than 0.80–0.90 [22].

However, less work has been performed to objectively assess the levels of carcass fatness and the results have not been as satisfactory. Fat is spread widely over beef carcasses and the distribution tends to be variable, which has been mentioned as one of the reasons for the poor results obtained up to date [19,24].

The work reported here employed an image analysis, a method that is simple yet quite objective, quantifiable and repeatable. It can provide accurate information on the degree of carcass fatness, the second of the beef carcass classification parameters. The method, as developed in this work, allows a shift from a discontinuous fatness scale of just five points (or expanded into a 15-point scale) to a continuous percentage scale from 0 to 100, which is much more comprehensive and therefore more accurate.

The results of the work reported here may have been conditioned by the type of animals included in the study. All the carcasses harvested were assigned SEUROP classification fatness scores between 3 and 6 (on a scale of 1–15) (Table 1). That is, the carcasses had little fat in spite of coming from young bulls slaughtered at weights considered high for the usual endpoint in the traditional beef production systems in Spain (570 ± 52.5 kg live weight). These fatness results were consistent with the findings reported by Albertí et al. [29] and Soret et al. [31], who described the Asturiana and Pirenaica cattle breeds as low-marbled breeds. A similar percentage cutting fat values (Table 1) were obtained for carcasses from animals of these two breeds used in the current study. This value is much less than for other faster-growing Spanish breeds such as the Avileña, Morucha or Retinta breeds, which have cutting fat values greater than 10% [29].

An image analysis provides a simple way of predicting carcass fatness by measuring the mean carcass surface area color value by way of an initial approximation. Taking the gray level value as a basis over a scale ranging from 0 (pure black) to 255 (pure white), a high gray level value for a carcass as a whole is indicative of a high level of carcass fatness whereas a low value is indicative of a very lean carcass. This can be seen in Figure 3, which shows that the predictions of the amount of cutting fat based on the mean carcass gray level attained values of $R^2 = 0.61$ and RMSE = 2.01% ($p < 0.001$; Figure 3). Despite being statistically significant, these values were smaller than those obtained for the predictions of the amount of cutting fat as scored by official slaughterhouse graders based on visual assessments, which yielded values of $R^2 = 0.66$ (RSME = 1.86%; $p < 0.001$; Figure 2). In other words, this method of predicting carcass fatness based on the mean carcass gray level value was not superior to the visual grading by official graders.

Carcass fat area, i.e., the difference in the gray level values for carcass fat and muscle, is a second parameter or measure for predicting carcass fatness using an image analysis. In this study, carcass scoring according to the SEUROP scale yielded values between 3 and 6 compared with carcass fat area scores ranging from 25.0% to 48.4% on the scale for this second image analysis parameter (Table 1). That is, the image analysis scale was much broader and more accurate than the scale used in the SEUROP classification system.

The accuracy achieved in predicting the carcass fatness, measured in this study as the cutting fat, was greater for the image analysis method employed than for the official SEUROP classification system ($R^2 = 0.72$ vs. 0.66 and RMSE = 1.70% vs. 1.86%, respectively; Figures 2 and 4).

Accordingly, the image analysis results were better than the results obtained based on the fatness score from the visual grading and yielded a more objective assessment of the degree of carcass fatness. However, the sample size used in this study was limited and it would be appropriate to increase it in future studies.

Finally, a multiple regression model that included all the variables considered in this study yielded highly accurate predictions of beef carcass fatness, attaining values of $R^2 > 0.79$ and RMSE < 1.54%, showing that an image analysis is a method capable of providing quantifiable and objective predictions of beef carcass fatness.

5. Conclusions

The findings of this study suggest that measuring carcass fat area using an image analysis can be regarded as a suitable indicator of carcass fatness in young bulls of Spanish meat breeds. Furthermore, including this assessment method in the framework of the EU's SEUROP classification system could be worthwhile because it provides an objective measure of carcass fatness. Nevertheless, before applying an image analysis to other breeds or production systems, the method should be tested on the carcasses of fatter animals spanning the broadest possible range of fatness scores and, if it is feasible, spanning the entire interval from 1 to 5.

Author Contributions: Conceptualization, J.A.M., G.R. and A.A.; methodology, J.A.M., G.R., K.I. and B.S.; software, G.R. and O.U.; formal analysis, J.A.M., A.A. and B.S.; investigation, J.A.M., G.R. and K.I.; writing—original draft preparation, J.A.M. and G.R.; writing—review and editing, O.U., K.I., B.S. and A.A.; project administration, J.A.M. All authors have read and agreed to the published version of the manuscript.

Funding: This research received no external funding.

Institutional Review Board Statement: All procedures (transport and slaughter) employed in this study met ethical guidelines and complied with Spanish legal requirements [32].

Informed Consent Statement: Not applicable.

Data Availability Statement: The data presented in this study are available from the corresponding author on request.

Conflicts of Interest: The authors declare no conflict of interest.

References

1. Council Regulation (ECC) No 1026/91 of 22 April 1991 amending Regulation (EEC) No 1208/81 determining the Community scale for the classitication of carcases of adult bovine animals. *Off. J. Eur. Union* **1991**, *21*, L106/1–L106/4. [CrossRef]
2. Fisher, A. *Beef Carcass Classification in the EU: An Historical Perspective*; Wageningen Academic Publishers: Wageningen, The Netherlands, 2007; pp. 19–30.
3. Cross, H.R.; Whittaker, A.D. The role of instrument grading in a beef value-based marketing system. *J. Anim. Sci.* **1992**, *70*, 984–989. [CrossRef] [PubMed]
4. Delgado-Pando, G.; Allen, P.; Troy, D.J.; McDonnell, C.K. Objective carcass measurement technologies: Latest developments and future trends. *Trends Food Sci. Technol.* **2021**, *111*, 771–782. [CrossRef]

5. Allen, P. Recent developments in the objective measurement of carcass and meat quality for industrial application. *Meat Sci.* **2021**, *181*, 108601. [CrossRef]

6. Griffin, D.B.; Savell, J.W.; Recio, H.A.; Garrett, R.P.; Cross, H.R. Predicting carcass composition of beef cattle using ultrasound technology. *J. Anim. Sci.* **1999**, *77*, 889–892. [CrossRef] [PubMed]

7. Williams, A.R. Ultrasound applications in beef cattle carcass research and management. *J. Anim. Sci.* **2002**, *80*, E183–E188. [CrossRef]

8. Greiner, S.P.; Rouse, G.H.; Wilson, D.E.; Cundiff, L.V.; Wheeler, T.L. Prediction of retail product weigth and percentage using ultrasound and carcass measurements in beef cattle. *J. Anim. Sci.* **2003**, *81*, 1736–1742. [CrossRef] [PubMed]

9. Beriain, M.J.; Insausti, K.; Valera, M.; Indurain, G.; Purroy, A.; Carr, T.R.; Horcada, A. Effectiveness of using ultrasound readings to predict carcass traits and sensory quality in young bulls. *Comput. Electron. Agric.* **2021**, *183*, 106060. [CrossRef]

10. Prieto, N.; Navajas, E.A.; Richardson, R.I.; Ross, D.W.; Hyslop, J.J.; Simm, G.; Roehe, R. Predicting beef cuts composition, fatty acids and meat quality characteristics by spiral computed tomography. *Meat Sci.* **2010**, *86*, 770–779. [CrossRef] [PubMed]

11. Navajas, E.A.; Glasbey, C.A.; Fisher, A.V.; Ross, D.W.; Hyslop, J.J.; Richardson, R.I.; Simm, G.; Roehe, R. Assessing beef carcass tissue weights using computed tomography spirals of primal cuts. *Meat Sci.* **2010**, *84*, 30–38. [CrossRef] [PubMed]

12. Lopez-Campos, O.; Larsen, I.L.; Prieto, N.; Juarez, M.; Dugan, M.E.R.; Aalhus, J.L. Evaluation of Total Lean and Saleable Meat Yield Prediction Equations and Dual Energy X-Ray Absorptiometry for a Rapid, Non-Invasive Yield Prediction in Beef. *Meat Muscle Biol.* **2016**, *1*, 104. [CrossRef]

13. Calnan, H.; Williams, A.; Peterse, J.; Starling, S.; Cook, J.; Connaughton, S.; Gardner, G.E. A prototype rapid dual energy X-ray absorptiometry (DEXA) system can predict the CT composition of beef carcases. *Meat Sci.* **2021**, *173*, 108397. [CrossRef]

14. Zollinger, B.L.; Farrow, R.L.; Lawrence, T.E.; Latman, N.S. Prediction of beef carcass salable yield and trimmable fat using bioelectrical impedance analysis. *Meat Sci.* **2010**, *84*, 449–454. [CrossRef]

15. Shackelford, S.D.; Wheeler, T.L.; Koohmaraie, M. On-line classification of US Select beef carcasses for longissimus tenderness using visible and near-infrared reflectance spectroscopy. *Meat Sci.* **2005**, *69*, 409–415. [CrossRef] [PubMed]

16. Chapman, J.; Elbourne, A.; Truong, V.K.; Cozzolino, D. Shining light into meat—A review on the recent advances in in vivo and carcass applications of near infrared spectroscopy. *Int. J. Food Sci. Technol.* **2020**, *55*, 935–941. [CrossRef]

17 Marimuthu, J.; Loudon, K.M.W.; Gardner, G.E. Ultrawide band microwave system as a non invasive technology to predict beef carcase fat depth. *Meat Sci.* **2021**, *179*, 108455. [CrossRef]

18. Cross, H.R.; Gilliland, D.A.; Durland, P.R.; Seideman, S. Beef carcass evaluation by use of a video image analysis system. *J. Anim. Sci.* **1983**, *57*, 908–917. [CrossRef]

19. Craigie, C.R.; Navajas, E.A.; Purchas, R.W.; Maltin, C.A.; Bünger, L.; Hoskin, S.O.; Ross, D.W.; Morris, S.T.; Roehe, R. A review of the development and use of video image analysis (VIA) for beef carcass evaluation as an alternative to the current EUROP system and other subjective systems. *Meat Sci.* **2012**, *92*, 307–318. [CrossRef]

20. Alempijevic, A.; Vidal-Calleja, T.; Falque, R.; Quin, P.; Toohey, E.; Walmsley, B.; McPhee, M. Lean meat yield estimation using a prototype 3D imaging approach. *Meat Sci.* **2021**, *181*, 108470. [CrossRef] [PubMed]

21. Díez, J.; Albertí, P.; Ripoll, G.; Lahoz, F.; Fernández, I.; Olleta, J.L.; Panea, B.; Sañudo, C.; Bahamonde, A.; Goyache, F. Using machine learning procedures to ascertain the influence of beef carcass profiles on carcass conformation scores. *Meat Sci.* **2006**, *73*, 109–115. [CrossRef]

22. Oliver, A.; Mendizabal, J.A.; Ripoll, G.; Albertí, P.; Purroy, A. Predicting meat yields and commercial meat cuts from carcasses of young bulls of Spanish breeds by the SEUROP method and an image analysis system. *Meat Sci.* **2010**, *84*, 628–633. [CrossRef]

23. Vote, D.J.; Bowling, M.B.; Cunha, B.C.N.; Belk, K.E.; Tatum, J.D.; Montossi, F.; Smith, G.C. Video image analysis as a potential grading system for Uruguayan beef carcasses. *J. Anim. Sci.* **2009**, *87*, 2376–2390. [CrossRef]

24. Heggli, A.; Gangsei, L.E.; Røe, M.; Alveseike, O.; Vinje, H. Objective carcass grading for bovine animals based on carcass length. *Acta Agric. Scand. A Anim. Sci.* **2021**, *70*, 113–121. [CrossRef]

25. Directive 2010/63/EU of the European Parliament and of the Council of 22 September 2010 on the protection of animals used for scientific purposes. *Off. J. Eur. Union* **2010**, L276/33–L276/79.

26. Council Regulation (EC) No 1099/2009 of 24 September 2009 on the protection of animals at the time of killing. *Off. J. Eur. Union* **2009**, *53*, L303/1–L303/30.

27. Panea, B.; Ripoll, G.; Albertí, P.; Joy, M.; Teixeira, A. Atlas of dissection of ruminant's carcass. *ITEA* **2012**, *108*, 3–105.

28. IBM Corp. *IBM SPSS Statistic for Windows*; Version 27.0; IBM Corp.: Armonk, NY, USA, 2020.

29. Albertí, P.; Ripoll, G.; Goyache, F.; Lahoz, F.; Olleta, J.L.; Panea, B.; Sañudo, C. Carcass characterisation of seven Spanish beef breeds slaughtered at two commercial weights. *Meat Sci.* **2005**, *71*, 514–521. [CrossRef]

30. Commission Delegated Regulation (EU) 2017/1182—of 20 April 2017—Supplementing Regulation (EU) No 1308/2013 of the European Parliament and of the Council as regards the Union scales for the classification of beef, pig and sheep carcasses and as rega. *Off. J. Eur. Union* **2017**, *L171*, 74–99.

31. Soret, B.; Mendizabal, J.A.; Arana, A.; Alfonso, L. Expression of genes involved in adipogenesis and lipid metabolism in subcutaneous adipose tissue and longissimus muscle in low-marbled Pirenaica beef cattle. *Animal* **2016**, *10*, 2018–2026. [CrossRef]

32. Real Decreto 53/2013, de 1 de febrero, por el que se establecen las normas básicas aplicables para la protección de los animales utilizados en experimentación y otros fines científicos, incluyendo la docencia. *Off. Gaz. Spain* **2013**, *1337*, 1–49.

Article

Fatty Acid Composition of Salami Made by Meat from Different Commercial Categories of Indigenous Dairy Cattle

Marco Alabiso [1], Giuseppe Maniaci [1,*], Cristina Giosuè [2], Antonino Di Grigoli [1] and Adriana Bonanno [1]

[1] Dipartimento Scienze Agrarie, Alimentari e Forestali, University of Palermo, Viale delle Scienze, 90128 Palermo, Italy; marco.alabiso@unipa.it (M.A.); antonino.digrigoli@unipa.it (A.D.G.); adriana.bonanno@unipa.it (A.B.)

[2] Institute for Anthropic Impacts and Sustainability in the Marine Environment, National Council of Research (IAS-CNR), 90128 Palermo, Italy; cristina.giosue@ias.cnr.it

* Correspondence: giuseppe.maniaci@unipa.it

Simple Summary: The Cinisara is a Sicilian breed raised on pasture to produce the Caciocavallo Palermitano cheese. Even if it is penalized by competition with meat breeds, characterized by higher growth rate and yield, the production of fresh meat represents a considerable added value for the smallfarms. The meat of Cinisara is not appreciated, despite having a high content of iron, vitamin E, conjugated linoleic acid (CLA), and low content of lipids and cholesterol, above all due to incorrect management of the supply chain phases that negatively affects the quality of the final product. Alternative production such as bresaola and salami could contribute to the enhancement of Cinisara meat. The present study investigated the fatty acid profile of salamis produced by processing the meat of young bulls and adult cows with the addition of lard pork to provide additional information to that reported in a previous experiment on physicochemical and sensory properties. The results suggest the possibility of producing cured meats with Cinisara meat, even if the addition of pork lard mitigates some favorable effects deriving from the livestock system of this breed, based on grazing. Further studies should be conducted to investigate the possibility of making cured meats with beef only.

Abstract: In autochthonous dairy cattle farms, the production of salami could represent an alternative commercial opportunity. Therefore, a study was carried out to investigate the fatty acid (FA) composition of salami made using the meat from grazing (GB) or housed (HB) young bulls and grazing adult cows (AC) of Cinisara breed. The products were manufactured by adding 20% of pork lard. Animal category influenced the FA composition, although the addition of lard mitigated the differences found in fresh meat. The salami from GB showed higher polyunsaturated FA content ($p \leq 0.01$) and, in particular, a higher level of linoleic acid ($p \leq 0.05$), than from other animal categories. Salami made from AC meat showed lower polyunsaturated/saturated FA ratio ($p \leq 0.05$), but a better *n-6/n-3* ratio compared to HB ($p \leq 0.05$), due to the lower content of linoleic acid. Multivariate analysis showed an important influence of animal category on FA composition due to age, feeding system and meat fat content of animals, despite the addition of lard.

Keywords: cinisara breed; beef; cured meat; fat; fermented sausage

Citation: Alabiso, M.; Maniaci, G.; Giosuè, C.; Di Grigoli, A.; Bonanno, A. Fatty Acid Composition of Salami Made by Meat from Different Commercial Categories of Indigenous Dairy Cattle. *Animals* **2021**, *11*, 1060. https://doi.org/10.3390/ani11041060

Academic Editors: Guillermo Ripoll and Begoña Panea

Received: 14 March 2021
Accepted: 6 April 2021
Published: 8 April 2021

Publisher's Note: MDPI stays neutral with regard to jurisdictional claims in published maps and institutional affiliations.

1. Introduction

Cinisara cow, a Sicilian autochthonous dairy breed, is traditionally reared with a feeding system based on the prevalent exploitation of natural pasture, and its milk is used to produce the Caciocavallo Palermitano cheese [1–3]. The manufacture of processed products, such as salami and bresaola, an alternative to fresh meat, could increase the economic profit of farms, diversifying the offer in the market [4–7]; in particular, some cuts of the posterior quarter (semimembranosus, semitendinosus and quadriceps femorismuscles) and the anterior quarter (brachial biceps muscle) could be used for the production of bresaola and the rest of the meat for the production of salami.

In recent years, consumers' interest in products rich in protein and low in lipid has increased, promoting the consumption of alternative meat to pork. Traditionally, the salami is made with meat and fat of pork [8], and shows high-fat content [9]. Therefore, the reformulation of meat-based products is directed to the promotion of consumer health by reducing lipid content and improving its fatty acid (FA) profiles [10]. In particular, lipid reformulation is based on the use of low-fat meat and/or the replacing of the fat with one having better healthy characteristics. Interest in cured, fermented and dried products of other livestock animals, such as cattle, donkeys, mutton and poultry has greatly increased, representing alternative products in the market [11–13].

During fermentation and ripening of salami, several biochemical, microbiological, physical and sensorial changes occur in meat under defined conditions of temperature and relative humidity (RH). Specifically, carbohydrates, proteins and lipids are the main substrates of these reactions. Endogenous muscular enzymes (cathepsins) may determine proteolysis and lipolysis at the value of pH found in fermented sausages, as well as the microorganisms' enzymes [14–16]. Lipolytic phenomena, with the release of FA and carbonyl compounds, and proteolytic phenomena, with the release of non-protein nitrogen compounds, are important for the development of characteristic taste and flavor of the final product [17,18].

In meat, spontaneous fermentation is associated with the biochemical activity of indigenous lactic acid bacteria (LAB). This process is technically easy and inexpensive but determines a certain variability in quality products and increases the risk of contamination from pathogenic microorganisms [19]. Controlled fermentation is an alternative, and involves the use of selected bacteria (living or resting) as starters, developing the desired metabolic activity in the meat. Further, different strains of indigenous LAB from meat are also isolated and used as commercial starters to made salami [20,21]. Consequently, the starter cultures influence the lipolytic processes, even if their effects on the FA composition are not clear. For example, higher contents of saturated FA and linoleic acid were found in Pastirma (Turkish dry-cured sausage) made using starter cultures [22], while no effect was detected in Tunisian sausage [23].

The breed, the age, the gender and the feeding system can influence the FA composition of fresh meat [24–28]. As also found in bresaola, grazing improves health and organoleptic properties of processed meat products compared to products obtained with the meat of animals reared in confined systems, positively influencing the FA profiles [4]. Indeed, different studies showed the use of pasture increases the percentage of *n-3* and conjugated linoleic acid (CLA) in products, which are FA with beneficial effects on human health [27]. Otherwise, the products from animals that are concentrate-fed show generally high level of *n-6*, in particular linoleic acid, negatively affecting their healthy quality [29].

The diet administered to the animal is one of the most important factors influencing the composition of production in cattle and sheep. This effect is due to specific compounds, including FA, which is transferred directly from the feed to the final products or derives from the microbial activity of the rumen and/or from the metabolism of the animal under the effect of specific diets. These compounds can therefore serve as a marker of an animal's food background [30].

The present study investigated the FA profile of salami made by processing meat from carcasses of different commercial categories, such as grazing or housed young bulls and grazing adult cows of Cinisara breed, under different microbiological conditions due to spontaneous fermentation or the inoculum of a starter culture. This investigation provides additional information to that reported by Gaglio et al. [7] regarding the physicochemical and sensory properties of salami, with the aim to further characterize this product.

2. Materials and Methods

2.1. Meat and Salami Manufacturing

Carcasses of animals of Cinisara breed, selected on the basis of age and the livestock system (use or non-use of pasture) were used, and in the specific:

n. 4 grazing young bulls (GB–18 months old), fed pasture-based diets from a 6-month age until slaughtering, with hay and concentrate supplements in the final phase (16–18 months);

n. 4 housed young bulls (HB–18 months old), fed pasture-based diets from a 6-month age, whereas housed and fed a diet of only hay and concentrate in the finishing phase (16–18 months);

n. 4 adult cows at the end of their productive life (AC–10 years old), fed pasture-based diets from 6 months of age until slaughtering, receiving hay and concentrate supplements from the beginning of the reproductive career (24 months).

The animals were slaughtered at an EU-licensed abattoir, according to the standard handling procedures with respect to EU regulations [31] on the protection of animals at the time of slaughter. The carcasses were stored in a cooling room at 4–8 °C for a 7-day ageing period. After, they were dissected (day zero), and the meat from the forequarters of each carcass was minced separately with a 6 mm plate; 50 kg of minced meat of each carcass was used.Each batch of minced meat was supplemented with pork lard (20% *w/w*) from Nebrodi Black Pig (Sicilian native pig breed) and other ingredients (sodium chloride, 2.5%; sucrose, dextrose and maltodextrin,0.35%; sodium ascorbate, 0.016%; sodium nitrate, 0.01%; potassium nitrate, 0.051%),obtaining in a blender a raw mixture for each animal (4 for each animal category, 12 total mixtures). Subsequently, each raw mixture was separated into two batches; one part fermented spontaneously, while the other one was inoculated with freeze-dried cultures, containing *Staphylococcus xylosus* and *Pedioccocuspentosaceus* (Tec-AL. s.r.l., Traversetolo, Italy; final concentration of approximately 107 CFU g^{-1}).

The meat was separately stuffed into natural casings. Each salami was formed at approximately 35 cm in length, 7 cm in diameter and an initial weight of about 1.2 kg. Dissection, mincing, blending and stuffing operations were performed on the same day for all carcasses. Afterwards, the salamis were hung for 3 days at 20 °C and ambient RH, and then transferred to cold rooms with controlled temperature and RH following the protocol reported by Gaglio et al. [7], reaching a final weight of about 0.7 kg. In brief, the ripening protocol was as follows: day 1, 20–22 °C and 62–72% RH; day 2, 19–21 °C and 64–74% RH; day 3, 18–20 °C and 66–76% RH; day 4, 17–19 °C and 68–78% RH; day 5, 16–18 °C and 70–80% RH; day 6, 15–17 °C and 72–82% RH; day 7, 14–16 °C and 74–84% RH; day 8, 13–15 °C and 76–86% RH; days 9–10, 12–14 °C and 78–88% RH; days 11–55, 11–13 °C and 80–90% RH. All salami were made at the "Lipari salami factory" in Alcamo (Sicily, Italy).

2.2. Sampling

For each carcass, a sample of minced meat (before the addition of other ingredients) and pork fat was collected.

The salami were sampled at day 0 after mixing all ingredients (raw mixture) (12 samples, 1 for each animal) and at day 45 (end of the maturing process) (12 samples of uninoculated salami and 12 samples of inoculated salami). The samples, placed in sterile vacuum containers, were refrigerated at 8°C for the transfer to laboratory where they were homogenized (2 min at maximum speed) by stomacher (LAB Blender 400, Seward Medical, London, UK) and then freeze-dried for successive analysis as described by Alabiso et al. [10] (SCANVAC Coolsafe 55-9, Labogene Aps, Lynge, Denmark).

2.3. Fatty Acids Composition

Fatty acids (FA) were extracted according to the method developed by O'Fallon et al. [32]. The total FA quantification was performed using C23:0 (Sigma-Aldrich, Darmstadt, Germany) as internal standard (0.5 mg/g freeze-dried sample). Each sample (1 µL) was injected by autosampler into an HP 6890 gas chromatography system equipped with a flameionization detector (Agilent Technologies Inc., Santa Clara, CA, USA). For separation of FA methyl esters from each sample, a capillary column (100 m length, 0.25 mm i.d., 0.25 μm; CP-Sil 88; Chrompack, Middelburg, the Netherlands) was used. Temperatures of the injector and the detector were kept at 255 °C and 250 °C, respectively, using H_2 flow of 40 mL/min, air flow of 400 mL/min and He flow of 45 mL/min. The oven temperature was held for 1 min at 70 °C, increased at 5 °C/min to 100 °C, held for 2 min, increased at 10 °C/min to 175 °C, held for 40 min and finally increased at 5 °C/min to 225 °C, held for 45 min. The carrier gas was He, used with a head pressure of 158.6 kPa and a flow rate of 0.7 mL/min (linear velocity of 14 cm/s). The identification of each FA was performed as described by Alabiso et al. [4].

The health-promoting index (HPI) was calculated as suggested by Ashkezary et al. [33]: (*n-3* PUFA + *n-6* PUFA + MUFA)/(C12:0 + (4 × C14:0) + C16:0). The thrombogenic index (TI) was calculated according to Ulbricht and Southgate [34] as follows: TI = (C14:0 + C16:0 + C18:0)/((0.5 × ΣMUFA) + (0.5 × ΣPUFA *n-6*) + (3 × ΣPUFA *n-3*) + (*n-3/n-6*)).

2.4. Statistical and Explorative Multivariate Analysis

Data were statistically processed using the SAS 9.2 software [35]. FA composition was analyzed according to a MIXED model including the fixed effects of animal category (A, with three levels: GB, HB and AC), product (P, with three levels: raw mixture RM, inoculated salami IS and uninoculated salami US) and the animal within category as a random effect. The interactions between the effects A × P was removed from the model since it was always not significant. Tukey's test was used to compare means when the effects were significant ($p < 0.05$).

The principal components analysis (PCA), performed using the PRINCOMP SAS procedure, was based on FA composition in order to assess its specific contribution in explaining the differences among salami type, due to the different animal category. The variables used in the analysis were identified on the basis of a stepwise selection using the STEPDISC SAS procedure, after they were standardized by multiplying them by the inverse of standard deviation (1/SD). The number of main components was selected according to Kaiser's criterion and only those with Eigen values above 1.00 were retained.

3. Results and Discussion

3.1. Fatty Acids Composition

Table 1 shows the FA composition of meat of each animal category and the pork lard. AC is characterized by a higher percentage content of monounsaturated fatty acids (MUFA) and a lower presence of polyunsaturated fatty acids (PUFA) compared to the levels found in GB, similar to what was found in bresaola obtained from cows at the end of their career [4]. Intermediate features were found in HB.

Pork lard showed a high MUFA content, consisting of 87% of oleic acid (OA, C18:1 *n-9*), and a low PUFA content possibly related to direct deposition of MUFA from acorns present in the diet to the fat depots. The FA composition of pork lard is comparable to that found by Pugliese et al. [36] in outdoor rearing system. The FA composition of pork lard used suggests that it derived from animals that evidently were reared outdoors with a feeding integration to the natural pasture, according to the farming system typically used for the Nebrodi Black Pig.

In general, the processed products show fat quantity and profile depending on the ingredients of the mixture. In this study, the same receipt was used for all products, changing only the kind of meat. Therefore, the differences in fat content and FA composition should be mainly attributable to beef rather than to lard and other ingredients.

Table 1. Fatty acids profile (% of total FA) and health indexes of fresh meat and fresh pork lard.

	Animal Category (A) [a]			
	GB	HB	AC	PL
Fat	3.30	4.65	8.80	76.12
Total FA,% DM [b]	2.94	4.19	7.92	68.52
Others SFA [c]	0.71	1.43	1.87	n.d.
C14:0	2.32	2.02	2.23	n.d.
C14:1	0.03	0.08	0.06	2.12
C15:0	1.89	1.30	1.80	0.05
C16:0	16.06	19.24	20.84	26.18
C16:1	2.24	2.4	1.91	3.51
C17:0	1.57	1.13	1.42	0.39
C17:1	0.50	0.76	1,00	0.16
C18:0	19.19	18.55	16.91	9.28
C18:1 c9	17.42	24.38	35.51	43.29
C18:1 t11	1.14	1.23	1.59	n.d.
Other C18:1	1.41	1.51	1.76	0.67
Other C18:2	0.32	0.24	0.64	0.07
C18:2 *n-6* LA	19.83	13.21	4.07	11.27
CLA C18:2 c9t11 RA	0.16	0.21	0.22	0.12
Other CLA isomers	0.09	0.03	0.07	0.05
C18:3 *n-3* ALA	2.35	1.25	1.22	0.61
C18:3 *n-6* GLA	0.16	0.18	0.19	0.17
C20:0	0.16	0.12	0.11	0.10
C20:2 *n-6*	0.15	0.17	0.13	0.04
C20:3 *n-6*	0.42	0.40	0.43	0.08
C20:4 *n-6* AA	6.97	5.83	2.61	0.93
C20:5 *n-3* EPA	0.92	0.59	0.41	0.28
C22:0	1.13	1.08	1.01	0.03
C22:2 *n-6*	0.22	0.15	0.05	n.d.
C22:4 *n-6*	0.37	0.32	0.21	n.d.
C22:5 *n3* DPA	2.03	1.49	0.85	n.d.
SFA	43.03	44.87	46.19	36.03
MUFA [d]	22.74	30.36	41.83	49.75
PUFA [e]	33.99	24.07	11.10	13.62
MUFA/SFA	0.53	0.68	0.91	1.38
PUFA/SFA	0.79	0.54	0.24	0.38
n-6	28.12	20.26	7.69	12.49
n-3	5.30	3.33	2.48	0.89
n-6/n-3	5.31	6.08	3.10	14.03
HPI [f]	2.23	1.98	1.76	2.42
TI [g]	0.90	1.09	1.22	1.04

The results indicate mean values of three measurements performed on each of the four carcasses for animal category and for each sample of lard. [a] Animal category: GB = grazing young bull; HB = housed young bull; AC = adult cow; PL = pork lard. [b] DM = dry matter; [c] SFA = saturated fatty acids; [d] MUFA = monounsaturated fatty acids; [e] PUFA = polyunsaturated fatty acids; [f] HPI = health promoting index; [g] TI = thrombogenic index.

Table 2 shows FA profile in relation to the animal category and the product type.

The total FA content (Table 2) was higher in AC than in GB and HB ($p < 0.05$), confirming that meat of adult cows is characterized by a major fat content compared to young animals [5] and that fat and proportions of FA change depending on the age of the animal and its diet [4].

Table 2. Effects of animal category and product on fatty acids profile (% of total FA) and health indexes of salami fat.

	Product (P) [b]		Animal Category (A) [a]			SEM [h]	Significance	
			GB	HB	AC		A	P
Total FA,% DM	RM	34.86	30.76	33.40	40.40	1.569	0.042	0.538
	IS	35.79	31.17	33.79	42.42			
	US	36.86	34.24	35.11	41.23			
	Total		31.38 [b]	34.10 [b]	41.35 [a]			
SFA [c]	RM	39.76	39.03	39.39	40.85	2.872	0.650	0.363
	IS	39.98	38.92	39.85	41.17			
	US	39.80	38.68	40.26	40.46			
	Total		38.88	39.83	40.82			
MUFA [d]	RM	39.60	34.01	39.33	45.45	2.085	0.042	0.469
	IS	40.96	35.85	40.27	46.75			
	US	40.85	35.66	40.12	46.77			
	Total		35.17 [b]	39.90 [a, b]	46.32 [a]			
PUFA [e]	RM	20.18	26.42	21.08	13.04	0.989	0.008	0.780
	IS	18.84	25.16	19.30	12.07			
	US	18.86	25.18	19.48	11.93			
	Total	.	25.59 [a]	19.95 [b]	12.35 [c]			
MUFA/SFA	RM	0.99	0.87	1.00	1.11	0.196	0.407	0.358
	IS	1.02	0.92	1.01	1.14			
	US	1.03	0.93	1.00	1.15			
	Total		0.91	1.00	1.14			
PUFA/SFA	RM	0.51	0.68	0.53	0.32	0.146	0.031	0.376
	IS	0.48	0.65	0.49	0.29			
	US	0.47	0.65	0.48	0.29			
	Total		0.66 [a]	0.51 [a]	0.30 [b]			
n-6	RM	17.33	22.78	18.50	10.77	0.997	0.007	0.258
	IS	15.98	21.48	16.63	9.79			
	US	15.91	21.40	16.75	9.57			
	Total		21.88 [a]	17.29 [b]	10.04 [c]			
n-3	RM	2.62	3.51	2.39	1.97	0.095	0.008	0.668
	IS	2.53	3.45	2.30	1.84			
	US	2.61	3.57	2.40	1.86			
	Total		3.51 [a]	2.36 [b]	1.89 [b]			
n-6/n-3	RM	6.57	6.49	7.74	5.47	0.752	0.031	0.529
	IS	6.26	6.23	7.23	5.32			
	US	6.04	5.99	6.98	5.15			
	Total		6.24 [a, b]	7.31 [a]	5.31 [b]			
HPI [f]	RM	2.05	2.21	2.06	1.87	0.054	0.041	0.843
	IS	2.03	2.35	1.93	1.82			
	US	2.04	2.29	2.01	1.81			
	Total		2.28 [a]	2.00 [a, b]	1.83 [b]			
TI [g]	RM	1.08	0.98	1.10	1.16	0.096	0.576	0.809
	IS	1.06	0.97	1.13	1.17			
	US	1.08	0.97	1.12	1.18			
	Total		0.98	1.12	1.17			

The results indicate mean values of three measurements performed on each trial of each carcass. [a] Animal category: GB = grazing young bull; HB = housed young bull; AC = adult cow. [b] Product: RM= raw mixture (day 0), IS= inoculated salami (day 45), US = uninoculated salami (day 45).[c] SFA = saturated fatty acids; [d] MUFA = monounsaturated fatty acids; [e] PUFA = polyunsaturated fatty acids; [f] HPI = health promoting index; [g] TI = thrombogenic index. [h] SEM= standard error of the means. The interactions among the effects were removed from the model because they were not significant ($p > 0.05$). Different letters ([a, b, c]) on horizontal rows indicate $p < 0.05$.

The saturated fatty acids (SFA) (Table 2) did not significantly change among animal categories, while MUFA were higher in AC ($p \leq 0.05$) compared to GB. HB showed intermediate values (differences not statistically significant). PUFA were higher in GB than in HB, which yet registered higher values compared with AC ($p \leq 0.01$).

The animal category influenced the *n-6* and *n-3* contents, the PUFA/SFA and *n-6/n-3* ratio, as well as HPI (Table 2). GB showed higher values of both *n-6* ($p \leq 0.01$) and *n-3*

($p \leq 0.05$) compared to AC and HB, and higher HPI ($p \leq 0.05$) compared to AC. The ratio PUFA/SFA was lower in AC compared to GB and HB ($p \leq 0.05$), while the ratio *n-6/n-3* was higher in HB compared to AC ($p \leq 0.05$). The FAO/WHO report indicates an optimal *n-6/n-3* ratio of 4.5–5 [37]. A diet contrasting various "lifestyle diseases", such as coronary heart diseases and cancers, requires a ratio of PUFA/SFA above about 0.45 and a ratio of *n-6/n-3* below 4.0 [38]. Therefore, AC show an unfavorable PUFA/SFA ratio compared to GB and HB. However, despite the lower PUFA content, AC showed a health favorable *n-6/n-3* ratio, especially due to the lower linoleic acid (LA) level.

Table 3 shows the SFA composition of raw mixtures and ripened salami. Significant differences were found for palmitic acid (C16:0), higher for HB and AC compared to GB ($p \leq 0.01$), for C15:0 and C17:0, lower in GB, and for stearic acid (C18:0) higher in GB compared to AC ($p \leq 0.05$), as found in bresaola by Alabiso et al. [4]. The effect of the product type emerged for C17:0, lower in the raw mixtures than in salami.

Table 3. Effects of animal category and product on saturated fatty acids (% of total FA) of salami fat.

	Product (P) [b]		Animal Category (A) [a]			SEM [d]	Significance	
			GB	HB	AC		A	P
Others SFA [c]	RM	0.77	0.23	0.66	1.40	0.328	0.419	0.685
	IS	0.79	0.83	0.3	1.22			
	US	0.54	0.24	0.79	0.61			
	Total		0.44	0.58	1.08			
C14:0	RM	1.62	1.64	1.51	1.71	0.136	0.089	0.653
	IS	1.71	1.48	1.84	1.81			
	US	1.64	1.5	1.55	1.89			
	Total		1.54	1.63	1.80			
C15:0	RM	0.24	0.20	0.16	0.37	0.071	0.021	0.752
	IS	0.31	0.23	0.33	0.39			
	US	0.30	0.15	0.24	0.51			
	Total		0.19 [b]	0.25 [b]	0.42 [a]			
C16:0	RM	22.81	20.80	23.25	24.39	0.983	0.009	0.462
	IS	22.80	20.04	23.39	24.98			
	US	22.89	20.52	23.36	24.78			
	Total		20.50 [b]	23.35 [a]	26.71 [a]			
C17:0	RM	0.46 [b]	0.47	0.35	0.58	0.091	0.029	0.019
	IS	0.65 [a]	0.53	0.67	0.74			
	US	0.65 [a]	0.43	0.56	0.94			
	Total		0.48 [b]	0.53 [b]	0.76 [a]			
C18:0	RM	13.01	14.82	12.61	11.60	1.084	0.026	0.291
	IS	12.97	15.06	12.55	11.29			
	US	12.74	14.18	12.97	11.07			
	Total		14.67 [a]	12.71 [a, b]	11.32 [b]			
C20:0	RM	0.10	0.10	0.10	0.10	0.019	0.299	0.738
	IS	0.08	0.10	0.09	0.08			
	US	0.10	0.09	0.10	0.10			
	Total		0.10	0.10	0.10			
C22:0	RM	0.73	0.77	0.75	0.70	0.080	0.938	0.093
	IS	0.66	0.65	0.68	0.66			
	US	0.67	0.76	0.69	0.56			
	Total		0.72	0.71	0.65			

The results indicate mean values of three measurements performed on each trial of each carcass. [a] Animal category: GB = grazing young bull; HB = housed young bull; AC = adult cow. [b] Product: RM= raw mixture (day 0), IS= inoculated salami (day 45), US = uninoculated salami (day 45). [c] SFA = saturated fatty acids. [d] SEM= standard error of the means. The interactions among the effects were removed from the model because they were not significant ($p > 0.05$). Different letters ([a, b]) on horizontal and vertical rows indicate $p < 0.05$.

Table 4 shows the MUFA composition of raw mixtures and ripened salami. In particular, the C14:1 and especially OA were higher in AC than in GB ($p \leq 0.05$), while HB showed intermediate values (not statistically significant); as in bresaola, OA was the most representative of the MUFA and its content was influenced by the age of the animals [4,29]. The raw mixtures showed the lower level of C17:1.

Table 4. Effects of animal category and product on monounsaturated fatty acids (% of total FA) of salami fat.

	Product (P) [b]		Animal Category (A) [a]			SEM [c]	Significance	
			GB	HB	AC		A	P
C14:1	RM	0.08	0.07	0.06	0.12	0.025	0.006	0.997
	IS	0.09	0.04	0.11	0.14			
	US	0.10	0.03	0.08	0.18			
	Total		0.04 [b]	0.08 [b]	0.18 [a]			
C16:1	RM	2.41	2.55	2.24	2.46	0.203	0.066	0.062
	IS	2.60	2.21	2.70	2.81			
	US	2.67	2.47	2.53	3.01			
	Total		2.43	2.49	2.76			
C17:1	RM	0.27 [b]	0.27	0.25	0.29	0.022	0.218	0.041
	IS	0.36 [a]	0.32	0.37	0.43			
	US	0.36 [a]	0.30	0.36	0.53			
	Total		0.30	0.33	0.37			
C18:1 c9 OA	RM	34.58	29.33	34.46	39.94	2.440	0.031	0.598
	IS	34.68	30.41	33.83	39.80			
	US	34.38	30.16	33.41	39.57			
	Total		32.80 [b]	37.56 [a, b]	40.44 [a]			
Other C18:1	RM	2.25	1.79	2.32	2.64	0.764	0.241	0.439
	IS	3.90	3.87	4.26	3.57			
	US	3.64	3.70	3.74	3.48			
	Total		2.79	3.44	3.23			

The results indicate mean values of three measurements performed on each trial of each carcass. [a] Animal category: GB = grazing young bull; HB = housed young bull; AC = adult cow. [b] Product: RM= raw mixture (day 0), IS= inoculated salami (day 45), US = uninoculated salami (day 45). [c] SEM= standard error of the means. The interactions among the effects were removed from the model because not significant ($p > 0.05$). Different letters ([a, b]) on horizontal and vertical rows indicate $p < 0.05$.

Table 5 shows the PUFA composition of raw mixtures and ripened salami. In all animal categories, the most represented among PUFA was linoleic acid (LA, 18:2 *n-6*), and was lower in AC compared to HB and GB ($p < 0.01$). In accordance with Wood et al. [29], the LA content was negatively related to the trend of salami fat. In fact, total FA content (Table 2) was higher in AC than in HB and GB. Furthermore, the arachidonic acid (AA, C20:4 *n-6*), derived from LA through the action of $\Delta 5$ and $\Delta 6$ desaturase enzymes and elongase [4], was on average lower in AC compared to HB and GB ($p < 0.05$), similar to its precursor. Moreover, eicosapentaenoic (EPA, C20:5 *n-3*) and docosapentaenoic (DPA, C22:5 *n-3*) acids were on average higher in GB than HB and AC ($p < 0.05$ and $p < 0.01$, respectively). Feeding systems based on grazing have been known to have a positive effect on PUFA content of beef, and in general, on products obtained from ruminants [39–41]. Indeed, the use of grazing until slaughter resulted in a higher PUFA content of GB compared to HB.AC, despite being grazed until slaughter, showed lower PUFA concentrations than GB.These trends can be linked to the older age of AC, confirming the results obtained in bresaola, in which beef had higher concentrations of fat and lower PUFA levels than those of younger subjects, fed under the same conditions. Analogous differences were found between HB and GB which differed in the feeding pattern [4].

Table 5. Effects of animal category and product on polyunsaturated fatty acids (% of total FA) of salami fat.

	Product (P) [b]		Animal Categories (A) [a]			SEM [c]	Significance	
			GB	HB	AC		A	P
Other C18:2	RM	0.06 [b]	0.03	0.03	0.12	0.024	0.027	0.011
	IS	0.16 [a]	0.08	0.17	0.22			
	US	0.15 [a]	0.06	0.14	0.26			
	Total		0.05 [c]	0.11 [b]	0.20 [a]			
C18:2 *n-6* LA	RM	13.76	18.29	14.47	8.52	0.704	0.015	0.281
	IS	12.71	17.69	12.87	7.56			
	US	12.65	17.61	13.02	7.33			
	Total		17.86 [a]	13.45 [b]	7.80 [c]			
CLA C18:2 c9t11 RA	RM	0.11	0.12	0.10	0.12	0.013	0.039	0.256
	IS	0.11	0.11	0.07	0.14			
	US	0.11	0.10	0.08	0.15			
	Total		0.11 [b]	0.11 [b]	0.14 [a]			
Other CLA isomers	RM	0.04	0.08	0.04	0.06	0.015	0.483	0.111
	IS	0.09	0.08	0.09	0.08			
	US	0.07	0.07	0.09	0.09			
	Total		0.07	0.07	0.07			
C18:3 *n-3* ALA	RM	1.42	1.78	1.24	1.24	0.073	0.059	0.361
	IS	1.33	1.73	1.12	1.14			
	US	1.37	1.79	1.18	1.15			
	Total		1.77	1.18	1.18			
C18:3 *n-6* GLA	RM	0.16	0.15	0.16	0.15	0.017	0.118	0.128
	IS	0.18	0.21	0.17	0.17			
	US	0.18	0.20	0.17	0.16			
	Total		0.19	0.17	0.16			
C20:2 *n-6*	RM	0.10	0.11	0.10	0.09	0.027	0.420	0.230
	IS	0.12	0.13	0.15	0.08			
	US	0.10	0.10	0.09	0.11			
	Total		0.11	0.11	0.09			
C20:3 *n-6*	RM	0.06	0.07	0.06	0.05	0.012	0.110	0.273
	IS	0.07	0.06	0.06	0.09			
	US	0.02	0.01	0.01	0.04			
	Total		0.05	0.04	0.06			
C20:4 *n-6* AA	RM	3.03	3.86	3.42	1.81	0.013	0.041	0.651
	IS	2.67	3.13	3.11	1.77			
	US	2.73	3.22	3.19	1.78			
	Total		3.40 [a]	3.24 [a]	1.79 [b]			
C20:5 *n-3* EPA	RM	0.48	0.69	0.41	0.35	0.031	0.039	0.628
	IS	0.48	0.67	0.43	0.34			
	US	0.51	0.72	0.47	0.34			
	Total		0.69 [a]	0.44 [a, b]	0.34 [b]			
C22:2 *n-6*	RM	0.11	0.15	0.14	0.05	0.003	0.036	0.361
	IS	0.10	0.13	0.14	0.03			
	US	0.10	0.13	0.13	0.03			
	Total		0.14 [a]	0.14 [a]	0.04 [b]			
C22:4 *n-6*	RM	0.13	0.15	0.15	0.10	0.009	0.305	0.306
	IS	0.12	0.13	0.13	0.09			
	US	0.13	0.13	0.14	0.12			
	Total		0.14	0.14	0.10			
C22:5 *n-3* DPA	RM	0.72	1.04	0.74	0.38	0.007	0.009	0.242
	IS	0.72	1.05	0.75	0.36			
	US	0.73	1.06	0.75	0.37			
	Total		1.05 [a]	0.75 [b]	0.37 [c]			

The results indicate mean values of three measurements performed on each trial of each carcass. [a] Animal category: GB = grazing young bull; HB = housed young bull; AC = adult cow. [b] Product: RM= raw mixture (day 0), IS = inoculated salami (day 45), US = uninoculated salami (day 45). [c] SEM = standard error of the means. The interactions among the effects were removed from the model because they were not significant ($p>0.05$). Different letters ([a, b]) on horizontal and vertical rows indicate $p < 0.05$. LA = linoleic acid; ALA = α-linolenic acid; RA = rumenic acid; CLA = conjugated linoleic acid; AA = arachidonic acid; EPA = eicosapentaenoic acid; DPA = docosapentaenoic acid.

The overall concentration of rumenic acid (RA), the main of CLA isomers, in ripened salami was higher in AC ($p < 0.05$), although the differences between the commercial categories of salami were smaller with respect to those found in bresaola [4], probably due to the mitigating effect of pork lard addition.

Since the addition of pork lard did not induce any positive effects due to the diet and farming system of pigs, the FA composition was better than that found on salami produced with only Nebrodi Black Pig [42].

The fermentation processes in both inoculated and uninoculated salami did not show appreciable differences in FA composition at the end of the maturation. A significant increase in the raw mixture in comparison with the ripened salami was found only for C17:0, C17:1 and others, C18:2 ($p < 0.05$). As found in Gaglio et al. [7], the microbiological evolution may have been comparable between inoculated and uninoculated samples, especially with regard toits role in modifying the FA profile, probably as a consequence of microbial contamination of raw meat during slaughter and subsequent processing stages.

3.2. Multivariate Analysis

The plot generated by PCA is shown in Figure 1, where the length of each vector measures the contribution of each variable to the main components.

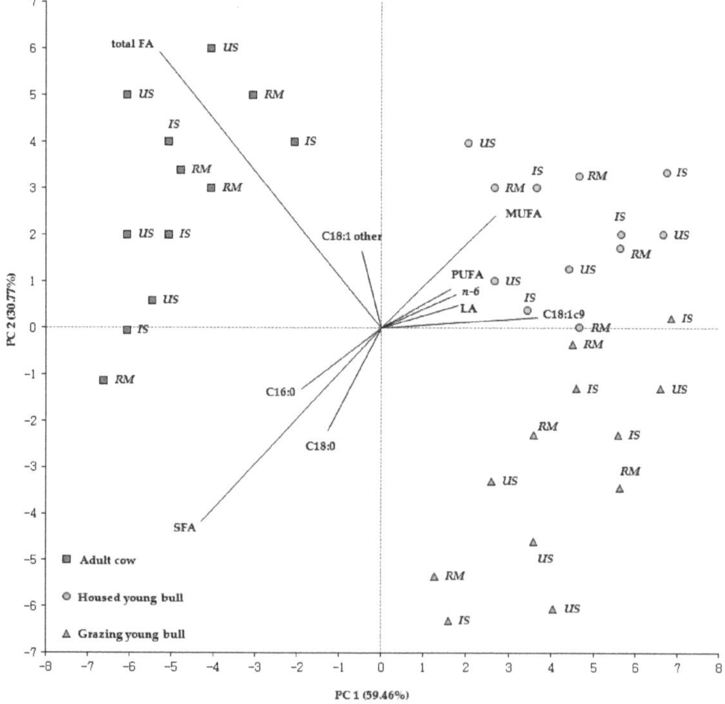

Figure 1. Score plot and loading plot from principal component analysis (PCA) of raw mixture and ripened salami. Abbreviations: RM = raw mixture; IS = inoculated salami; US = uninoculated salami; SFA = saturated fatty acids; MUFA = monounsaturated fatty acids; PUFA = polyunsaturated fatty acids; LA = linoleic acid.

The first two principal components accounted for 90.23% of the total variance. With 59.46% of the total variance, the first principal component was able to discriminate the AC salami from those of GB and HB on the basis of the main contributions of total FA, SFA, C18:1*n-9* and MUFA.Instead, with 30.77% of the total variance, the second main component was able to discriminate GB salami from HB salami, especially for the contribution of total FA, SFA, C18:0, MUFA and C16:0. Overall, the animal category, which associates the effect of age and feeding system, influenced the qualitative traits of salami more markedly than the inoculated starter culture, since the effect of the inoculation did not allow a clear distinction of the salami within the animal category.

4. Conclusions

The origin of meat, with regard to the animal category, differing for age and feeding system, influenced the FA composition of salami. The addition of pork lard reduced the initial differences between the animal categories.

In bulls, the use of pasture up to slaughter improved the FA profile, determining an increase in *n-3* PUFA content. Adult cows, despite benefiting from the pasture until slaughter, showed lower PUFA contents than young bulls, due to the higher fat content. However, the lower LA level resulted in a more favorable *n-6/n-3* ratio in AC than HB. The higher age also determined an increase of OA in AC.

On the basis of these results, salami made with Cinisara meat could be an alternative product, even if the addition of pork lard mitigated the favorable effect of the livestock system based on grazing in increasing the PUFA content and thus in improving the FA profile.

In salami, fermentation processes and the addition of pork lard would allow one to appreciate only significant differences linked to the animal categories. Further studies should be conducted in which salami is made only with beef, considering also the consumers who do not eat pork.

Author Contributions: Conceptualization, M.A. and G.M.; methodology, M.A. and G.M.; software, M.A. and G.M.; formal analysis, G.M.; investigation, G.M., C.G. and A.B.; resources, G.M.; data curation, M.A., G.M. and A.D.G.; writing—original draft preparation, M.A. and G.M.; writing—review and editing, M.A., G.M. and A.B.; visualization, C.G. and A.D.G.; supervision, A.B.; funding acquisition, A.B. All authors have read and agreed to the published version of the manuscript.

Funding: This research was funded by the Sicilian Region (Italy) under Grant PO FESR 2007–2013, Linea 4.1.1.2., "Applicazione biotecnologiche e bioinformatiche per la tracciabilità delle produzioni carnee siciliane".

Institutional Review Board Statement: Not applicable.

Data Availability Statement: All data included in this study are available upon request by contacting the corresponding author.

Conflicts of Interest: The authors declare that there is no conflict of interest regarding the publication of this article.

References

1. Alabiso, M.; Giosuè, C.; Alicata, M.L.; Parrino, G. The milk yield by Cinisara cows in different management systems: Note 2. Effects of season production. In Proceedings of the XX International Grassland Congress, Satellite Workshop, Glasgow, UK, 3 July 2005; p. 201.
2. Bonanno, A.; Tornambè, G.; Bellina, V.; De Pasquale, C.; Mazza, F.; Maniaci, G.; Di Grigoli, A. Effect of farming system and cheesemaking technology on the physicochemical characteristics, fatty acid profile, and sensory properties of Caciocavallo Palermitano cheese. *J. Dairy Sci.* **2013**, *96*, 710–724. [CrossRef]
3. Di Gregorio, P.; Di Grigoli, A.; Di Trana, A.; Alabiso, M.; Maniaci, G.; Rando, A.; Vallucci, C.; Finizio, D.; Bonanno, A. Effects of different genotypes at the CSN3 and LGB loci on milk and cheese-making characteristics of the bovine Cinisara breed. *Int. Dairy J.* **2017**, *71*, 1–5. [CrossRef]
4. Alabiso, M.; Maniaci, G.; Giosuè, C.; Gaglio, R.; Francesca, N.; Di Grigoli, A.; Portolano, B.; Bonanno, A. Effect of muscle type and animal category on fatty acid composition of bresaola made from meat of Cinisara cattle: Preliminary investigation. *CyTA-J. Food* **2020**, *18*, 734–741. [CrossRef]
5. Maniaci, G.; Alabiso, M.; Francesca, N.; Giosuè, C.; Di Grigoli, A.; Corona, O.; Cardamone, C.; Graci, G.; Portolano, B.; Bonanno, A. Bresaola made from Cinisara cattle: Effect of muscle type and animal category on physicochemical and sensory traits. *CyTA-J Food* **2020**, *18*, 383–391. [CrossRef]
6. Liotta, L.; Trentacoste, F.; Magazzù, G.; D'Angelo, G.; Chiofalo, V. Technological and quality characteristics of Bresaola from Cinisara cattle breed. Proceedings of the 21th ASPA Congress, 2015 June 9-12, Milano, Italy. *Ital. J. Anim. Sci.* **2015**, *14*, 93.

7. Gaglio, R.; Francesca, N.; Maniaci, G.; Corona, O.; Alfonso, A.; Di Noto, A.M.; Cardamone, C.; Sardina, M.T.; Portolano, B.; Alabiso, M. Valorization of indigenous dairy cattle breed through salami production. *Meat Sci.* **2016**, *114*, 58–68. [CrossRef] [PubMed]

8. Talon, R.; Leroy, S.; Lebert, I. Microbial ecosystems of traditional fermented meat products: The importance of indigenous starters. *Meat Sci.* **2007**, *77*, 55–62. [CrossRef] [PubMed]

9. Biesalski, H.K. Meat as a component of a healthy diet – are there any risks or benefits if meat is avoided in the diet? *Meat Sci.* **2005**, *70*, 509–524. [CrossRef]

10. Jiménez-Colmenero, F.; Ventanas, J.; Toldrá, F. Nutritional composition of dry-cured ham and its role in a healthy diet. *Meat Sci.* **2010**, *84*, 585–593. [CrossRef]

11. Paleari, M.A.; Moretti, V.M.; Beretta, G.; Mentasti, T.; Bersani, C. Cured products from different animal species. *Meat Sci.* **2003**, *63*, 485–489. [CrossRef]

12. Omer, M.K.; Hauge, S.J.; Østensvik, Ø.; Moen, B.; Alvseike, O.; Røtterud, O.J.; Prieto, M.; Dommersnes, S.; Nesteng, O.H.; Nesbakken, T. Effects of hygienic treatments during slaughtering on microbial dynamics and contamination of sheep meat. *Int. J. Food Microbiol.* **2015**, *194*, 7–14. [CrossRef] [PubMed]

13. Todorov, S.D.; Vaz-Velho, M.; De Melo Franco, B.D.G.; Holzapfel, W.H. Partial characterization of bacteriocins produced by three strains of Lactobacillus sakei, isolated from salpicao, a fermented meat product from North-West of Portugal. *Food Control* **2013**, *30*, 111–121. [CrossRef]

14. Lücke, F.K. Microbiological processes in the manufacture of dry sausage and raw ham. *Fleischwirtschaft* **1986**, *66*, 1505–1509.

15. Johansson, G.; Berdagué, J.L.; Larsson, M.; Tran, N.; Borch, E. Lipolysis, proteolysis and formation of volatile components during ripening of a fermented sausage with Pediococcus pentosaceus and Staphylococcus xylosus as starter cultures. *Meat Sci.* **1994**, *38*, 203–218. [CrossRef]

16. Toldra, F. Proteolysis and lipolysis in flavour development of dry-cured meat products. *Meat Sci.* **1998**, *4 9*, S101–S110. [CrossRef]

17. Molly, K.; Demeyer, D.; Johansson, G.; Raemaekers, M.; Ghistelinck, M.; Geenen, I. The importance of meat enzymes in ripening and flavour generation in dry fermented sausages. First results of a European project. *Food Chem.* **1997**, *59*, 539–545. [CrossRef]

18. Fernández, M.; De La Hoz, L.; Díaz, O.; Cambero, M.I.; Ordóñez, J.A. Effect of the addition of pancreatic lipase on the ripening of dry-fermented sausages—part 2. Free fatty acids, short-chain fatty acids, carbonyls and sensory quality. *Meat Sci.* **1995**, *40*, 351–362. [CrossRef]

19. Antara, N.S.; Sujaya, I.N.; Yokota, A.; Asano, K.; Tomita, F. Effects of indigenous starter cultures on the microbial and physico-chemical characteristics of Urutan, a Balinese fermented sausage. *J. Biosci. Bioeng.* **2004**, *98*, 92–98. [CrossRef]

20. Erkkilä, S.; Petäjä, E. Screening of commercial meat starter cultures at low pH and in the presence of bile salts for potential probiotic use. *Meat Sci.* **2000**, *55*, 297–300. [CrossRef]

21. Pennacchia, C.; Vaughan, E.E.; Villani, F. Potential probiotic Lactobacillus strains from fermented sausages: Further investigations on their probiotic properties. *Meat Sci.* **2006**, *73*, 90–101. [CrossRef] [PubMed]

22. Aksu, M.I.; Kaya, M. Effect of commercial starter cultures on the fatty acid composition of pastirma (Turkish dry meat product). *J. Food Sci.* **2002**, *67*, 2342–2345. [CrossRef]

23. El Adab, S.; Essid, I.; Hassouna, M. Microbiological, Biochemical and Textural Characteristics of a Tunisian Dry Fermented Poultry Meat Sausage Inoculated With Selected Starter Cultures. *J. Food Saf.* **2015**, *35*, 75–85. [CrossRef]

24. De Smet, S.; Raes, K.; Demeyer, D. Meat fatty acid composition as affected by fatness and genetic factors: A review. *Anim. Res.* **2004**, *53*, 81–98. [CrossRef]

25. Lengyela, Z.; Husve, F.; Polga, P.; Szabo, F.; Magyarb, L. Fatty acid composition of intramuscular lipids in various muscles of Holstein-Friesian bulls slaughtered at different ages. *Meat Sci.* **2003**, *65*, 593–598. [CrossRef]

26. Malau-Aduli, A.E.O.; Siebert, B.D.; Bottema, C.D.; Pitchford, W.S. Breed comparison of the fatty acid composition of muscle phospholipids in Jersey and Limousin cattle. *Ital. J. Anim. Sci.* **1998**, *76*, 766–773. [CrossRef] [PubMed]

27. Nuernberg, K.; Dannenberger, D.; Nuernberg, G.; Ender, K.; Voigt, J.; Scollan, N.D.; Wood, J.D.; Nutec, G.R.; Richardson, R.I. Effect of a grass-based and a concentrate feeding system on meat quality characteristics and fatty acid composition of longissimus muscle in different cattle breeds. *Livest. Prod. Sci.* **2005**, *94*, 137–147. [CrossRef]

28. Steen, R.W.J.; Porter, M.G. The effects of high-concentrate diets and pasture on the concentration of conjugated linoleic acid in beef muscle and subcutaneous fat. *Grass Forage Sci.* **2003**, *58*, 50–57. [CrossRef]

29. Wood, J.D.; Enser, M.; Fisher, A.V.; Nute, G.R.; Sheard, P.R.; Richardson, R.I.; Hughes, S.I.; Whittington, F.M. Fat deposition, fatty acid composition and meat quality. *Meat Sci.* **2008**, *78*, 343–358. [CrossRef]

30. Prache, S.; Martin, B.; Coppa, M. Authentication of grass-fed meat and dairy products from cattle and sheep. *Animal* **2020**, *14*, 854–863. [CrossRef] [PubMed]

31. The Council of the European Union. Council Regulation (EC) No1099/2009 of 24 September 2009 on the protection of animal sat the time of killing. *Off. J. Eur. Union* **2009**, *7*, 1–30.

32. O'Fallon, J.V.; Busboom, J.R.; Nelson, M.L.; Gaskins, C.T. A direct method for fatty acid methyl ester synthesis: Application to wet meat tissues, oils, and feedstuffs. *J. Anim. Sci.* **2007**, *85*, 1511–1521. [CrossRef]

33. Ashkezary, M.R.; Bonanno, A.; Todaro, M.; Settanni, L.; Gaglio, R.; Todaro, A.; Alabiso, M.; Maniaci, G.; Mazza, F.; Di Grigoli, A. Effects of adding solid and molten chocolate on the physicochemical, antioxidant, microbiological, and sensory properties of ewe's milk cheese. *J. Food Sci.* **2020**, *85*, 556–566. [CrossRef] [PubMed]

34. Ulbritch, T.L.V.; Southgate, D.A.T. Coronary Heart Disease, Seven Dietary Factors. *Lancet* **1991**, *338*, 985–992. [CrossRef]
35. *SAS/STAT Qualification Tools User's Guide in SAS 9*; Version 9.2; SAS Institute Inc.: Campus Drive Cary, NC, USA, 2010.
36. Pugliese, C.; Calagna, G.; Chiofalo, V.; Moretti, V.M.; Margiotta, S.; Franci, O.; Gandini, G. Comparison of the performances of Nero Siciliano pigs reared indoors and outdoors: 2. Joints composition, meat and fat traits. *Meat Sci.* **2004**, *68*, 523–528. [CrossRef] [PubMed]
37. FAO/WHO. *Fats and Fatty Acids in Human Nutrition: Report of an Expert Consultation*; FAO Food and Nutrition Paper Series 91; FAO: Rome, Italy, 2010.
38. Simopoulos, A.P. The importance of the ratio of omega-6/omega-3 essential fatty acids. *Biomed. Pharmacother.* **2002**, *56*, 353–379. [CrossRef]
39. Di Grigoli, A.; Di Trana, A.; Alabiso, M.; Maniaci, G.; Giorgio, D.; Bonanno, A. Effects of grazing on the behaviour, oxidative and immune status, and production of organic dairy cows. *Animals* **2019**, *9*, 371. [CrossRef]
40. French, P.; O'riordan, E.G.; Monahan, F.J.; Caffrey, P.J.; Moloney, A.P. Fatty acid composition of intra-muscular triacylglycerols of steers fed autumn grass and concentrates. *Livest. Prod. Sci.* **2003**, *81*, 307–317. [CrossRef]
41. Raes, K.; Balcaen, A.; Dirinck, P.; Winne, A.; Claeys, E.; Demeyer, D.; De Smet, S. Meat quality, fatty acid composition and flavor analysis in Belgian retail beef. *Meat Sci.* **2003**, *65*, 1237–1246. [CrossRef]
42. Chiofalo, B.; Liotta, L.; Sanzarello, L.; Piccolo, D.; Chiofalo, V. Vitamin E administration in nero siciliano pigs: Effet on the oxidative stability of "nebrodi" cured sausage. In Proceedings of the 6th International Symposium on the Mediterranean Pig, Messina, Italy, 11–13 October 2007; pp. 183–186.

MDPI

St. Alban-Anlage 66

4052 Basel

Switzerland

Tel. +41 61 683 77 34

Fax +41 61 302 89 18

www.mdpi.com

Animals Editorial Office

E-mail: animals@mdpi.com

www.mdpi.com/journal/animals